Blockchain Security and Its Application in Internet of Things

Blockchain Security and Its Application in Internet of Things

Guest Editor

Chun-Ta Li

MDPI

Basel • Beijing • Wuhan • Barcelona • Belgrade • Novi Sad • Cluj • Manchester

Guest Editor
Chun-Ta Li
Bachelor's Program of
Artificial Intelligence and
Information Security
Fu Jen Catholic University
New Taipei City
Taiwan

Editorial Office
MDPI AG
Grosspeteranlage 5
4052 Basel, Switzerland

This is a reprint of the Special Issue, published open access by the journal *Sensors* (ISSN 1424-8220), freely accessible at: https://www.mdpi.com/journal/sensors/special_issues/Blockchain_Security_IoT.

For citation purposes, cite each article independently as indicated on the article page online and as indicated below:

Lastname, A.A.; Lastname, B.B. Article Title. *Journal Name* **Year**, *Volume Number*, Page Range.

ISBN 978-3-7258-4277-3 (Hbk)
ISBN 978-3-7258-4278-0 (PDF)
https://doi.org/10.3390/books978-3-7258-4278-0

Contents

About the Editor

Chun-Ta Li

Chun-Ta Li received a Ph.D. degree in Computer Science and Engineering from the National Chung Hsing University, Taiwan, in 2008. Dr. Li is currently a full-time Professor in the Bachelor's Program of Artificial Intelligence and Information Security, at Fu Jen Catholic University, Taiwan. Dr. Li received the 2011 IJICIC Most Cited Paper Award from the International Journal of Innovative Computing, Information and Control. Dr. Li is included in the list of the World's Top 2% of Scientists published by Standford University from 2020 to 2024. Dr. Li is a senior member of IEEE, a member of the Chinese Information Security Association (CCISA), a member of IFIP WG 11.3, a member of Machine Intelligence Research Labs (MIR Labs), and an editorial board member of the International Journal of Network Security (IJNS). His research interests include information security, wireless sensor networks, mobile computing, and security protocols for IoT and cloud-based applications. Dr. Li has published eight Taiwan patents and more than 80 SCI/SSCI-indexed research papers in international journals on the above research fields. He also served as a reviewer and a guest editor for many SCI-index journals.

sensors MDPI

Article

Authorized Shared Electronic Medical Record System with Proxy Re-Encryption and Blockchain Technology

Weizhe Chen [1], Shunzhi Zhu [1,*], Jianmin Li [1], Jiaxin Wu [1], Chin-Ling Chen [1,2,3,*] and Yong-Yuan Deng [3]

1 School of Computer and Information Engineering, Xiamen University of Technology, Xiamen 361024, China; wzchen@stu.xmut.edu.cn (W.C.); lijm@xmut.edu.cn (J.L.); wujiaxin1996@stu.xmut.edu.cn (J.W.)
2 School of Information Engineering, Changchun Sci-Tech University, Changchun 130600, China
3 Department of Computer Science and Information Engineering, Chaoyang University of Technology, Taichung 41349, Taiwan; allendeng@cyut.edu.tw
* Correspondence: szzhu@xmut.edu.cn (S.Z.); clc@mail.cyut.edu.tw (C.-L.C.)

Abstract: With the popularity of the internet 5G network, the network constructions of hospitals have also rapidly developed. Operations management in the healthcare system is becoming paperless, for example, via a shared electronic medical record (EMR) system. A shared electronic medical record system plays an important role in reducing diagnosis costs and improving diagnostic accuracy. In the traditional electronic medical record system, centralized database storage is typically used. Once there is a problem with the data storage, it could cause data privacy disclosure and security risks. Blockchain is tamper-proof and data traceable. It can ensure the security and correctness of data. Proxy re-encryption technology can ensure the safe sharing and transmission of relatively sensitive data. Based on the above situation, we propose an electronic medical record system based on consortium blockchain and proxy re-encryption to solve the problem of EMR security sharing. Electronic equipment in this process is connected to the blockchain network, and the security of data access is ensured through the automatic execution of blockchain chaincodes; the attribute-based access control method ensures fine-grained access to the data and improves the system security. Compared with the existing electronic medical records based on cloud storage, the system not only realizes the sharing of electronic medical records, but it also has advantages in privacy protection, access control, data security, etc.

Keywords: EMR sharing; consortium blockchain; proxy re-encryption; privacy protection; IoT; BCoT

1. Introduction

1.1. Background

E-health appeared in the early 21st century; it refers to the use of modern information and communication technology, to provide medical services in the health sector [1]. Electronic medical records (EMRs) is a hot topic in the field of e-health [2]. According to one survey [3], as of March 2017, 86% of provincial hospitals and 75.6% of municipal hospitals have established EMR databases; only 32% of provincial hospitals and 35.2% of municipal hospitals have established electronic medical record information platforms. Although the goal of establishing electronic medical record systems has been achieved throughout the world, there is not enough medical record data sharing.

Electronic medical records (EMRs) involve systematic collections of patient- and population-based health data, which are stored electronically in digital format [4,5]. Effective EMR implementation and networking can save more than USD 81 billion annually, by improving medical efficiency and security [6].

At present, there are problems with the use of EMRs [7,8]. Firstly, data stored in the centralized system databases face many risks, such as hacker intrusions, data stealing, and artificial tampering of data that will endanger data security. Secondly, thousands of medical record data are saved by each regional hospital, separately, and unsystematic storage modes (e.g., the isolation of medical information systems) could result in the database information not being "connected". Without a unified system for the integration of resources, the sharing of medical data assets cannot be brought into full play, and the efficiency of the entire medical diagnosis is greatly reduced.

In a traditional EMR management system [9,10], the data are invisible, unmanageable, and uncontrollable for patients. Patients do not know which electronic medical record data are stored at a hospital, whether the electronic medical records stored in the cloud are used by a hospital, or shared by other hospitals, or whether the data are leaked. Moreover, even when electronic medical record data are leaked, they cannot be traced.

In January 2009, Satoshi Nakamoto invented bitcoin [11] and proposed a blockchain technology; then, Ethereum [12] expanded the concept of a smart contract, a consensus mechanism [13], and a point-to-point network [14]. Blockchain technology has the characteristics of peer-to-peer computing that is open and transparent, as well as communication security [15] that is difficult to tamper with and has multi-party consensus. It proposed a new solution to the disadvantages of the centralized network, and provided a new feasible paradigm for sharing electronic medical records [16]. Blaze et al. designed a proxy re-encryption scheme [17] in 1998, for the first time, to provide access control to data, to share encrypted data in a third-party server. These technologies bring new hope on how to achieve secure access control for EMR sharing.

1.2. EMR Sharing Advantages, threat Models, and Knowing Attacks

The EMR sharing system has the following advantages [18]:

1. The electronic medical record sharing system integrates electronic medical record data between hospitals so that it can be used and browsed across different hospitals, ensuring the availability and accuracy of electronic medical record data.

2. The establishment of a reasonable and effective electronic medical record sharing system facilitates a doctor's access to a patient's medical history, significantly reducing the costs associated with a patient's repeat examinations, and improving the efficiency of treatment.

3. Patients benefit from the sharing of electronic medical records because they can directly query their medical records, examination reports, and drug use in the hospital on the relevant networks.

4. The sharing of electronic medical records contributes to public health safety. The sharing of electronic medical records aids in the monitoring of the epidemic situation, allowing for early prevention and treatment of the epidemic situation, preventing the spread of large-scale infection, and preventing the occurrence of public health emergencies.

However, we may come across some potential threats and attacks while using the system. To make the proposed scheme more effective and safe, we analyze potential threats and attacks. The threat models and knowing attacks are as follows:

1. Data integrity issue [19].

In an insecure network environment, any information transmitted is vulnerable to tampering attacks, so that the data received by the receiver is not the original data. Data integrity is threatened. Therefore, it is necessary to ensure the integrity of the transmission data and protect it from tampering during transportation.

2. Illegal access issue [20].

Unauthorized access refers to the unauthorized use of network resources or the use of network resources in an unauthorized way. In this scheme, users are not allowed to operate other people's data in an unauthorized way.

3. Forgery and tampering [21].

If an attacker forges or tampers with the data stored in the shared electronic medical record system, it will have a significant impact on the entire system, resulting in massive data loss and errors. As a result, leveraging the non-tampering ability of the blockchain could significantly improve data security.

4. Replay Attack [22].

Replay attack means that the same information or data are repeatedly sent twice or more. If the receiver does not take relevant measures and continuously receives information, it would not be able to effectively identify that the data were received, which would lead to replay vulnerabilities.

5. Collusion Attack [23].

In the proxy re-encryption scheme, if the proxy colludes with the authorized party, the data and encryption key of the authorizer may be decoded. Therefore, in a proxy re-encryption scheme, we must verify whether the scheme can resist collision attacks.

Based on the above situation, this study proposes an EMR data sharing mechanism based on the advantages of anti-tampering and traceability of the consortium blockchain and the security authorization characteristics of proxy re-encryption. The proposed scheme, such that hospitals join the medical consortium blockchain, store the EMR data generated by patients in the consortium blockchain service center, and protect the data security according to the relevant national laws and regulations. At the same time, the chaincode functions are used to realize the EMR, and the proxy re-encryption is used for sharing and authorization. Furthermore, we write the attributes of users and devices into the digital certificate to provide different data access functions for users and devices with different attributes.

The rest of this paper is organized as follows. In Section 2, we review some preliminaries. The system model and detailed design are introduced in Sections 3 and 4, respectively. The analysis of the system is given in Section 5. Finally, Section 6 concludes the paper.

2. Preliminary

2.1. Blockchain and Smart Contract (Chaincode)

Blockchain is a kind of chain data structure that combines data blocks with time sequences [24]. A smart contract was a concept proposed by Nick Szabo in the 1990s [25], which is almost the same age as the internet. He defines a smart contract as a set of commitments defined in digital form, including agreements in which contract participants can execute these commitments automatically [26]. In this article, we refer to smart contracts as chaincodes, which are programs that are deployed and run on the blockchain network. The chaincode presets some conditions and rules to trigger the execution of the chaincode under certain events and conditions. The goal of the chaincode is to generate ledger data on the blockchain, which means that all operations on the blockchain data are completed by the chaincode. Moreover, security policies (including data encryption and decryption, data signature and signature verification, access control) will be automatically invoked through chaincodes.

2.2. Blockchain of Things

In ordinary internet of things devices [27], there are problems, such as poor data privacy and difficulty accessing data safely. The blockchain will have a significant impact on the internet of things due to its peer-to-peer, open and transparent communication, secure communication, difficulty to tamper with, and multi-party consensus. The encryption mechanism and data storage characteristics of blockchain just meet the security requirements of the internet of things. The integration of blockchain and the internet of things is called the blockchain of things (BCoT) in academia.

2.3. QR Code

A QR code [28], a kind of readable bar code, can identify the binary data recorded in it and obtain the information contained in it by scanning the QR code. A QR code's characteristics includes the following: it has large information capacity, high decoding reliability, and adopts certain security encryption measures. A QR code is widely used in near-field secure data exchange.

2.4. Elliptic Curve Digital Signature Algorithm (ECDSA)

In an elliptic curve system, we need to use a much shorter key than RSA to achieve the same security strength [29]. The elliptic curve digital signature algorithm (ECDSA) is the elliptic curve analog of the digital signature algorithm [30].

We assume that the sender sends M and signs M to a receiver, and the receiver needs to verify the signature to ensure the correctness of M. We assume that the sender has already generated their own key-pairs, $G(x, y)$ is the base point on the elliptic curve E/F_n (based on P256 curve), and satisfies $nG = 0$, n is a big prime number and is also the order of G.

Select a random number $Sk_A \in [1, n-1]$ as a private key, and a public key is $Pk_A = Sk_A G$.

Suppose the sender's key pair is (Sk_A, Pk_A).

2.4.1. Signature Generation

(1) Select a random number $r \in [1, n-1]$.
(2) Calculate $R = rG = (x_1, y_1)$.
(3) Calculate $S = (SHA(M) + Sk_A x_1) \cdot r^{-1}$ according to random number r, private key Sk_A, and $SHA(M)$, which is the secure hash value of message M.
(4) Send message M, and signature (R, S) to the receiver.

2.4.2. Signature Verification

(1) Receiver receives M and signature (R, S).
(2) Hash $SHA(M)$ according to message M.
(3) Use the sender's public key Pk_A to calculate $SHA(M)G/S + x_1 Pk_A/S$ and compare it, if it is equal to R:

$$\begin{aligned}
&SHA(M)G/S + x_1 Pk_A/S \\
&= SHA(M)G/S + x_1 Sk_A G/S \\
&= (SHA(M) + x_1 Sk_A)G/((SHA(M) + Sk_A x_1) \cdot r^{-1}) \\
&= rG \\
&= R
\end{aligned}$$

If it is equal, the signature verification is successful.

For reading convenience, in the following sections, we use $Sig_{SkX}(\cdot)$ to represent **the signature generation** and $Sign_X$ to represent the signature (R, S); we use $Ver_{PkX}(\cdot)$ to represent **the signature verification**.

2.5. The Algorithm Elliptic Curve Based Proxy Re-Encryption

In elliptic curve based proxy re-encryption [31], let E be an elliptic curve over a limited field F_q, where q is a large prime number, and G is a point on the elliptic curve E of order n. Let G_1, G_2 be two multiplicative cyclic groups of prime modulo n. Let $e : G_1 \times G_1 \to G_2$ be a bilinear map [32], $z = e(G_1, G_1) \in G_2$.

Key-Generation: let private key $a \in Z_n^*$ and a public key $Pk_a = aG \in$ point on E.

Encryption: generate an arbitrary number $r \in Z_n^*$, and output $C_a = (A, B) = (r \cdot Pk_a, z^r G + P_m)$, where $P_m = f(m)$ [33].

Re-Key-Generation: the re-encryption key $rk_{a \to b} = a^{-1}bG$ is generated by private key a and public key bG.

Re-Encryption: given $rk_{a \to b}$ and message C_a computes $C_b = (A', B') = (e(raG, a^{-1}bG), z^r G + P_m) = (z^{rb}, z^r G + P_m)$.

Decryption: (1) given ciphertext C_a and private key a, output $P_m = B - (e(A, Sk_a^{-1}G))G$; (2) given C_b and private key Sk_b, compute $P_m = B' - (A')^{1/b}G$; (3) compute $m = f^{-1}(P_m)$.

For reading convenience, in the next section, we use $Enc_{PkX}(\cdot)$ for **Encryption**, $reKeyGen(\cdot)$ for **Re-Key-Generation**, $reEnc_{A \to B}(\cdot)$ for **Re-Encryption**, and $Dec_{SkX}(\cdot)$ for **Decryption**.

3. The System Model

3.1. The Application Scenario

Figure 1 is the application scenario of the proposed scheme.

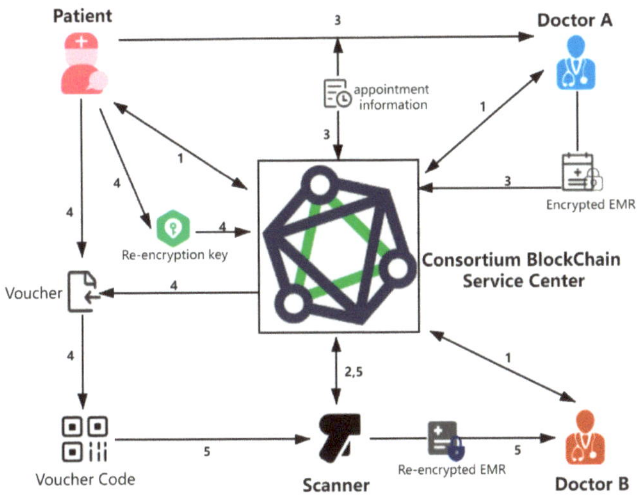

Figure 1. The System Framework.

In the proposed framework, there are four roles in the scheme.

(1) Consortium blockchain service center (CBSC): the distributed cloud storage and the peer-to-peer network of each hospital organization node constitute the consortium blockchain service center. In fact, the consortium blockchain service center is a decentralized network service. For the convenience of understanding, we call it CBSC. The consortium blockchain service center implements identity management and digital certificate issuance, and chaincode operation, data storage, and data legitimacy verification.

(2) Patient (P): the patient can make an appointment with a doctor. The patient uses the voucher and the scanner to authorize the doctor to view the patient's historical medical records. The patient authorizes the doctor to view his/her historical medical record through the combination of vouchers generated by proxy re-encryption and scanner.

(3) Doctor (D): the doctor obtains the patient's appointment information through the CBSC and generates the EMR with the encrypted fields by the patient's public key after diagnosis and stores it in CBSC.

(4) Scanner (SC): the scanner is used to scan the voucher code presented by the patient and request CBSC to obtain the re-encrypted ciphertext data through the voucher, which CBSC will convert the ciphertext through the re-encryption key.

1. **User registration phase:** all users (including patients and doctors) must register in the CBSC and obtain the private key and the corresponding X.509 digital certificate. The digital certificate contains the user's role attribute information (doctors, patients) and the user's public key.
2. **Device registration phase:** the scanner has to register in CBS to obtain the private key and the corresponding X.509 digital certificate. The digital certificate contains the hospital information and public key of the equipment.
3. **Appointment and EMR generation phase:** the patient makes an appointment to see doctor A and chains up the appointment information to CBSC. Then doctor A

obtains the appointment information from CBSC and makes a diagnosis. After the diagnosis, doctor A generates and chains up the encrypted EMR to CBSC.

4. **The generation of re-encryption key and access voucher phase:** the patient authorizes doctor B to view his/her historical EMR through proxy re-encryption and blockchain-network scanners. Firstly, the patient calculates the re-encryption key through his/her private key and doctor B's public key and submits it to CBSC. CBSC returns an authorization voucher to the patient. Then, the patient displays the voucher to doctor B in the form of a QR code.
5. **Scanning of access voucher and acquisition of re-encrypted EMR phase:** the scanner scans the QR code to obtain the voucher, and requests the re-encrypted EMR through the voucher from CBSC, which uses the re-encryption key to convert the historical EMR into the re-encrypted EMR. Finally, doctor B obtains the re-encrypted EMR through the scanner and decrypts it through the private key.

3.2. The Consortium Blockchain Service Center Architecture

CBSC refers to embedding the blockchain framework into the cloud-computing platform, making use of the deployment and management advantages of cloud service infrastructure to provide users with a convenient and high-performance blockchain ecological environment and ecological supporting services. In this paper, our CBSC architecture is shown in Figure 2, which is divided into the application layer, service layer, and consortium blockchain layer.

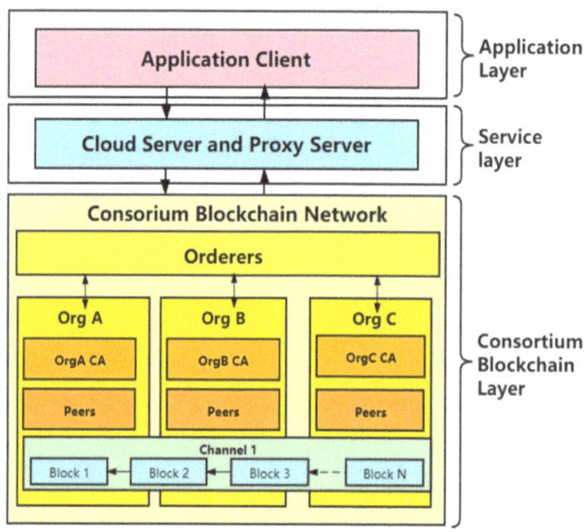

Figure 2. The consortium blockchain service center architecture.

Application layer and service layer: the application layer provides the EMR function service for patients, doctors, and scanners. It can interact with the background blockchain network via the API provided by the HTTPS server from the service layer. The service layer plays the role of middleware. Besides receiving and processing HTTPS requests from applications, the service layer must interact with the consortium blockchain layer directly to achieve the business logic by invoking a specific chaincode. In this way, the service layer can decouple applications and the data layer.

Consortium blockchain Layer: the consortium blockchain layer takes the Hyperledger Fabric technology as the core to provide blockchain services to users. Hyperledger Fabric technology is a new distributed infrastructure and computing mode, which uses chained data structure to verify and store data, uses distributed node consensus algorithm to generate and update data, uses cryptography to ensure the security of data transmission and access, and uses chaincode composed of automatic script code to program and operate data. In fabric, the order nodes use the consensus algorithm to sort the data and packages into blocks. Organizations that join the alliance chain verify and store data through peers.

4. The Proposed Scheme

4.1. X.509 Digital Certificate

CBSC uses X.509 digital certificate to identify the identity of users and devices in CBSC. The most basic information of the X.509 digital certificate includes the public key, the owner information of the public key, and the digital signature of CBSC. The "role" attributes in the digital certificate indicates the registered user or device attribute. If the value of "role" is "D", it is the doctor user certificate; if the value of "role" is "P", it is the patient certificate; if the "role" value is "S", it is the scanner device certificate. The scanner certificate is uniquely identified by the "Sid" attribute of the scanner; in the user certificate, the "Uid" attribute is the user's identification number and is used as the unique identifier of the use. For the patient, "Uid" means "Pid"; for the doctor, "Uid" means "Did". Patients can use the public key in the doctor's certificate combined with their own private key to realize the proxy re-encryption of data and then realize the access control of data. The examples of the digital certificate are as Figures 3 and 4 shown below.

```
Certificate:
  Data:
    Version: 3 (0x2)
    Serial Number:
      56:01:fb:3e:31:4b:79:0d:32:af:be:62:44:69:25:ba:47:5b:c2:b9
    Signature Algorithm: ecdsa-with-SHA256
    Issuer: C=CN, ST= Fujian, L = Xiamen, O = orgxmut.example.com, CN = ca.orgxmut.example.com
    Subject: OU = client + OU = orgxmut + OU = department1, CN = scanner1
    Subject Public Key Info:
      Public Key Algorithm: id-ecPublicKey
        Public-Key: (256 bit)
        pub:
          04:6b:b1:f1:b2:35:fd:ed:9f:5f:64:0e:a1:e3:a5:
          ca:ad:c0:51:3a:12:b8:75:b6:e8:2a:9f:8c:3b:c4:
          ad:7c:c3:dd:7a:5a:03:ab:f8:e4:dd:5a:61:71:12
        ASN1 OID: prime256v1
        NIST CURVE: P-256
     X509v3 extensions:
      X509v3 Subject Key Identifier:
        AC:72:C3:6A:96:60:E5:FB:76:2C:36:A7:8A:82:BA:D9:A5:DE:74:5F
      X509v3 Authority Key Identifier:
        keyid:04:A3:8D:BE:E5:50:3E:7A:3D:29:FC:49:A4:8D:2B:25:F5:7D:81:1E:4D:C8:08:C6:BE:96:85:48:CA:10:CE:AA
        1.2.3.4.5.6.7.8.1: {"attrs":{"role":"S" ,"Sid":"123454554" ,"hf.Type":"client"}}
   Signature Algorithm: ecdsa-with-SHA256
        30:44:02:20:0c:ab:c1:ee:20:f8:a0:52:0e:6f:d1:16:38:81:
        2f:b4:0e:6a:cc:6a:d4:45:2f:0b:f8:b3:d2:78:14:8e:7a:92:
        02:20:54:4e:a7:47:02:2b:ef:d8:9e:25:bf:4b:d3:60:db:51
```

Figure 3. The example of the scanner device digital certificate.

```
Certificate:
  Data:
    Version: 3 (0x2)
    Serial Number:
      76:f1:a2:7b:6c:5d:8f:a3:7a:af:be:62:54:0e:99:e3:5d:b2:94:4a
    Signature Algorithm: ecdsa-with-SHA256
    Issuer: C=CN, ST= Fujian, L = Xiamen, O = orgxmut.example.com, CN = ca.orgxmut.example.com
    Subject: OU = client + OU = orgxmut + OU = department1, CN = patient1
    Subject Public Key Info:
      Public Key Algorithm: id-ecPublicKey
        Public-Key: (256 bits)
        pub:
          07:5a:c2:e6:c9:92:af:bc:0e:3b:64:7e:b9:f2:b7:
          cc:cd:c0:55:3c:52:b8:75:b6:e8:2c:9f:8c:3b:c4:
          cd:7c:c3:dd:7c:5c:03:cb:f8:e4:dd:5c:65:75:52
        ASN1 OID: prime256v1
        NIST CURVE: P-256
     X509v3 extensions:
      X509v3 Subject Key Identifier:
        AC:72:C3:6A:96:60:E5:FB:76:2C:36:A7:8A:82:BA:D9:A5:DE:74:5F
      X509v3 Authority Key Identifier:
        keyid:04:A3:8D:BE:E5:50:3E:7A:3F:29:F9:49:A4:8D:2B:25:F5:7D:81:1E:4D:C8:08:C6:BE:96:85:48:CA:10:CE:AA
        1.2.3.4.5.6.7.8.1: {"attrs":{"role":"P" ,"Uid":"12345667QWZ","hf.Type":"client"}}
   Signature Algorithm: ecdsa-with-SHA256
        15:22:52:25:5a:ab:a1:ee:25:f8:a5:41:5b:6a:e1:16:38:81:
        2f:b2:5e:6a:aa:6a:e2:25:2f:5b:f8:b3:e2:78:12:8e:7a:73:
        52:25:52:2e:a7:27:52:2b:ef:e8:9e:25:bf:2b:e3:65:ea:28
```

Figure 4. The example of a user's digital certificate.

4.2. Deployment and Initialization of the Chaincode

The chaincode is event-driven, with state storage and programs running on the blockchain. The user realizes data access on CBSC through the chaincode. In this scheme, the chaincode data structure and function of the key information for the proposed architecture we define as Figure 5 follows, and Table 1 shows the detailed introduction of chaincode data structure. In particular, the field "state" of the appointment is divided into two states, namely "CREATE" and "FINISH". When an appointment is created, the patient needs to sign it and set the state to "CREATE". After the doctor's diagnosis, he/she needs to sign and set the state to "FINISH". In EMR, "EncryptedSickDetail" and "EncryptedDrugDetail" are encrypted fields. When authorizing EMR by using proxy re-encryption, these two fields are re-encrypted. We define the access data structure for access control, in which the "ReEncryptionkey" field is used to re-encrypt the EMR. Similarly, the "state" of access has three states. When the patient creates an access, the state is "CREATE". When the scanner scans the voucher and obtains the re-encrypted EMR, the state is "SCAN". When the doctor obtains the EMR, the state is "FINISH".

Table 1. Data Structure in CBSC.

Data Structure	Meaning
Doctor	The doctor data structure represents the role of the doctor user, where *Did* is the doctor's ID, the field *DName* is the doctor's name, the *Phone* field refers to the doctor's mobile phone number, the field *HospitalName* refers to the name of the doctor's Hospital, and the *HospitalId* field refers to the code of the doctor's hospital. These fields form the basic information of the doctor.
Patient	Patient data structure represents the role of the patient-user, where *Pid* is the patient's ID, the field *PName* is the patient's name, the field *Phone* refers to the patient's mobile phone number, the field *Address* refers to the name of the patient's home address.
Scanner	Scanner data structure refers to the basic information of the scanner equipment connected to the network, the *Sid* field refers to the equipment code, the *belongto* field refers to the organization, the *Sname* field refers to the equipment model name, and the field *activationTime* refers to the activation time of the scanner equipment.
Appointment	*Appointment* data structure stores the information of the patient's appointment with the doctor. The *BeginTime* field refers to the creation time of the appointment, *EndTime* refers to the end time of the appointment, the *Pid* field refers to the patient's ID, *Did* is the doctor's ID, *PatientSignature* refers to the patient's signature, *DoctorSignature* refers to the doctor's signature, the *state* field refers to the current state of the appointment
EMR	*EMR* data structure refers to the electronic medical record, which *Pid* refers to the patient's ID, *Did* refers to the doctor's ID, *createTime* refers to the creation time of the *EMR*, the *EncryptedSickDetail* field refers to the encrypted sick detail, the *EncryptedDrugDetail* field refers to the encrypted drug detail, and *doctorSignature* refers to the signature of the diagnostic doctor
Access	*Access* data structure records the information that the patient authorizes the doctor to view the historical EMR. *Pid* refers to the patient's ID, *Did* refers to the doctor's ID, *BeginTime* refers to the authorization start time, *EndTime* refers to the authorization end time, *reEncryptionKey* refers to the re-encryption key, *PatientSignature* refers to the patient's signature, *scannerSignature* refers to the scanner's device signature, *DoctorSignature* refers to the doctor's signature, and *state* refers to the current status of access.

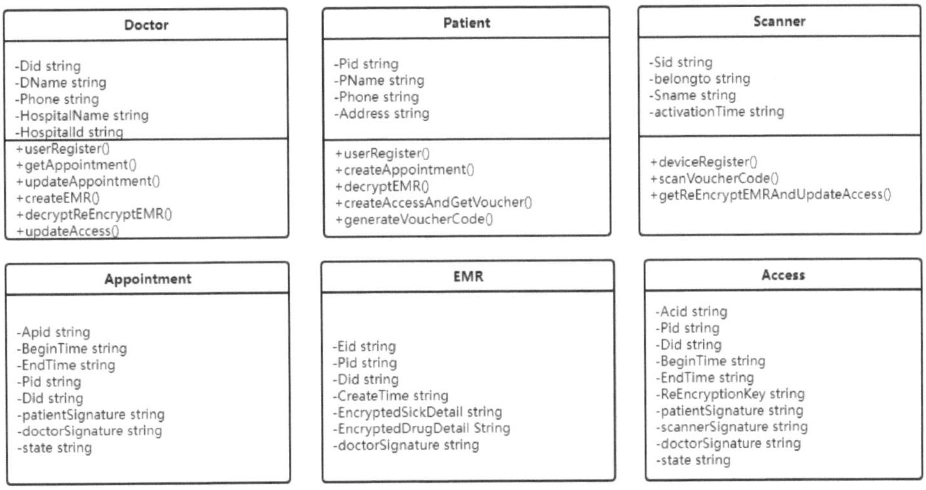

Figure 5. The chaincode data structure.

4.3. Notation

Table 2 shows the notations and their meaning.

Table 2. Notations.

Notation	Meaning
G	A generator for the elliptic group
Pk_X	X's public key
Sk_X	X's private key
$role$	the *role* is an attribute stored in the digital certificate of the user or device
$Cert_X$	X's X.509 digital certificate
M	X's original text
C_X	X's ciphertext
$timestamp$	The timestamp of the current time
$voucher$	Data voucher requesting re-encrypted data
$SickDetail$	The original detail of sick in EMR
$DrugDetail$	The original detail of drug in EMR
$reEncryptedSickDetail$	Re-encrypted sick detail
$reEncryptedDrugDetail$	Re-encrypted drug detail
$reEMR$	Re-encrypted EMR
$SHA(\cdot)$	Secure hash algorithm function
$f(m)$	The elliptic curve function $f(\cdot)$ of the embedding message m
P_m	Data embedded in the elliptic curve function $f(\cdot)$
$f^{-1}(P_m)$	The inverse function of elliptic curve function $f(\cdot)$, P_m is a point on elliptic curve
$info_X$	X's basic information (include name, phone number, and id, etc.)
$rk_{A\rightarrow B}$	Proxy re-encryption key generated by A and B

Table 2. *Cont.*

Notation	Meaning
$A \overset{?}{=} B$	Determine A if equal to B
$A\|\|B$	Concatenate A and B
$Sign_X$	X's signature
$Sig_{SkX}(M)$	The signature function, use the X's private key Sk_X to sign the message M.
$Ver_{PkX}(Sign_X)$	The verification function, use the X's public key Pk_X to verify the correctness of the signature $Sign_X$
$Enc_{PkX}(\cdot)$	The function of encryption with Pk_X
$reKeyGen(\cdot)$	The generation of re-encryption keys from A to B
$reEnc_{A\to B}(\cdot)$	The function of re-encrypting the ciphertext of A into the ciphertext that B can decrypt.
$Dec_{SkX}(\cdot)$	The function of decryption with the key Sk_X

4.4. User Registration Phase

Users (including patients and doctors) should register with CBSC. CBSC will generate a private key and digital certificate (including public key) for users, sign the private key and digital certificate, and issue them to users. After the user obtains the private key and digital certificate, the user will verify the correctness of the signature. After the signature is verified successfully, the user will call the chaincode to store the user details in CBSC. The flow chart of the user registration phase is shown in Figure 6.

- Step 1: the user chooses the attributes *role* and *Uid* to request for registration in CBSC.
- Step 2: when the CBSC receives the request, it selects a random number as the private key Sk_U and uses it with the generator for the elliptic group G to compute a public key $Pk_U = Sk_U G$.

  Then the CBSC signs the signature $Sign_{CBS}$ as follows:

$$Sign_{CBSC} = Sig_{SkCBSC}(role||Uid||Pk_U||timestamp) \quad (1)$$

- Step 3: CBSC generates digital certificates $Cert_U$ for users and sends Sk_U and $Cert_U$ to users.

$$Cert_U = (role||Uid||Pk_U||timestamp||Sign_{CBSC}) \quad (2)$$

- Step 4: after receiving the data, the user will verify the correctness of the signature. If it is correct, the certificate is legal.

$$(role\Big|\Big|Uid\Big|\Big|Pk_U\Big|\Big|timestamp) \overset{?}{=} Ver_{PkCBSC}(Sign_{CBSC}) \quad (3)$$

  and the user will store the private key Sk_U and certificate $Cert_U$.
- Step 5: the user selects the user information $Info_U$. For a patient, $Info_U$ means $(PName||Phone||Pid||Address||Phone||CreateTime)$, as shown in the above patient data structure; for doctors, $Info_U = (DName||Phone||DId||HospitalName||HospitalId|| CreateTime)$ as shown in the above doctor data structure. Then the user signs $(Info_U||timestamp)$

$$Sign_{U1} = Sig_{SkU}(Info_U||timestamp) \quad (4)$$

  and chain up $(Info_U||timestamp||Sign_{U1})$ to the CBSC.
- Step 6: when the CBSC receives the data from the user, it will verify the signature $Sign_{U1}$

$$(Info_U||timestamp) \stackrel{?}{=} Ver_{PkU}(Sign_{U1}) \quad (5)$$

If it holds, then the CBSC saves the data.

The chaincode of the pseudo-code of user registration is shown in Algorithm 1.

Algorithm 1. The chaincode pseudo-code of the user registration information.

```
user_cert, sk_user = CBSC.CA.CreateIndenity(role,Uid)
if user_cert != NULL
role←user_cert
if (role == "P")
Info←Patient
else if(role == "D")
Info←Doctor
userSign = ecdsa.sign(sk_user, Info, timestamp)
UserRegister(Info, timestamp, userSign)
```

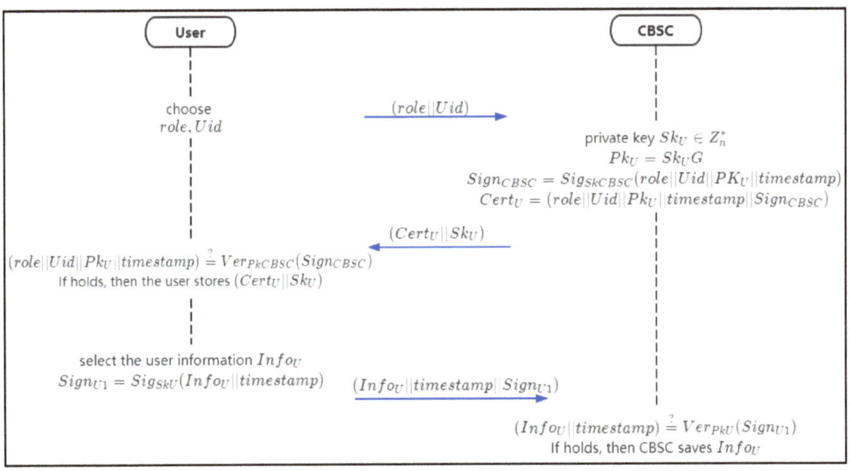

Figure 6. The flow chart of user registration.

4.5. Device Registration Phase

The scanner is registered in CBSC under the setting of person. Scanner devices are also registered in CBSC. CBSC will generate a private key and device digital certificate for the scanner, sign the private key and digital certificate, and issue them to devices. After the scanner obtains the private key and digital certificate, it will verify the correctness of the signature. After the signature is verified successfully, the scanner will call the chaincode automatically to store the device details in CBSC. The flow chart of the device registration phase is shown in Figure 7.

- Step 1: the scanner sets the attributes *role* and *Sid* to request for registration in CBSC.
- Step 2: when the CBSC receives the request, it selects a random number as the private key Sk_{SC} and uses it with the generator for the elliptic group G to compute a public key $Pk_{SC} = Sk_{SC}G$ for the scanner.

  Then the CBSC signs the signature $Sign_{CBSC2}$ as follows:

$$Sign_{CBSC2} = Sig_{SkCBSC}(role||Sid||Pk_{SC}||timestamp) \quad (6)$$

- Step 3: CBSC generates device digital certificates $Cert_{SC}$ and sends Sk_{SC} and $Cert_{SC}$ to users.

$$Cert_{SC} = (role||Sid||Pk_{SC}||timestamp||Sign_{CBSC2}) \quad (7)$$

- Step 4: after receiving the data, the scanner will verify the correctness of the signature. If it is correct, the certificate is legal.

$$(role||Sid||Pk_{SC}||timestamp) \overset{?}{=} Ver_{PkCBSC}(Sign_{CBSC2}) \quad (8)$$

  and the scanner will store the private key Sk_{SC} and certificate $Cert_{SC}$.
- Step 5: the scanner set up the basic information $Info_{SC}$ and signs $Info_{SC}$ with Sk_{SC},

$$Sign_{SC1} = Sig_{SkSC}(Info_{SC}||timestamp) \quad (9)$$

  and chain up $(Info_{SC}||timestamp||Sign_{SC1})$ to the CBSC.
- Step 6: when the CBSC receives the data from the scanner, it will verify the signature $Sign_{SC1}$

$$(Info_{SC}||timestamp) \overset{?}{=} Ver_{PkSC}(Sign_{SC1}) \quad (10)$$

  If it holds, then the CBSC saves the data.
  The chaincode of the pseudo-code of device registration is shown in Algorithm 2.

Algorithm 2. The chaincode pseudo-code of the device registration.

```
role,sid←Input()
(device_cert, sk_sc, timestamp,cbsSign) = CBSC.CA.CreateIndenity(role,sid)
pk_p = device_cert.getPublickey()
if ecdsa.verify(cbsSign,{role,sid,pk_p,timestamp})
scSign = ecdsa.sign(sk_sc, Info, timestamp)
deviceRegister(Info, timestamp, scSign)
```

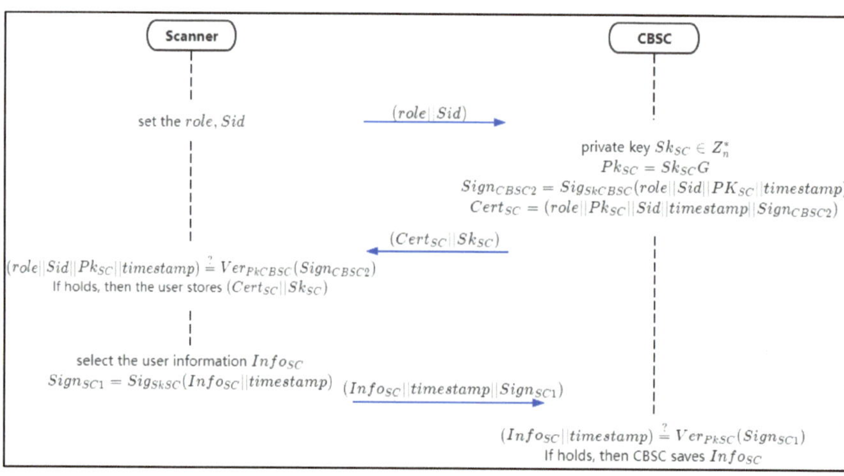

Figure 7. The flow chart of scanner device registration.

4.6. Appointment and EMR Generation Phase

In this phase, the patient makes an appointment to see the doctor and chains up the appointment information to the CBSC. Then doctor A obtains the appointment information from CBSC and makes a diagnosis. After diagnosis, doctor A generates and stores an encrypted EMR in CBSC. The flow chart of appointment and EMR generation is shown in Figure 8.

- Step 1: the patient sets the appointment data fields and sets the field "state" = "CREATE", then signs them with the patient's private key Sk_P,

$$Sign_{P1} = Sig_{SkP}(BeginTime||EndTime||Pid||Did||state||timestamp) \tag{11}$$

$$appointment = (BeginTime||EndTime||Pid||Did||state||Sign_{P1}) \tag{12}$$

  Then sends $(appointment||Sign_{P1}||timestamp)$ to CBSC.
- Step 2: when CBSC receives the data, it will get the data fields from *appointment* and verify the data signature,

$$(BeginTime\Big|\Big|EndTime\Big|\Big|Pid\Big|\Big|Did\Big|\Big|state\Big|\Big|timestamp) \overset{?}{=} Ver_{PkP}(Sign_{P1}) \tag{13}$$

  If holds, CBSC will save the data.
- Step 3: Doctor A requests the appointment form; the state field value is "CREATE" in CBSC,

$$Sign_{D1} = Sig_{SkD}(Did||state||timestamp) \tag{14}$$

  Then requests to CBSC with parameters $(Sign_{D1}||Did||state||timestamp)$
- Step 4: upon receiving the request, CBSC will verify the correctness of the signature,

$$(Did\Big|\Big|state\Big|\Big|timestamp) \overset{?}{=} Ver_{PkD}(Sign_{D1}) \tag{15}$$

If it holds, then the appointment will be found according to $(Did||state)$, and the appointment will be signed.

$$Sign_{CBSC3} = Sig_{SkCBSC}(appointment||timestamp) \tag{16}$$

and sent to doctor A with $(appointment||timestamp||Sign_{CBSC3})$.

- Step 5: after receiving the data, doctor A will verify the correctness of the data,

$$(appointment\Big|\Big|timestamp) \overset{?}{=} Ver_{PkCBSC}(Sign_{CBSC3}) \tag{17}$$

  If it is held, the patient's certificate is requested according to the patient's *Pid*.
- Step 6: when CBSC receives doctor A's request, it sends the patient's digital certificate $Cert_P$ to doctor A, and doctor A obtains the patient's public key Pk_P from the patient's digital certificate.
- Step 7: after the diagnosis, doctor A will generate the EMR for the patient and encrypt the *EncryptedSickDetail* (the encrypted sick detail filed) and *EncryptedDrugDetail* (the encrypted drug detail filed) fields in the EMR with the patient's public key.

$$EncryptedSickDetail = Enc_{PkP}(SickDetail) \tag{18}$$

$$EncryptedDrugDetail = Enc_{PkP}(DrugDetail) \tag{19}$$

  $Sign_{D2} = Sig_{SkD}(Eid||Pid||Did||CreateTime||EncryptedDetail||EncryptedDrugDetail)$
  $EMR = (Eid||Pid||Did||CreateTime||EncryptedSickDetail||EncryptedDrugDetail||Sign_{D2})$
  Then, set the "state" filed = "FINISH" in *appointment* and sign it,

$$Sign_{D3} = Sig_{SkD}(state||timestamp) \tag{20}$$

and doctor A sends $(EMR||Sign_{D2}||state||timestamp||Sign_{D3})$ to CBSC.

- Step 8: when CBSC receives doctor A's data, it will verify the correctness of the signature.

$$(Eid||Pid||Did||CreateTime||EncryptedDetail||EncryptedDrugDetail) \stackrel{?}{=} Ver_{PkD}(Sign_{D2}) \tag{21}$$

$$(state||timestamp) \stackrel{?}{=} Ver_{PkD}(Sign_{D3}) \tag{22}$$

If the signature is correct, it will store the EMR and update the "state" of the appointment.

Algorithms 3 and 4 show the appointment generation and EMR generation respectively.

Algorithm 3. The chaincode pseudo-code of appointment generation.

```
Did←Doctor.Did
Pid←Patient.Pid
BeginTime,EndTime←Input()
state←"CREATE"
timestamp←System.currentTime
if patient_cert != NULL
  pSign = ecdsa.sign(sk_p,Did,Pid,BeginTime,EndTime,timestamp)
  Appointment={Apid,BeginTime,EndTime,Pid,Did,pSign,state}
  createAppointment(Appointment, timestamp, pSign)
```

Algorithm 4. The chaincode of the pseudo-code of EMR generation.

```
Did←Doctor.Did
state←"CREATE"
timestamp←System.currentTime
if doctor_cert != NULL
    dSign1 = ecdsa.sign(sk_d,Did,state,timestamp)
    (Appointment,timestamp,cbsSign)=getAppointment(Did,state,,timestamp,dSign1)
    if ecdsa.verify(cbsSign,Appointment,timestamp)
        Pid←Appointment.Pid
        patient_cert = CBSC.CA.GetIndentityByAttribute(Pid)
        pk_p←patient_cert.getPublicKey
        EncryptedSickDetail = ECC.Encrypt(pk_p, sickdetail)
        EncryptedDrigDetail = ECC.Encrypt(pk_p, drugdetail)
        Did←Appointment.Did
        CreateTime←System.currentTime
        dSign2 = ecdsa.sign(sk_d,Pid,Did,CreateTime,EncryptedSickDetail,
                            EncryptedDrugDetail)
        EMR={Eid,Pid,Did,CreateTime,EncryptedSickDetail,EncryptedDrugDetail,dSign2}
        State←"FINISH"
        dSign3 = ecdsa.sign(sk_d,state,timestamp)
        createEMR(EMR)
        updateAppointment(state,timestamp,dSign3)
```

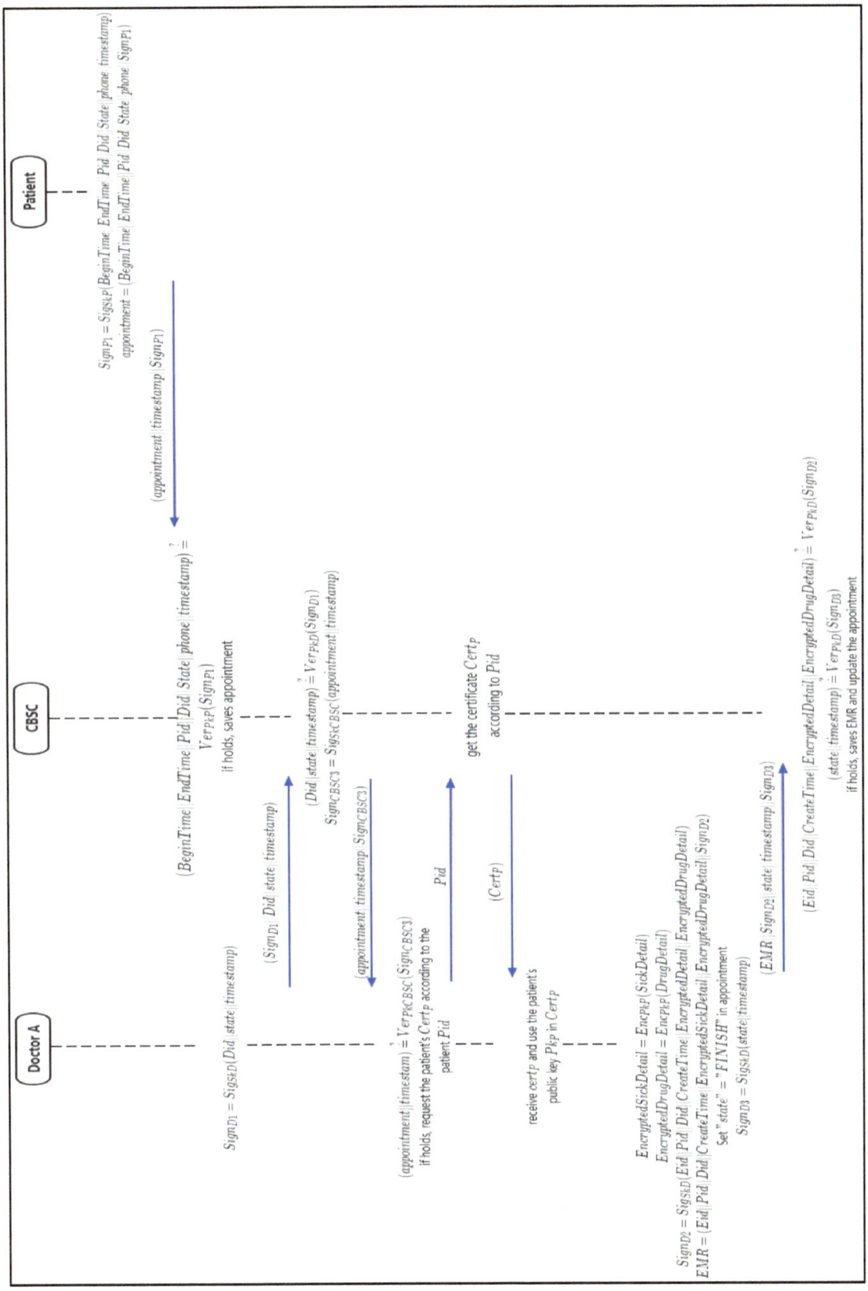

Figure 8. The flow chart of appointment and EMR generation.

4.7. The Generation of Re-Encryption Key and Access Voucher Phase

In this phase, the patient will combine his private key with the public key of doctor B to generate the re-encryption key and send it to CBSC. After verification, CBSC will return the access voucher to the patient. The patient displays the access voucher to doctor B in the form of a QR code. Figure 9 shows the data flow in this phase.

Step 1: The patient requests doctor B's certificate $Cert_D$ from CBSC through doctor B's ID *Did*.

$$Sign_{P2} = Sig_{SkP}(Did||timestamp) \tag{23}$$

Request with parameters $(Did||timestamp||Sign_{P2})$.

Step 2: when CBSC receives the request, it will verify the signature $Sign_{P3}$,

$$(Did||timestamp) \overset{?}{=} Ver_{PkP}(Sign_{P2}) \tag{24}$$

If it holds, then return doctor B's certificate $Cert_D$.

Step 3: the patient gets doctor B's public key Pk_D from $Cert_D$, then generates the re-encryption key $rk_{P\rightarrow D}$.

$$rk_{P\rightarrow D} = reKeyGen(Sk_P, Pk_D) \tag{25}$$

Then the patient sets the "state" = "CREATE" in *access*, signs the fields of *access*, and forms access data.

$$Sign_{P3} = Sig_{SkP}(Acid||Pid||Did||BeginTime||EndTime||rk_{P\rightarrow D}||state||timestamp) \tag{26}$$

$$access = (Acid||Pid||Did||BeginTime||EndTime||rk_{P\rightarrow D}||state||Sign_{P3}) \tag{27}$$

Then sends $(access||timestamp||Sign_{P3})$ to CBSC.

Step 4: when CBSC receives the data, it will verify the correctness of $Sign_{P3}$ firstly,

$$(Acid||Pid||Did||BeginTime||EndTime||rk_{P\rightarrow D}||state||timestamp) = Ver_{PkP}(Sign_{P3}) \tag{28}$$

If it is held, then save the *access* and generates the data digest and forms the data voucher,

$$digest = SHA(access) \tag{29}$$

$$voucher = (Acid||digest) \tag{30}$$

$$Sign_{CBSC4} = Sig_{SkCBSC}(voucher||timestamp) \tag{31}$$

Then sends $(voucher||timestamp||Sign_{CBSC4})$ to the patient.

Step 5: when the patient receives the data, the signature will be verified,

$$(voucher||timestamp) \overset{?}{=} Ver_{PkCBSC}(Sign_{CBSC4}) \tag{32}$$

If it holds, the voucher QR code will be generated and be shown to doctor B.

Algorithm 5 shows the generation of the registration of the re-encryption key and access voucher.

Algorithm 5. The chaincode of the pseudo-code of the generation of re-encryption key and access voucher.

```
Did←Input()
Pid←Input()
dortor_cert←CBSC.CA.GetIndentityByAttribute(Did)
pk_d←doctor_cert.getPublicKey
rkey = reKeyGen(sk_p,Pk_d)
BeginTime,EndTime←Input()
state = "CREATE"
timestamp = System.currentTime
if patient_cert != NULL
    pSign = ecdsa.sign(sk_p,{Pid,Did,BeginTime,EndTime,rkey,state,timestamp})
    access = {Acid, Pid,Did,BeginTime,EndTime,rkey,state,pSign}
    (voucher,timestamp,cbsSign) = createAccessAndGetVoucher(access,timestamp,pSign)
    If ecdsa.verify(cbsSign,voucher,timestamp)
        generateVoucherCode(voucher)
```

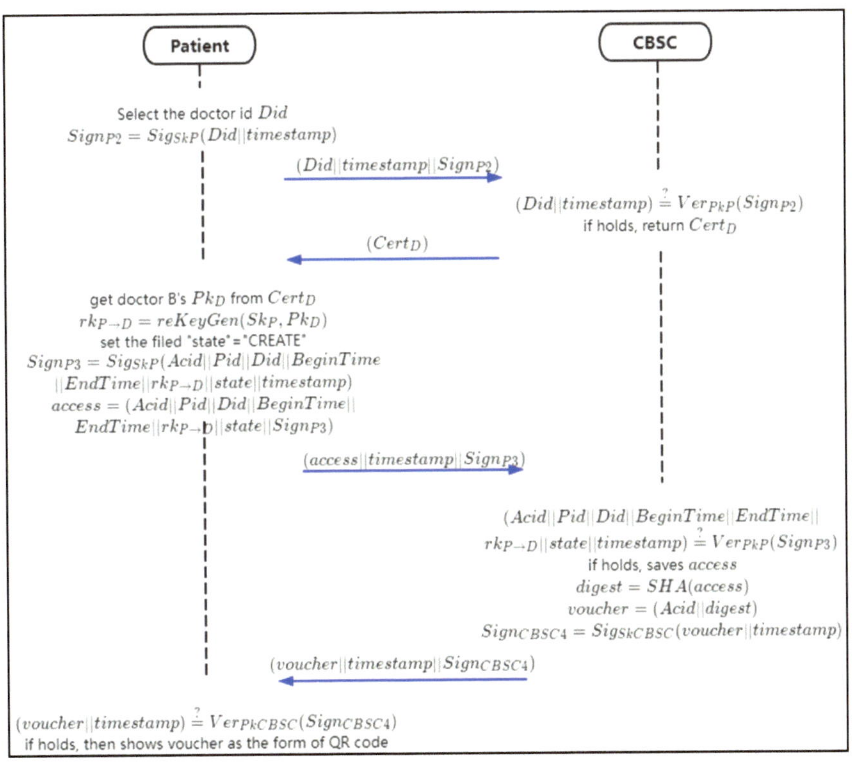

Figure 9. The flow chart of re-encryption key and access voucher generation.

4.8. Scanning of Access Voucher and Acquisition of Re-Encrypted EMR Phase

In this phase, the patient authorizes doctor B to view their historical EMR through proxy re-encryption and scanner.

Step 1: doctor B scans the QR code with the scanner and generates the signature

$$Sign_{SC} = Sig_{SkSC}(voucher||timestamp) \tag{33}$$

The scanner requests the data in CBSC $(voucher||timestamp||Sign_{SC})$.

Step 2: after receiving the request, CBSC first verifies the correctness of the signature,

$$(voucher||timestamp)\overset{?}{=}Ver_{PkSC}(Sign_{SC}) \quad (34)$$

If it holds, CBSC queries the *access* according to acid, hashes *access*, and compares it with the digest, $digest\overset{?}{=}SHA(access)$. If it is correct, update the *state* field *access* to "SCAN" and add the scanner's signature to *access*.

$$access = (Acid||Pid||Did||BeginTime||EndTime||rk_{P\to D}||state||Sign_{P3}||Sign_{SC}) \quad (35)$$

Then CBSC re-encrypts the encrypted field in EMR and sends it to doctor B through the scanner.

$$reEncryptedSickDetail = reEnc_{P\to D}(rk_{P\to D}, EncryptedSickDetail) \quad (36)$$

$$reEncryptedDrugDetail = reEnc_{P\to D}(rk_{P\to D}, EncryptedDrugDetail) \quad (37)$$

$$reEMR = (Pid||reEncryptedSickDetail||reEncryptedDrugDetail) \quad (38)$$

$$Sign_{CBSC5} = Sig_{SkCBSC}(reEMR||timestamp) \quad (39)$$

Then sends $(reEMR||timestamp||Sign_{CBSC5})$ to doctor.

Step 3: after receiving the data, doctor B will verify the signature.

$$(reEMR||timestamp)\overset{?}{=}Ver_{PkCBSC}(Sign_{CBSC5}) \quad (40)$$

If it is correct, decrypt the data, update the "state" file to "FINISH" in access and signs,

$$SickDetail = Dec_{SkD}(reEncrypyedSickDetail) \quad (41)$$

$$DrugDetail = Dec_{SkD}(reEncryptdDrugDetail) \quad (42)$$

$$Sign_{D4} = Sig_{SkD}(state||timestamp) \quad (43)$$

Request to update "state" field and add doctor B's signature into access to CBSC.

Step 4: when CBSC receives the data, it will verify the signature $Sign_{D4}$,

$$(state||timestamp)\overset{?}{=}Ver_{PkD}(Sign_{D}) \quad (44)$$

if holds, it will update the "state" field of access to "FINISH", and add doctor B's signature to access.

$$access = (Acid||Pid||Did||BeginTime||EndTime||rk_{P\to D}||state||Sign_{P3}||Sign_{SC}||Sign_{D4}) \quad (45)$$

Figure 10 shows the data flow; Algorithm 6 shows the chaincode in this phase.

Algorithm 6. The chaincode of the pseudo-code of scanning of access voucher and acquisition of re-encrypted EMR.

```
voucher←scanVoucherCode()
timestamp←System.CurrentTime
scSign = ecdsa.sign(sk_sc,voucher,timestamp)
(reEMR,timestamp,cbsSign) = getReEncryptEMRAndUpdateAccess(voucher,timestamp,scSign)
if ecdsa.verify(cbsSign,reEMR,timestamp)
    (sickDetail,drugDetail)=decryptReEncryptEMR(reEMR)
    state = "FINISH"
    dSign = ecdsa.sign(sk_d,state,timestamp)
    updateAccess(state,timestamp,dSign)
```

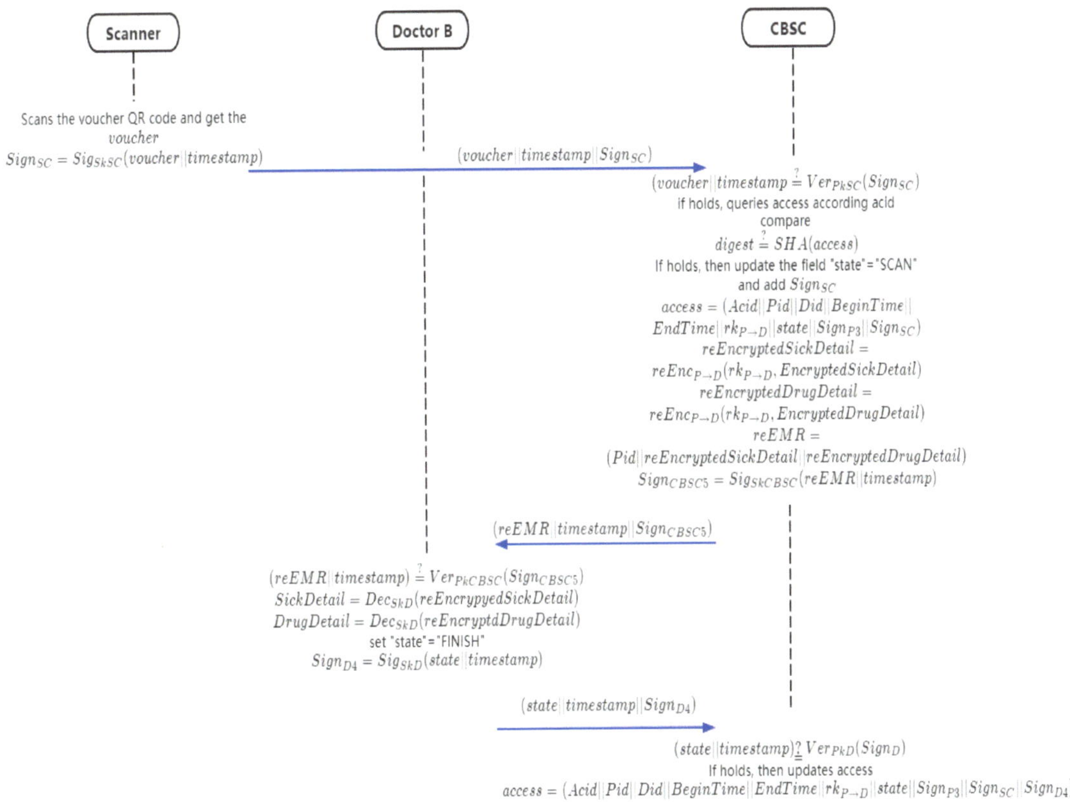

Figure 10. The flow chart of scanning of access voucher and acquisition of re-encrypted EMR.

5. Analysis

5.1. Data Integrity Analysis

In order to protect the integrity and security of the data, this paper uses an elliptic curve encryption algorithm (ECDSA) to sign the data.

Taking the user registration phase's signature as an example, the verification process of the signature $Sign_{CBSC}$ is as follows:

Because $Sign_{CBSC} = (R_{CBSC}, S_{CBSC}) = (rG, r + hash(role||Uid||Pk_U||timestamp||R_{CBSC})Sk_{CBSC})$; therefore, the verification is as follows:

$$\begin{aligned} E1: R_{CBSC} + hash(role||Uid||Pk_U||timestamp||R_{CBSC})Pk_{CBSC} \\ = rG + hash(role||Uid||Pk_U||timestamp||rG)Pk_{CBSC} \end{aligned} \quad (46)$$

$$\begin{aligned} E2: S_{CBSC}G &= (r + hash(role||Uid||Pk_U||timestamp||rP)Sk_{CBSC})G \\ &= rG + hash(role||Uid||Pk_U||timestamp||rG)Sk_{CBSC}G \\ &= rG + hash(role||Uid||Pk_U||timestamp||rG)Pk_{CBSC} \end{aligned} \quad (47)$$

When E1 equals E2, the signature verification is correct, which can prove the integrity of the data. Once the data are tampered with, then E1 will not match E2. In this way, the integrity of the data are guaranteed.

Scene: the malicious attacker intercepts the information transmitted from CBSC to the user and sends the modified information to the user.

Analysis: the attacker will not succeed. The user will verify the integrity of the data:

$$(role||Uid||Pk_U||timestamp) \overset{?}{=} Ver_{PkCBSC}(Sign_{CBSC}) \quad (48)$$

Because the attacker cannot obtain CBSC's private key, and if the data are modified, the signature verification will be incorrect, so the attacker will not be able to achieve the purpose of sending the modified data to the user.

5.2. Tamper-Resistant

Consortium blockchain technology can ensure that the chain-up information will not be tampered with. All of the chained data stored in a block will be constructed into a binary tree structure of the Merkle tree structure. As shown in Figure 11 below, the hash value between two data records in the Merkle tree will be directly concatenated as the input of the next binary tree. In this way, if an attacker attempts to change any of the data records, the root node of the Merkle tree will change greatly due to the characteristics of the SHA-256 encryption hash, so that other participants will find that the content has been changed when they verify the block information.

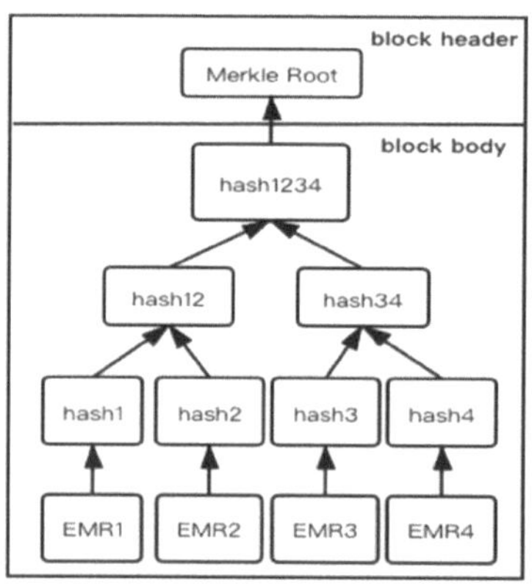

Figure 11. Block structure and Merkle tree in the proposed scheme.

5.3. Data Security Sharing and Access Control

In the process of authorizing the patient to share the historical EMR with the doctor, the proxy re-encryption algorithm is used to convert the original ciphertext into a ciphertext that can be decrypted by the doctor's private key. When the doctor wants to view the patient's historical medical record, the patient will generate the re-encryption key $rk_{P\to D}$ generated by the patient's private key sk_P and the doctor's public key Pk_D, and generate access data.

$$access = (Acid||Pid||Did||BeginTime||EndTime||rk_{P\to D}||state||Sign_{P3}) \quad (49)$$

Access specifies the usage time ($BeginTime$ and $EndTime$) of the re-encryption key $rk_{P\to D}$ and uses signature and $state$ to ensure the authenticity and usage record of the data.

Besides, when the users (including patients and doctors) and devices are registered, we write the role attribute into the user or device's digital certificate (where the patient's role data are "P", the doctor's role attribute is "D", and the scanner device's role attribute is "S"). When the user or scanner calls the chaincode to access data, the chaincode will obtain the attribute value in the user or device's digital certificate firstly, Different chaincode functions and data access are provided according to different attribute values. Table 3 shows the attribute-based access control in this paper.

Table 3. The attribute-based access control in the proposed scheme.

Role \ Function	User Register	Device Register	Appointment Generation	EMR Generation	Re-Encryption Key and Access Voucher Generation	Scanning of Access Voucher and Acquisition of Re-Encrypted EMR
P	✔		✔		✔	
D	✔			✔		✔
S		✔				✔

5.4. Blockchain of Things (BCoT)

In this paper, the voucher scanner will be connected to the blockchain. The blockchain-networking scanner will realize the data interaction with the blockchain network through the chaincode (smart contract). Once the chaincode reaches the trigger condition, it will be automatically executed and cannot be tampered with; and the attribute access control is used to specify the chaincode functions that can be accessed by the blockchain-networking scanner. It ensures the device's secure access to blockchain data.

5.5. Known Attacks

5.5.1. Resisting Replay Attack

Scene: the information transmitted between sender and receiver might be intercepted by malicious attackers. The attacker mimics the legitimate sender and then sends the same message to the target receiver again.

Analysis: because all information transmitted between sender and receiver is protected by ECDSA, and timestamp verification is added, the attacker cannot accurately timestamp parameters, so the attack will fail because the signature verification will fail. Since the information sent after each round will be changed, the same information cannot be sent twice. Therefore, a replay attack cannot succeed in this scheme.

5.5.2. Resisting Collusion Attack

Scene: suppose the doctor and the blockchain center (proxy) conspire to obtain the patient's private key.

Analysis: in this scheme, we use the proxy re-encryption scheme, which is collusion resistant. In the phase when the patient authorizes the doctor to view the patient's historical medical record, doctor B's public key Pk_D is used to calculate the re-encryption key $rk_{P\to D}$ through Pk_D and the patient's private key Sk_P.

$$rk_{P\to D} = Sk_P^{-1} Pk_D \tag{50}$$

CBS will convert the encrypted fields in EMR into data that can be decrypted by Sk_D through $rk_{P\to D}$. In the whole process, unless the patient exposes his private key, the doctor and the blockchain center (proxy) will not be able to obtain the patient's private key in collusion.

5.5.3. Man-in-the-Middle Attack

Scene: the attacker intercepts the transmitted data and then modifies the intercepted message and sends the modified message to the destination.

Analysis: all signatures in the proposed scheme contain a timestamp, and the scheme uses public-key cryptography as well as public and private keys. Therefore, the public key is used to encrypt the data, and the private key is used to sign the data. When the signature involves the private key, the attacker cannot modify the signature or the timestamp. Therefore, they cannot proceed with a man-in-the-middle attack because it is impossible to successfully modify the message.

6. Discussion

We test the performance of the blockchain service through the experimental simulation of the mentioned scheme in the following cluster host, as shown in Table 4:

Table 4. Experimental environment configuration.

Configuration	Detail
CPU	4-core CPU Intel® Xeon® Skylake 6133
Memory	8G
Network	4 Gbit/s
SSD	60 GB

The consortium blockchain service configuration is shown in Figure 12:

Figure 12. CBSC Configuration.

6.1. Send Rate

Caliper is a blockchain performance-testing framework that allows users to test different blockchain solutions using custom use cases, obtaining a set of performance test results. In this scheme, we use the caliper to test the performance of chaincode in five phases, and the results are shown in the figure below. We use 5665 transactions to test, and the sending rate is shown in Figure 13.

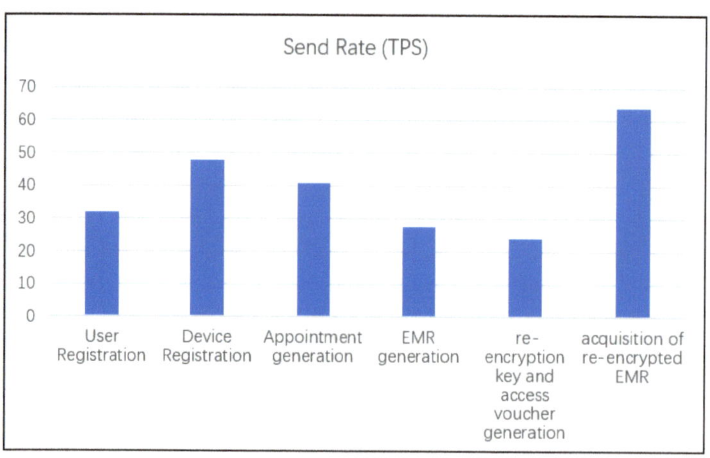

Figure 13. Send rate (TPS).

6.2. System Resource Consumption

The consumption of system resources is as follows. In the simulation experiment of this scheme, we set up two organization nodes, and each organization node consists of a peer node. At the same time, we set the order node, and its system resource consumption is shown in Table 5 below.

Table 5. System resource consumption.

Name	CPU% (max)	CPU% (avg)	Memory (max) (MB)	Memory (avg) (MB)	Traffic In (MB)	Traffic Out (MB)	Disc Write (KB)	Disc Read (KB)
peer0.org1	38.26	18	110	105	11.8	18.2	292	856
peer0.org2	3.19	1.86	54.6	48.5	0.19	0.133	292	68
orderer	1.68	0.26	29.4	27.9	0.1	0.193	288	236

6.3. The Function Comparison with Other Works

On the subject of patient data confidentiality, Yup et al. [34] investigated the use of blockchain technology in healthcare intelligence. The healthcare data gateway was created to ensure privacy and data access controls were proposed. Liang et al. [35] used blockchain technology to develop a mobile-based healthcare record sharing system, proposing a secure user-centric approach for access control and privacy via a channel formation scheme. Using blockchain, Sun et al. [36] proposed a distributed attribute-based signature scheme for medical systems and a record sharing protocol based on blockchain with supporting algorithms. Using distributed ledger technology, Yang and Li [37] developed an electronic medical record security architecture that improved interoperability between different organizations. The proposed scheme aims to establish a secure electronic medical record sharing system using blockchain smart contracts and cryptography algorithms. Table 6 below compares this work to other related works.

Table 6. The function comparison with other works.

Scheme	1	2	3	4	5
Yup et al. [34]	Yes	Yes	Yes	No	No
Liang et al. [35]	Yes	No	Yes	Yes	No
Sun et al. [36]	No	Yes	Yes	No	No
Yang and Li [37]	Yes	Yes	Yes	No	No
proposed scheme	Yes	Yes	Yes	Yes	Yes

1. Architecture. 2. Encryption key. 3. Access control. 4. Authorization sharing. 5. Traceability of access.

6.4. Computation Cost and Communication Cost

6.4.1. Computation Cost

The computation cost of the proposed scheme is shown in Table 7.

6.4.2. Communication Costs

The communication performance of the proposed scheme in the different networks is shown in Table 8.

L_{Cert} is the length of the certificate (5312 bits), L_{InfoU} is the length of $Info_U$ (192 bits), L_{InfoSC} is the length of $Info_{SC}$ (128 bits), L_{Sign} is the length of the signature (576 bits), L_{Sk} is the length of the private key (125 bits), L_{Ap} is the length of the appointment data structure (736 bits), L_{EMR} is the length of the electronic medical record data (768 bits), L_{Ac} is the length of access (800 bits), and L_{Other} is the length of other message data (32 bits).

Table 7. The computation cost of the proposed scheme.

Phase	Party	Cost
User registration	User	$T_{Cmp} + T_{Sig}$
	CBS	$T_{Sig} + T_{Cmp}$
Device registration	Scanner	$T_{Sig} + T_{Cmp}$
	CBS	$T_{Sig} + T_{Cmp}$
Appointment and EMR generation	Doctor A	$2T_{Enc} + 2T_{Sig} + T_{Cmp}$
	CBS	$T_{Sig} + 4T_{Cmp}$
	Patient	T_{Sig}
The generation of re-encryption key and access voucher	Patient	$2T_{Sig} + T_{Cmp} + T_{RkGen}$
	CBS	$2T_{Cmp} + T_{Sig} + T_H$
Scanning of access voucher and acquisition of re-encrypted EMR	Scanner	T_{Sig}
	Doctor B	$2T_{Dec} + T_{Sig}$
	CBS	$2T_{RkEnc} + 3T_{Cmp} + T_{Sig}$

Notes: T_P: polynomial function operation; T_{Cmp}: comparison operation; T_{Enc}: symmetric encryption operation; T_{Dec}: symmetric decryption operation; T_{Sig}: signature operation; T_{RkGen}: re-encrypt key operation; T_{RkEnc}: re-encryption operation.

Table 8. The communication performance of the proposed scheme in different network.

Phase \ Party	Message Length	4G (100 Mps)	5G (20 Gps)
1	$L_{Cert} + L_{InfoU} + L_{Sign} + L_{Sk} + 3L_{other}$	$6301/102{,}400 \approx 0.062$ ms	$6301/20{,}480{,}000 \approx 0.308$ us
2	$L_{Sign} + L_{Cert} + L_{InfoSC} + L_{Sk} + 3L_{Other}$	$6237/102{,}400 \approx 0.061$ ms	$6237/20{,}480{,}000 \approx 0.305$ us
3	$5L_{Sign} + 2L_{Ap} + L_{Cert} + L_{EMR} + 8L_{Other}$	$10{,}688/102{,}400 \approx 0.104$ ms	$10{,}688/20{,}480{,}000 \approx 0.522$ us
4	$3L_{Sign} + L_{Cert} + L_{Ac} + 7L_{Other}$	$8064/102{,}400 \approx 0.079$ ms	$8064/20{,}480{,}000 \approx 0.394$ us
5	$3L_{Sign} + 9L_{Other}$	$2016/102{,}400 \approx 0.02$ ms	$2016/20{,}480{,}000 \approx 0.098$ us

Notes: 1: User registration. 2: Device registration. 3: Appointment and EMR generation. 4: The generation of re-encryption key and access voucher. 5: Scanning of access voucher and acquisition of re-encrypted EMR.

7. Conclusions

Blockchain has brought about new ideas to internet medicine. Based on the consortium blockchain technology, this paper implements a sharing EMR system, realizing the following advantages and contributions:

1. The ECDSA signature algorithm and proxy re-encryption algorithm based on ECC were analyzed. Combined with attribute access control, the overall hierarchical architecture of sharing an EMR system based on consortium blockchain with secure access was designed and implemented.

2. According to different role attributes, different chaincodes were designed, and the data access control at the chaincode level was realized through attribute access control.

3. Through the proxy re-encryption algorithm, the data security sharing was realized. The sharing of privacy fields of electronic medical records could be used only with the authorization of patients, which greatly improves the control of patients over their own data.

4. The scanner device was connected to the blockchain network, and the blockchain-networking scanner interacted with the blockchain data through the chaincode, which was executed digitally and automatically. The blockchain-networking scanner used a specific chaincode according to its attributes to realize the device's secure access to blockchain data.

In future work, we will conduct additional research on the encryption and authorized access of electronic medical records, as well as investigate a more general solution in the form of a security pattern, particularly in fine-grained access to encrypted electronic medical records.

Author Contributions: Supervision and methodology, S.Z., J.L. and C.-L.C.; writing—original draft, W.C. and C.-L.C.; validation, C.-L.C. and Y.-Y.D.; surveyed related work, J.L., J.W. and W.C. All authors have read and agreed to the published version of the manuscript.

Funding: This research was funded by the Science and Technology Planning Project of Fujian Province, Young Teacher Education Research Project of FuJian (No. 2020H0023, 2019J05123, JAT190679, JT180435).

Institutional Review Board Statement: Not applicable.

Informed Consent Statement: Not applicable.

Data Availability Statement: The data used to support the findings of this study are available from the corresponding author upon request.

Conflicts of Interest: The authors declare no conflict of interest.

References

1. Albahri, O.S.; Albahri, A.S.; Mohammed, K.I.; Zaidan, A.A.; Zaidan, B.B.; Hashim, M.; Salman, O.H. Systematic Review of Real-time Remote Health Monitoring System in Triage and Priority-Based Sensor Technology: Taxonomy, Open Challenges, Motivation and Recommendations. *J. Med. Syst.* **2018**, *42*, 80. [CrossRef] [PubMed]
2. Enaizan, O.; Zaidan, A.A.; Alwi, N.H.M. Electronic medical record systems: Decision support examination framework for individual, security and privacy concerns using multi-perspective analysis. *Health Technol.* **2020**, *10*, 795–822. [CrossRef]
3. Li, C.; Xu, X.; Zhou, G.; He, K.; Qi, T.; Zhang, W.; Tian, F.; Zheng, Q.; Hu, J. Implementation of National Health Informatization in China: Survey About the Status Quo. *JMIR Med. Inform.* **2019**, *7*, 12238. [CrossRef] [PubMed]
4. Mackey, T.K.; Kuo, T.T.; Gummadi, B.; Clauson, K.A.; Church, G.; Grishin, D.; Obbad, K.; Barkovich, R.; Palombini, M. 'Fit-for-purpose?'—Challenges and opportunities for applications of blockchain technology in the future of healthcare. *BMC Med.* **2019**, *17*, 68. [CrossRef]
5. Halamka, J.D.; Alterovitz, G.; Buchanan, W.J.; Cenaj, T.; Clauson, K.A.; Dhillon, V.; Hudson, F.D.; Mokhtari, M.; Porto, D.A.; Rutschman, A.; et al. Top 10 Blockchain Predictions for the (Near) Future of Healthcare. *Blockchain Healthc. Today* **2019**, *2*. [CrossRef]
6. Hillestad, R.; Bigelow, J.; Bower, A.; Girosi, F.; Meili, R.; Scoville, R.; Taylor, R. Can electronic medical record systems transform health care? Potential health benefits, savings, and costs. *Health Aff.* **2005**, *24*, 1103–1117.
7. Sood, S.P.; Nwabueze, S.N.; Mbarika, V.W.; Prakash, N.; Chatterjee, S.; Ray, P.; Mishra, S. Electronic medical records: A review comparing the challenges in developed and developing countries. In Proceedings of the 41st Annual Hawaii International Conference on System Sciences, Waikoloa, HI, USA, 7–10 January 2008; p. 248.
8. Stafford, T.F.; Treiblmaier, H. Characteristics of a blockchain ecosystem for secure and sharable electronic medical records. *IEEE Trans. Eng. Manag.* **2020**, *67*, 1340–1362. [CrossRef]
9. Souther, E. Implementation of the electronic medical record: The team approach. *Comput. Nurs.* **2001**, *19*, 47–55.
10. Jung, E.Y.; Kim, J.; Chung, K.Y.; Park, D.K. Mobile healthcare application with EMR interoperability for diabetes patients. *Clust. Comput.* **2014**, *17*, 871–880. [CrossRef]
11. Nakamoto, S. A Peer-To-Peer Electronic Cash System. Bitcoin. 2008. Available online: https://bitcoin.org/bitcoin (accessed on 6 July 2021).
12. Wood, G. Ethereum: A secure decentralised generalised transaction ledger. *Ethereum Proj. Yellow Pap.* **2014**, *151*, 1–32.
13. Ongaro, D.; Ousterhout, J. In search of an understandable consensus algorithm. In Proceedings of the 2014 {USENIX} Annual Technical Conference ({USENIX}{ATC} 14), Philadelphia, PA, USA, 19–20 June 2014; pp. 305–319.
14. Schollmeier, R. A definition of peer-to-peer networking for the classification of peer-to-peer architectures and applications. In Proceedings of the First International Conference on Peer-to-Peer Computing, Linköping, Sweden, 27–29 August 2001; pp. 101–102.
15. Zyskind, G.; Nathan, O. Decentralizing privacy: Using blockchain to protect personal data. In Proceedings of the 2015 IEEE Security and Privacy Workshops, San Jose, CA, USA, 21–22 May 2015; pp. 180–184.
16. Azaria, A.; Ekblaw, A.; Vieira, T.; Lippman, A. Medrec: Using blockchain for medical data access and permission management. In Proceedings of the 2016 2nd International Conference on Open and Big Data (OBD), Vienna, Austria, 22–24 August 2016; pp. 25–30.
17. Blaze, M.; Bleumer, G.; Strauss, M. Divertible protocols and atomic proxy cryptography. In Proceedings of the International Conference on the Theory and Applications of Cryptographic Techniques, Espoo, Finland, 31 May–4 June 1998; pp. 127–144.
18. Zhang, X.X.; Zhang, L. Construction of Platform for Decision-making Management and Data Center in Hospitals. *Chin. Med. Equip. J.* **2012**, *33*, 79–81.

19. Sattarova Feruza, Y.; Kim, T.H. IT security review: Privacy, protection, access control, assurance and system security. *Int. J. Multimed. Ubiquitous Eng.* **2007**, *2*, 17–32.
20. Ding, S.; Cao, J.; Li, C.; Fan, K.; Li, H. A novel attribute-based access control scheme using blockchain for IoT. *IEEE Access* **2019**, *7*, 38431–38441. [CrossRef]
21. Lee, Y.; Rathore, S.; Park, J.H.; Park, J.H. A blockchain-based smart home gateway architecture for preventing data forgery. *Hum.-Cent. Comput. Inf. Sci.* **2020**, *10*, 9. [CrossRef]
22. Miao, F.; Pajic, M.; Pappas, G.J. Stochastic game approach for replay attack detection. In Proceedings of the 52nd IEEE Conference on Decision and Control, Firenze, Italy, 10–13 December 2013; pp. 1854–1859.
23. Nuñez, D.; Agudo, I.; Lopez, J. Proxy re-encryption: Analysis of constructions and its application to secure access delegation. *J. Netw. Comput. Appl.* **2017**, *87*, 193–209. [CrossRef]
24. Swan, M. *Blockchain: Blueprint for a New Economy*; O'Reilly Media, Inc.: Sebastopol, CA, USA, 2015.
25. Szabo, N. Formalizing and securing relationships on public networks. *First Monday* **1997**, *2*, 9. [CrossRef]
26. Wang, S.; Yuan, Y.; Wang, X.; Li, J.; Qin, R.; Wang, F.Y. An overview of smart contract: Architecture, applications, and future trends. In Proceedings of the 2018 IEEE Intelligent Vehicles Symposium (IV), Changshu, China, 22 October 2018; pp. 108–113.
27. Dai, H.N.; Zheng, Z.; Zhang, Y. Blockchain for Internet of Things: A survey. *IEEE Internet Things J.* **2019**, *6*, 8076–8094. [CrossRef]
28. Soon, T.J. QR code. *Synth. J.* **2008**, 59–78. Available online: https://foxdesignsstudio.com/uploads/pdf/Three_QR_Code.pdf (accessed on 6 July 2021).
29. Johnson, D.; Menezes, A.; Vanstone, S. The elliptic curve digital signature algorithm (ECDSA). *Int. J. Inf. Secur.* **2001**, *1*, 36–63. [CrossRef]
30. Chandrakar, P.; Om, H. A secure and robust anonymous three-factor remote user authentication scheme for multi-server environment using ECC. *Comput. Commun.* **2017**, *110*, 26–34. [CrossRef]
31. Thangam, V.; Chandrasekaran, K. Elliptic curve based proxy re-encryption. In Proceedings of the Second International Conference on Information and Communication Technology for Competitive Strategies, Udaipur, India, 4–5 March 2016; pp. 1–6.
32. Zhang, F.; Safavi-Naini, R.; Susilo, W. An efficient signature scheme from bilinear pairings and its applications. In Proceedings of the International Workshop on Public Key Cryptography, Singapore, 1–4 March 2004; pp. 277–290.
33. Udin, M.N.; Abd Halim, S.; Jayes, M.I.; Kamarulhaili, H. Application of message embedding technique in ElGamal elliptic curve cryptosystem. In Proceedings of the 2012 International Conference on Statistics in Science, Business and Engineering (ICSSBE), Langkawi, Malaysia, 31 December 2012; pp. 1–6.
34. Yue, X.; Wang, H.; Jin, D.; Li, M.; Jiang, W. Healthcare data gateways: Found health care intelligence on blockchain with novel privacy risk control. *J. Med. Syst.* **2016**, *40*, 218. [CrossRef]
35. Liang, X.; Zhao, J.; Shetty, S.; Liu, J.; Li, D. Integrating blockchain for data sharing and collaboration in mobile healthcare applications. In Proceedings of the 2017 IEEE 28th Annual International Symposium on Personal, Indoor, and Mobile Radio Communications (PIMRC), Montreal, QC, Canada, 8–13 October 2017; pp. 1–5.
36. Sun, Y.; Zhang, R.; Wang, X.; Gao, K.; Liu, L. A decentralizing attribute-based signature for healthcare blockchain. In Proceedings of the 2018 27th International Conference on Computer Communication and Networks (ICCCN), Hangzhou, China, 30 July–2 August 2018; pp. 1–9.
37. Yang, G.; Li, C. A design of blockchain-based architecture for the security of electronic health record (EHR) systems. In Proceedings of the 2018 IEEE International Conference on Cloud Computing Technology and Science (CloudCom), Nicosia, Cyprus, 10–13 December 2018; pp. 261–265.

sensors

Article

Enterprise Data Sharing with Privacy-Preserved Based on Hyperledger Fabric Blockchain in IIOT's Application

Chin-Ling Chen [1,2,3], Jiaxin Yang [1,*], Woei-Jiunn Tsaur [4,*], Wei Weng [1], Chih-Ming Wu [5] and Xiaojun Wei [1]

1 School of Computer and Information Engineering, Xiamen University of Technology, Xiamen 361024, China; clc@mail.cyut.edu.tw (C.-L.C.); wwweng@xmut.edu.cn (W.W.); xjwei@xmut.edu.cn (X.W.)
2 School of Information Engineering, Changchun Sci-Tech University, Changchun 130600, China
3 Department of Computer Science and Information Engineering, Chaoyang University of Technology, Taichung 41349, Taiwan
4 Computer Center, National Taipei University, New Taipei City 237303, Taiwan
5 School of Civil Engineering and Architecture, Xiamen University of Technology, Xiamen 361024, China; chihmingwu@xmut.edu.cn
* Correspondence: 2022031448@stu.xmut.edu.cn (J.Y.); wjtsaur@mail.ntpu.edu.tw (W.-J.T.)

Abstract: Internet of Things (IoT) technology is now widely used in energy, healthcare, services, transportation, and other fields. With the increase in industrial equipment (e.g., smart mobile terminals, sensors, and other embedded devices) in the Internet of Things and the advent of Industry 4.0, there has been an explosion of data generated that is characterized by a high volume but small size. How to manage and protect sensitive private data in data sharing has become an urgent issue for enterprises. Traditional data sharing and storage relies on trusted third-party platforms or distributed cloud storage, but these approaches run the risk of single-node failure, and third parties and cloud storage providers can be vulnerable to attacks that can lead to data theft. To solve these problems, this paper proposes a Hyperledger Fabric blockchain-based secure data transfer scheme for enterprises in the Industrial Internet of Things (IIOT). We store raw data in the IIoT in the InterPlanetary File System (IPFS) network after encryption and store the Keyword-index table we designed in Hyperledger Fabric blockchain, and enterprises share the data by querying the Keyword-index table. We use Fabric's channel mechanism combined with our designed Chaincode to achieve privacy protection and efficient data transmission while using the Elliptic Curve Digital Signature Algorithm (ECDSA) to ensure data integrity. Finally, we performed security analysis and experiments on the proposed scheme, and the results show that overall the data transfer performance in the IPFS network is generally better than the traditional network, In the case of transferring 5 MB file size data, the transmission speed and latency of IPFS are 19.23 mb/s and 0.26 s, respectively, and the IPFS network is almost 4 times faster than the TCP/IP network while taking only a quarter of the time, which is more advantageous when transferring small files, such as data in the IIOT. In addition, our scheme outperforms the blockchain systems mainly used today in terms of both throughput, latency, and system overhead. The average throughput of our solution can reach 110 tps (transactions are executed per second), and the minimum throughput in experimental tests can reach 101 tps.

Keywords: Chaincode; data security sharing; IPFS; Industrial Internet of Things (IIoT); Hyperledger Fabric blockchain; privacy-preserved

1. Introduction

1.1. Background

In recent years, with the rapid development of the Industrial Internet of Things (IIoT), the increase in productivity has also resulted in a significant challenge-data explosion. Enterprises in the industrial IoT use smart portable mobile terminals (e.g., drones, smartphones, electronic watches), sensors (e.g., infrared sensors, laser scanners, gyroscopes), and other large embedded devices (e.g., magnetic resonance imaging devices, traffic lights, avionics) to collect data, which is mostly unstructured and difficult to store and is maintained in traditional relational databases. Moreover, the significant amount of data also poses a challenge to IIOT terminal devices with limited computing and storage capacity. Unfortunately, the current industrial IoT still lacks a unified data management service due to the adoption of different data management systems among enterprises. In addition, the replicable and easily disseminated nature of data makes it difficult to trace the data shared among enterprises [1]. Moreover, enterprises store a large amount of data in third-party cloud storage platforms. This approach is at risk of single-node failure, and once the cloud storage server is attacked, there is a risk of data leakage, which brings serious asset loss to enterprises.

Several incidents related to the loss of stored data have already occurred in 2021 alone. Examples include the database breach of Ubiquiti, one of the world's largest IoT technology providers [2], the database breach of Société Internationale de Télécommunications Aéronautiques (SITA) [3], the data breach of the healthcare system IT company CapturеRx [4], and the data breach of Volkswagen and Audi, a famous car brand [5].

From the above events, the future needs a decentralized storage approach to provide data storage and sharing services for the enterprise. Fortunately, the nature of blockchain technology can provide a good solution for such decentralized storage systems. The blockchain consists of individual blocks connected by a hash function, and each block contains the hash value of the previous block, a timestamp, transaction data, etc. [6]. The blockchain can be considered a distributed ledger database, which is decentralized, open and transparent, tamperproof, and traceable, and it provides a safe and reliable storage method for enterprise data. However, each client of a blockchain system must maintain a complete copy of the block data [7], and storing a large amount of data directly in the blockchain can impose a high overhead on the client. Secondly, blocks are added to the blockchain on a time-based basis [8], and data for a product in industrial systems are often contributed by multiple participants at different points in time, and uploading them directly can place a significant load on the blockchain and may make the system congested.

Therefore, this paper proposes an enterprise data sharing scheme based on the Hyperledger Fabric [9] blockchain. Sensitive raw data collected by enterprises in IIOT are encrypted with Advanced Encryption Standard (AES) [10] and stored in the InterPlanetary File System (IPFS) [11], a peer-to-peer distributed file system that provides a high-throughput content-addressable block storage system. Then, we construct the data hash address returned by IPFS into a Keyword-index table to upload to the Hyperledger Fabric blockchain and share the data between enterprises through the Keyword-index table, which can effectively reduce the load on the blockchain network. In addition, we use Chaincode deployed in the blockchain to achieve a high degree of automation in the invocation of data, and the Elliptic Curve Digital Signature Algorithm (ECDSA) [12] to sign the messages transmitted by all parties to ensure data integrity.

In summary, our contributions are as follows.

(1) We designed a data security sharing and privacy protection framework to solve the blockchain load problem and achieve enterprise privacy protection of sensitive data while improving the scalability of the system.
(2) We designed a Keyword-index table for data sharing between enterprises and designed a Chaincode to realize the automatic call of data.
(3) Our scheme realizes mutual authentication of all parties and protection of data integrity.

1.2. Related Works

Few studies have focused on the use of the blockchain to share data between companies or organizations. We outlined the trends in related research, focusing on discussions that combine blockchain technology, as shown in Table 1.

Table 1. Comparison between the proposed and existing enterprise data sharing solutions.

Authors	Year	Objective	Technologies	Merits	Demerits
Teslya et al. [13]	2017	Proposed a blockchain-based IIOT trust information sharing platform	Blockchain, Smart Contracts, Smart-M3	It can single out the search for information in the detachment.	Leading to a slower rate of information entering the smart space
Wang et al. [14]	2018	To use blockchain double-link structure combined with proxy re-encryption for data sharing	Blockchain, Proxy re-encryption	The two chains store original data and transaction data separately, combined with proxy re-encryption to achieve reliable data sharing	There is no detailed experimental process to prove the actual effect of the program, and the safety analysis is not detailed enough
Zhang et al. [15]	2018	To realize data sharing in the electronic medical system through alliance chain	Blockchain, Bilinear maps and complexity assumptions	A detailed description of the sharing of medical data through the alliance chain, and detailed experimental analysis	Security analysis is not complete enough
Ra Lee et al. [16]	2019	To use blockchain registry and FHIR to share healthcare data	Blockchain, FHIR	Improve query efficiency by storing the registry in the blockchain and storing the original data in the database	There is no analysis process, and the plan is not complete enough
Kumar S et al. [17]	2020	To provide controlled access and secure transmission of patient health information between various healthcare organizations	Hyperledger Fabric	Investigated related literature and provided detailed algorithms and steps	No comparison with other programs, no experiments to demonstrate the actual effect of the program

Teslya et al. [13] proposed a blockchain-based IIOT trust information sharing platform, and such a combination made it possible to use the mechanisms implemented in the blockchain to solve the problems identified in the platforms for IoT. Wang et al. [14] proposed a blockchain dual-chain structure, where one chain stores the original data and the other chain stores the transaction data, combined with proxy re-encryption for reliable data sharing. The scheme proposed by Zhang et al. [15] describes in detail the implementation of data sharing in eHealth systems through federated chains, where multiple hospitals form a federated chain and use bilinear mapping to ensure secure data sharing, with a very detailed evaluation of the efficiency and cost. Ra Lee et al. [16] proposed a healthcare data-sharing framework using blockchain registries and Fast Healthcare Interoperability Resources (FHIR) technology to improve operability by storing registries on the blockchain while storing the raw data in a database. Kumar et al. [17] proposed a method for health data sharing using Hyperledger Fabric by calling chain codes and listing the specific algorithmic steps. However, the above schemes are still not perfect in terms of identity authentication and data traceability, and the communication parties do not have complete trust, and there is still the risk of data leakage.

Our scheme focuses on proposing a secure data sharing and privacy protection scheme based on blockchain and smart contract technology that allows data to be shared between authorized enterprises. We ensure that the entire process from data submission to data transfer is fully recorded in the blockchain and that ECDSA is used for data integrity protection. We use data stored independently of each other to increase the scalability of the blockchain network, reduce latency and energy costs, and improve the transmission effectiveness of the network. The perfect authentication and access control mechanism can ensure that the sensitive data of enterprises will not be leaked out and effectively protect the privacy of enterprises.

The contents of the rest of the paper are as follows: Section 2 presents some related knowledge of our study. Section 3 describes our proposed architecture and the detailed workflow. In Section 4, we analyze the security of the scheme. In Section 5, we evaluate the performance of the scheme. In Section 6, we perform an experimental test of the proposed scheme. Finally, Section 7 concludes the paper.

2. Preliminary

2.1. Elliptic Curve Digital Signature Algorithm (ECDSA)

Elliptic Curve Cryptography (ECC) [18] is a public key encryption algorithm based on elliptic curve mathematics. The main advantage of ECC is that it uses a smaller key length and provides a comparable level of security compared to the Rivest–Shamir–Adleman (RSA) encryption algorithm. ECDSA is a combination of ECC and DSA (Digital Signature Algorithm). Compared with RSA, the public key length of ECDSA is shorter and the encrypted message will be smaller, so the computation and processing time will be shorter, and the memory and bandwidth requirements will be smaller. The following is the signature and verification process of ECDSA:

Signing process: Suppose Alice wants to sign a message m, the elliptic curve parameter used is $D = (p, a, b, G, n, h)$, Alice needs to choose a random number between [1, N−1], d_A as Alice's private key, and generate a public key $Q_A = d_A G$. Alice will sign according to the following steps: First, Alice needs to generate a random number k between [1, N−1]; then calculate $(x_1, y_1) = kG$, $z = h(m)$, $r = x_1 \ mod \ n$, $s = (z + d_A r)k^{-1} \ mod \ n$. Finally, Alice sends the ECDSA signature result (*r*,*s*) to Bob.

Verification process: Bob needs to verify after receiving the signature. The verification steps are as follows: First, verify whether (*r*,*s*) is between [1, N−1]; then, calculate the following parameters: $z' = h(m)$, $u = 1z's^{-1} \ mod \ n$, $u = 2rs^{-1} \ mod \ n$, $(x_1', y_1') = u_1 G + u_2 Q_A$. Finally, check whether the equation $x_1' \ mod \ n = r$ is Equality: If they are equal, Bob confirms that the signature and message sent by Alice are correct.

2.2. Hyperledger Fabric

Hyperledger Fabric [19] is a platform for blockchain-based distributed ledger solutions that control transactions through chain codes, based on a modular architecture that provides a high degree of confidentiality, flexibility, and scalability. The transaction process is divided into the proposal phase, endorsement phase, sorting and packaging phase, and on-chain storage phase.

The Hyperledger Fabric architecture is mainly composed of the following parts: Client: the blockchain network used to connect members, through the SDK to call the proposal for transactions; Certificate Authorities (CA): Certificate and public and private key issuers, mainly responsible for the identity of the member's Management; Peers: can be divided into Leader Peer, Anchor Peer, Endorsing Peer, and Committing Peer, responsible for storing copies of the ledger and executing smart contracts (called Chaincode in Hyperledger Fabric) and approving transactions; Ordering Service (OS): responsible for collecting transaction of each channel and broadcasting to all Peers in the channel for storage on the chain. The specific workflow is shown in Figure 1:

(1) Proposal stage: The user sends the transaction to multiple Endorsing Peer through the Client.
(2) Endorsement stage: EP1, EP2, EP3 are Endorsing Peers. After receiving the proposal from the Client, it verifies and executes the endorsement, and then returns the endorsement result to the client.
(3) Sorting stage: The Client receives the endorsement results of all Endorsing Peers and compares whether they are consistent, and then sends the transaction to the Ordering Service, and the Ordering Service receives the transactions of all channels and sorts the transactions to form a block.
(4) On-chain stage: The Ordering Service broadcasts the packaged block to all Peers, and then the Peers verifies the transaction and uploads it to the blockchain.

2.3. Chaincode

The Chaincode in Hyperledger Fabric encapsulates the business logic used to create and modify business logic in the ledger, which can be written in different programming languages (e.g., Java, Go, and Node.js) [21]. Chaincode is created and executed by Peers to facilitate, authenticate, and enforce rules for reading, and the business logic of chain codes is defined by mutual agreement between members to read, execute, and update the current state of the ledger. When conditions are triggered, the chain code performs specific tasks, and the results of the transaction execution are submitted to the blockchain network and eventually attached to all Peers' copies of the ledger [22].

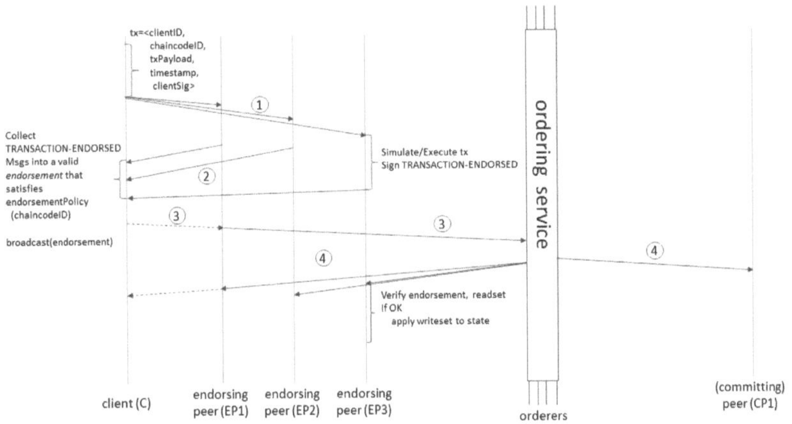

Figure 1. Hyperledger Fabric Transaction Flow [20].

2.4. *InterPlanetary File System (IPFS)*

The Interplanetary File System (IPFS) is a peer-to-peer distributed file system used as a distributed data storage service where the contents of the resources received by IPFS correspond to unique hashes 31. Any node in the IPFS network is independent and does not depend on other nodes, and the nodes do not need to trust each other, so there is no single point of failure as in traditional HTTP transfers. Data access will select the nearest node, greatly speeding up data transfer and reducing the storage footprint [23]. IPFS peer-to-peer transmission can effectively save network bandwidth, distributed files can effectively avoid potential DDoS attacks, and it has features, e.g., high throughput, content addressing, data anti-tampering, and de-duplication.

2.5. *BAN Logic*

BAN Logic [24] was first proposed by Burrows et al. It is a trust-based modal logic that is usually used to prove the correctness of a protocol or scheme. During the reasoning of BAN Logic, the trust of the subjects participating in the protocol changes and evolves as the message exchange evolves. When applying BAN Logic for analysis, it is divided into the following four steps:

(1) Describe the protocol messages that are not formally described in BAN Logic notation.
(2) Identify the initial assumptions from the protocol description and describe them in BAN Logic notation.
(3) List the goals to be achieved by the protocol.
(4) Using the messages, initial conditions, and inference rules in the communication, prove whether the protocol can achieve the goal.

2.6. *Threat Model*

The threat model is an important consideration for system security issues, and the following security issues are worth analyzing in our scenario.

(1) Mutual authentication of nodes [25]: Mutual authentication refers to two parties who authenticate each other simultaneously in an authentication protocol. To ensure data security, mutual authentication is the ideal solution among authentication schemes for transmitting sensitive data. The receiver/sender must be able to confirm the

legitimate identity of the sender/receiver of the message during the transmission of the message, and failure to do so will pose a great threat to data security.

(2) Data integrity [26]: Data integrity is the key to ensuring data accuracy and consistency, and to processing or retrieving data. Any accidental changes to data as a result of storage, retrieval, or processing operations can compromise data integrity. For messages transmitted in an unencrypted network environment that may be maliciously modified, data integrity may also be compromised.
(3) Data traceability [27]: Data loss due to malicious data theft by attackers, posing a serious threat to corporate assets.
(4) Non-repudiation [28]: Non-repudiation means that people cannot deny the act of sending a message and the content of the message due to the existence of some mechanism. The sender denies the message it sent, which can cause damage to the trust relationship between nodes.
(5) Resist known attack [29]: Cyber-attacks may cause data corruption or system paralysis, posing challenges to the stability and security of the system. Common attacks on blockchain networks are man-in-the-middle attacks, replay attacks, etc. For enterprises, cyber-attacks can disrupt critical infrastructure and lead to data leakage or corruption.

3. Proposed Scheme

3.1. System Architecture

In this article, we elaborate on the Hyperledger blockchain-based framework for enterprise data sharing and privacy protection, as shown in Figure 2. The framework is divided into three layers.

(1) Hyperledger Network Layer: This includes Peers, Ordering Service Node, Channels, and Certificate Authority (CA). The CA is responsible for issuing public and private keys and digital certificates. Administrators and Peers must be authenticated by the CA to become part of the blockchain network. The Channel is a private blockchain built based on data isolation and confidentiality. The data in the channel (e.g., Ledger information and member information) is known only to the members in the channel, and the data cannot be shared between different channels, and the channel mechanism ensures data sharing between different enterprises while protecting privacy. The Ordering Service Node only sorts and packs the transactions received in the channel and does not verify the legitimacy of the transactions, and then broadcasts the packaged transactions to all Peers in the channel. Peers are a network entity that maintains the ledger and runs the Chaincode to do read and write operations on the ledger.
(2) Client Layer: Each enterprise in the industrial IoT has an administrator who is responsible for interacting with the Hyperledger Blockchain Network. The administrator is connected to the blockchain network through the Client, which uses the SDK (Software Development Kit) to interact with the blockchain network and can access the ledger through Peers using the Chaincode, and the administrator needs to register through CA to participate in transactions in the system.
(3) Storage Layer: Enterprises that join the same channel will also join the channel's IPFS network, which is a distributed file system for storing and sharing data, and generating a hash address for storing data, which is a key component. The administrator stores the data encrypted using AES in IPFS while constructing a Keyword-index table of the hash addresses returned by IPFS to upload to the blockchain, which greatly increases the scalability of the system. Moreover, each data transaction carries a timestamp and is permanently stored in the blockchain.

The Hyperledger Fabric blockchain can be configured with multiple Channels, and multiple enterprises can join a single Channel or join different Channels for data sharing. Enterprise administrators create their own CA in the blockchain network and then apply for a public-private key and a digital certificate using the X.509 standard from the CA to provide signatures for transactions and to endorse the results of transactions. The digital certificate contains basic information, e.g., version number, serial number, business registration number, public key, enterprise tax number, and valid time.

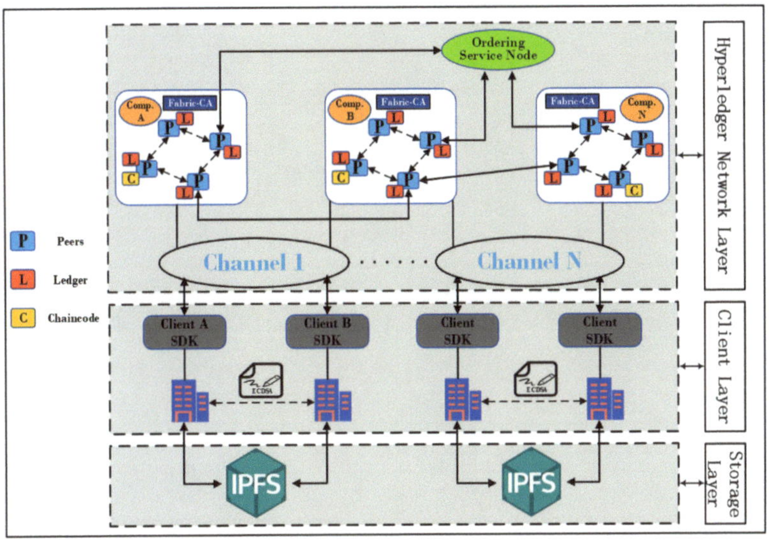

Figure 2. Hyperledger Fabric-based Framework for Enterprise Data Sharing and Privacy Protection.

3.2. Hyperledger Fabric Detailed Transaction Information Flow

Data sharing among industrial IoT companies is realized through Channel, and different companies' businesses may have crossover, so all parties can join the same Channel for data sharing. For example, Enterprise Administrator *A* (*A*) and Enterprise Administrator *B* (*B*) can join the same Channel for data sharing, which can be divided into four phases: registration phase, data storage phase, data query phase, and data transfer phase, and the workflow is shown in Figure 3.

Step 1. *A* and *B* need to register with the Fabric CA in Hyperledger Fabric Blockchain through the Client, and then the Fabric CA issues the public and private keys and digital certificates to the client of A and B, and the registration phase is completed.

Step 2. *B* uses the AES encryption algorithm to symmetrically encrypt and sign the sensitive and private data, and the encrypted data is saved to IPFS.

Step 3. IPFS returns the hash address of the encrypted data to the *B* Client.

Step 4. *B* Client receives the hash address and generates a Keyword-index table for the data keywords, and executes the Chaincode to add the Keyword-index table to the blockchain, and the data storage phase is completed.

Step 5. *A* sends a data access request containing keywords to the blockchain through the client.

Step 6. If the request initiated by *A* is legitimate and the queried data exists in the blockchain index directory, the blockchain network will return to *A* the required Keyword-index table containing the data hash address stored in IPFS, and the data query phase is completed.

Step 7. *A* initiates a data request to *B*, which contains *A*'s ID.

Step 8. *B* receives a request from *A*, requests *A*'s public key from the blockchain network, and verifies *A*'s request message, and then uses *A*'s public key to encrypt the AES key to form an encrypted key message and sends the message to *A*.

Step 9. After receiving the message, *A* uses its private key to decrypt it to obtain the AES key, obtains the encrypted data through the hash address provided by the

Keyword-index table in IPFS, and then uses the AES key to decrypt the encrypted data to obtain the original data. The transfer phase is completed.

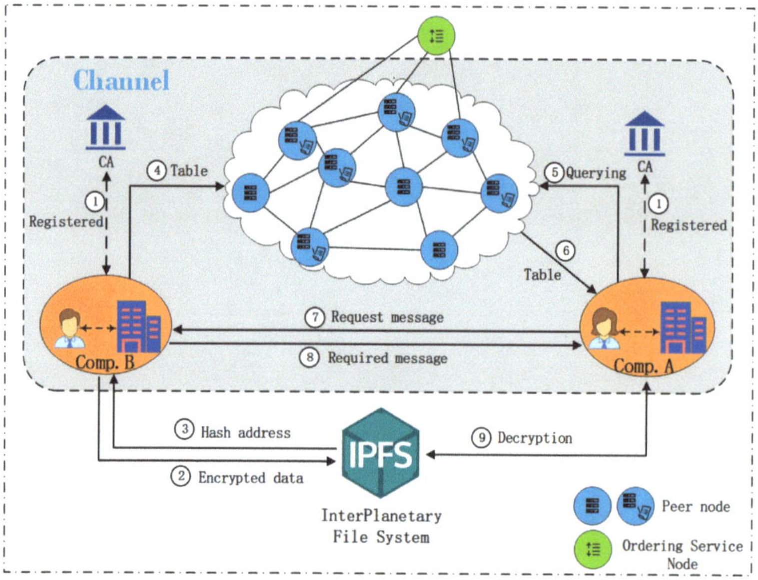

Figure 3. Enterprise data sharing process within Channel.

3.3. Registration Phase (Phase 1)

In this phase, enterprises joining the blockchain network for the first time need the administrator (X) to register with the CA via the client, and the registration phase proceeds as follows.

Step 1. X hashes the registration information to be submitted to obtain $h(M_{SUBMIT})$ and sends it to the CA.

Step 2. CA generates ECDSA private key d_X based on the X and calculates $Q_X = d_X \times G$. If the identity of the registered role is verified as legitimate, the CA sends (d_X, Q_X) and $Cert_X$ to the X Client, where $Cert_X$ contains a unique ID_X.

Step 3. X stores (d_X, Q_X) and $Cert_X$.

3.4. Data Storage Phase (Phase 2)

In this phase, B will store the original data in IPFS after AES encryption through the client, and at the same time, construct the hash address of the encrypted data returned from IPFS to generate a *Keyword-index table* (as shown in Figure 4) for uploading to Hyperledger Fabric Blockchain (*HFB*). The workflow is shown in Algorithm 1 and can be divided into four steps.

The Keyword-index table structure is as follows:

(1) "Holder": Name of the enterprise holding the data. "Signature": Signature of the enterprise administrator to ensure the integrity of the data. "ID": The unique identifier of the enterprise administrator, which is included in the certificate.

(2) "Hash_Address": The hash address of the data, the only basis for content addressing in an IPFS network. "Summary_Data": a brief description of the data content. "keyword": the search basis for the data requester to query this index table in the blockchain by keyword. "size": the size of the data. "type": the type of the data.

(3) "Timestamp": Indicates the time when this index table was added to the blockchain, added by Peers. "TXnumber": transaction serial number, which is the unique value of

the index table to search in the blockchain. "Version": including IPFS version number and Fabric version number.

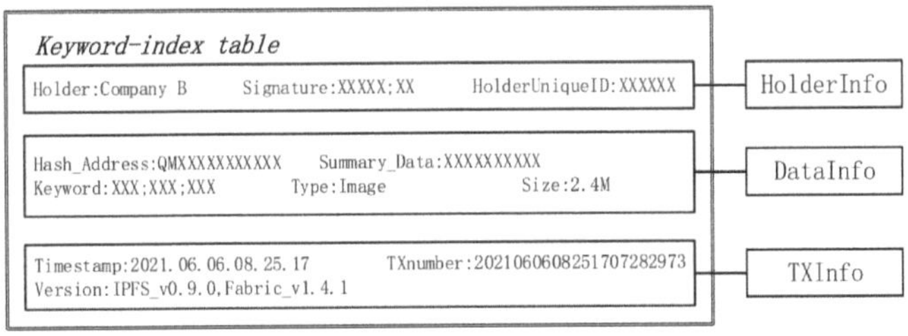

Figure 4. Keyword-index table structure.

Algorithm 1: Data Storing.

Input: DT_B;
Step 1: M_1;
B chooses a random number k_{B1};
$M_1 = (ID_B \parallel T_1 \parallel DT_B)$;
Sign M_1; call function $Sign(M_1, d_B, k_{B1})$, return (r_{B1}, s_{B1});
$C_{B1} \leftarrow E_{SK_B}(M_1)$;
Send C_{B1} to IPFS;
Step 2: hash address;
Upon receiving; check whether $T_{NOW} - T_1 \leq \tau$;
if $T_{NOW} - T_1 \leq \tau$ **then**
  store C_{B1} and generating *Hash_Data*;
  send *Hash_Data* to B;
end
Step 3: *Keyword-index table;*
Upon receiving; Generate $Keyword - index\ table$;
B chooses a random number k_{B2};
$M_2 = (ID_B \parallel T_2 \parallel Keyword - index\ table)$;
Sign M_2; call function $Sign(M_2, d_B, k_{B2})$, return (r_{B2}, s_{B2});
send M_2, (r_{B2}, s_{B2}) to *HFB*;
Step 4: Add to ledger;
Upon receiving; check whether $T_{NOW} - T_2 \leq \tau$;
call function $Verify(z_{B2}', r_{B2}, s_{B2})$, return result;
if $T_{NOW} - T_2 \leq \tau$ **then**
  if result = "valid" **then**
    call chaincode "Subfile", add $(M_2, Submit_B)$ to ledger;
  end
end

Step 1. B selects a random number k_{B1}, selects the data DT_B to be stored, and generates the message:

$$M_1 = (ID_B \parallel T_1 \parallel DT_B)$$

The function $Sign(M_1, d_B, k_{B1})$ is called to generate the signature (r_{B1}, s_{B1}) for M_1 (as shown in Algorithm 2) and then uses the AES encryption algorithm to symmetrically encrypt M_1 to get $C_{B1} = E_{SK_B}(M_1)$. C_{B1} and (r_{B1}, s_{B1}) are stored in IPFS.

Algorithm 2: Signature and Verification of the Scheme.

---

func *Sign* (M string, d string, k string)(r string, s string){
$(x, y) = k \times G$;
$z \leftarrow h(M)$;
$r \leftarrow x \bmod n$, $s \leftarrow k^{-1}(z + r \times d) \bmod n$;
return r, s
}
func *Verify* (z string, r string, s string, Q string) (result string){
$u_1 \leftarrow z \times s^{-1} \bmod n$
$u_2 \leftarrow r \times s^{-1} \bmod n$
$(x', y') = u_1 * G + u_2 * Q$
if $x' == r \bmod n$ **then**
return "valid"
else
return "invalid"
end
}

Step 2. IPFS first checks the validity of the timestamp to prevent replay attacks, then stores the message in the IPFS network and returns the hash address to *B*.

Step 3. *B* generates a *Keyword-index table* for data keywords, and then selects a random number k_{B2} to generate a message:

$$M_2 = (ID_B \parallel T_2 \parallel Keyword - index\ table)$$

The function $Sign(M_2, d_B, k_{B2})$ is called to generate the signature (r_{B2}, s_{B2}) for M_2 and then send M_2, (r_{B1}, s_{B1}) to *HFB*.

Step 4. *HFB* checks the validity of the timestamp $T_{NOW} - T_2 \leq \tau$ and then calls the function $Verify(z_{B2}{}', r_{B2}, s_{B2})$ (as shown in Algorithm 2) to verify the legitimacy of the signature. If $x_{B2}{}' = r_{B2}\ mod\ n$, the signature is legal. Then, the Chaincode "Subfile" (as shown in Figure 5) is executed to be added $(M_2, Submit_B)$ to the blockchain ledger, $Submit_B = h(r_{B2}, s_{B2})$. The data storage phase is completed.

```
type HashVal struct {
        Holder_Id       string
        Holder_Sig      string
        Holder_Name     string
        Hash_Data       string
}

func PutTable(stub shim.ChaincodeStubInterface, table HashVal) bool {
        tableJsonBytes, err := json.Marshal(table)                  if err != nil {
                return false }
        err = stub.PutState(table.Holder_Id, tableJsonBytes)     if err != nil {
                return false }
        return true
}

func (s *HashVal) add(stub shim.ChaincodeStubInterface, args []string)
pb.Response {
        if len(args) != 1 {
                return shim.Error("Incorrect number of arguments. Expecting 1")}
        var table HashVal
        err := json.Unmarshal([]byte(args[0]), &table)            if err != nil {
                return shim.Error("Unmarshal table failed")}
        _, exist := GetTableInfo(stub, table.Holder_Id)           if exist {
                return shim.Error("Id specified already exists")}
        flag := PutTable(stub, table)                              if !flag {
                return shim.Error("Add table failed")}
        return shim.Success([]byte("Add table succeed"))
}
```

Figure 5. Chaincode Subfile of the proposed scheme.

3.5. *Data Query Phase (Phase 3)*

Enterprises that need to query data, for example, *A*, need to submit a query request to *HFB*, and if the submitted request is legitimate, *HFB* will return the *Keyword-index table* to *A*. The workflow is shown in Algorithm 3 and can be divided into two steps.

Algorithm 3: Data Querying.

Input: M_{A-HFB};
Step 1: query
A chooses a random number k_{A1};
$M_{A-HFB} = (ID_A \parallel T_{A-HFB} \parallel Keywords)$;
call function $Sign(M_{A-HFB}, d_A, k_{A1})$, return (r_{A1}, s_{A1});
send M_{A-HFB}, (r_{A1}, s_{A1}) to *HFB*
Step 2: return
Upon receiving; check whether $T_{NOW} - T_{A-HFB} \leq \tau$;
call function $Verify(z_{A1}{}', r_{A1}, s_{A1})$, return result;
if $T_{NOW} - T_{A-HFB} \leq \tau$ **then**
  if result = "valid" **then**
    call chaincode "Querfile", return *Keyword-index table* to *A*;
  end
end

Step 1. *A* selects a random number k_{A1}, enters keywords, and draws up a query message:

$$M_{A-HFB} = (ID_A \parallel T_{A-HFB} \parallel Keyword)$$

and calls the function $Sign(M_{A-HFB}, d_A, k_{A1})$ to generate the signature (r_{A1}, s_{A1}) for M_{A-HFB} and then sends M_{A-HFB}, (r_{A1}, s_{A1}) to *HFB*.

Step 2. *HFB* checks the timestamp $T_{NOW} - T_{A-HFB} \leq \tau$ and calls the function $Verify(z_{A1}', r_{A1}, s_{A1})$ to verify the validity of the signature. If $x_{A1}' = r_{A1} \ mod \ n$, the signature is legal. Then, the Chaincode "Querfile" (as shown in Figure 6) is executed to be added $(M_{A-HFB}, Query_A)$ to the blockchain, $Query_A = h(r_{A1}, s_{A1})$. Moreover, then, the blockchain returns the *Keyword-index table* to *A*. The data query phase is completed.

```
type Query struct {
        Requester_Id        string
        Requester_Sig       string
        Requester_Name      string
        Keywords            string
}

func GetTable(stub shim.ChaincodeStubInterface, Requester_Id string, Requester_Sig)
(HashVal, bool) {
        var table HashVal
        tableJsonBytes, err := stub.GetState(Requester_Id)
                if err != nil || tableJsonBytes == nil {
                        return table, false }
                err = json.Unmarshal(tableJsonBytes, &table)
                if err != nil {
                        return table, false }
                return table, true
}

func (s *Query) query(stub shim.ChaincodeStubInterface, args []string) pb.Response {
                if len(args) != 1 {
                        return shim.Error("incorrect nums of args, expecting 1") }
                result, err := stub.GetState(args[0])
                if err != nil {
                        return shim.Error("query failed according to id") }
                if result == nil {
                        return shim.Error("get nothing according to id") }
                return shim.Success(result)
}
```

Figure 6. Chaincode Querfile of the proposed scheme.

3.6. Data Transfer Phase (Phase 4)

A request SK_B from *B*, and then the original data is obtained through SK_B. See the workflow Algorithm 4, which can be divided into 3 steps.

Algorithm 4: Data Transferring.

Input: M_{A-B};
Step 1: request;
A chooses a random number k_{A2};
$M_{A-B} = (ID_A \parallel T_{A-B} \parallel ID_B \parallel Hash\_Data \parallel Txnumber)$;
call function $Sign(M_{A-B}, d_A, k_{A2})$, return (r_{A2}, s_{A2});
$C_{A-B} \leftarrow E_{Puk_B}(M_{A-B})$
send C_{A-B}, (r_{A2}, s_{A2}) to *HFB*;
Step 2: return;
Upon receiving; $M_{A-B} = D_{Prk_B}(C_{A-B})$;
check whether $T_{NOW} - T_{A-B} \leq \tau$;
call function $Verify(z_{A2}', r_{A2, s_{A2}})$, return result;
if $T_{NOW} - T_{A-B} \leq \tau$ **then**
  if result = "valid" **then**
    B chooses a random number k_{B3};
    $M_{B-A} = (ID_B \parallel T_{B-A} \parallel ID_A \parallel SK_B)$
    call function $Sign(M_{B-A}, d_B, k_{B3})$, return (r_{B3}, s_{B3});
    $C_{B2} \leftarrow E_{Puk_A}(M_{B-A})$;
    send C_{B2}, (r_{B3}, s_{B3}), to *A*;
  end
end
Step 3: Descrypt data;
$M_{B-A} = D_{Prk_A}(C_{B-A})$; check whether $T_{NOW} - T_{B-A} \leq \tau$;
call function $Verify(z_{B3}', r_{B3}, s_{B3})$, return result;
if $T_{NOW} - T_{B-A} \leq \tau$ **then**
  if result = "valid" **then**
    store SK_B; get encrypted data in IPFS;
    $M_1 = D_{SK_B}(C_{B1})$;
  end
end

Step 1. A selects a random number k_{A2}, and draws up the requested message:

$$M_{A-B} = (ID_A \parallel ID_B \parallel T_{A-B} \parallel Hash\_Data \parallel TXnumber)$$

and calls the function $Sign(M_{A-B}, d_A, k_{A2})$ to generate the signature (r_{A1}, s_{A1}) for M_{A-B} and uses Puk_B to encrypt the M_{A-B} to obtain $C_{A-B} = E_{Puk_B}(M_{A-B})$. Then, send C_{A-B}, (r_{A2}, s_{A2}) to B.

Step 2. After receiving the requested message, B decrypts the message $M_{A-B} = D_{Prk_B}(C_{A-B})$ using Prk_B, and checks the validity of the timestamp $T_{NOW} - T_{A-B} \leq \tau$. Then, it calls the function $Verify(z_{A2}', r_{A2}, s_{A2})$ to verify the validity of the signature. If $x_{A2}' = r_{A2} \ mod \ n$, the signature is legal. Next, B selects the random number k_{B3} and adds SK_B to the message:

$$M_{B-A} = (ID_B \parallel ID_A \parallel T_{B-A} \parallel SK_B)$$

and calls the function $Sign(M_{B-A}, d_B, k_{B3})$ to generate the signature (r_{B3}, s_{B3}) for M_{B-A}. Afterward, B uses Puk_A to encrypt the M_{B-A} to obtain $C_{B-A} = E_{Puk_A}(M_{B-A})$. Then, send C_{B-A}, (r_{B3}, s_{B3}) to A.

Step 3. A decrypts the message $M_{B-A} = D_{Prk_A}(C_{B-A})$ using Prk_A to obtain SK_B and checks the validity of the timestamp $T_{NOW} - T_{B-A} \leq \tau$. Then, it calls the function $Verify(z_{B3}', r_{B3}, s_{B3})$ to verify the validity of the signature. If $x_{B3}' = r_{B3} \ mod \ n$, the signature is legal. Afterward, A obtains the encrypted data C_{B1} using the *Keyword-index table* in the IPFS network and decrypts the C_{B1} with SK_B, $M_1 = D_{SK_B}(C_{B1})$, $M_1 = (ID_B \parallel T_1 \parallel DT_B)$. The data transfer phase is completed.

4. Security Analysis

4.1. Mutual Authentication

In this article, we use BAN Logic to demonstrate the mutual authentication of the two parties in the data transmission process, mainly to ensure that the data is not tampered with during the transfer phase. Table 2 shows syntax and semantics are associated with BAN Logic.

Table 2. BAN Logic.

Symbol	Description
$P \mid\equiv X$	P trusts X or P is qualified to trust X
$P \triangleleft X$	P received a message containing X
$P \mid\sim X$	P has sent a message containing X
$P \mid\Rightarrow X$	P has jurisdiction over X
$\#(X)$	X is the latest
$P \overset{K}{\leftrightarrow} Q$	The shared key K is used for communication by P and Q.
$\overset{K}{\rightarrow} P$	P has X as a public key
$\{X\}_K$	The message X is encrypted by K
$< X >_Y$	This indicates that X combined with Y

In the data transfer phase, the scheme mainly authenticates the legitimacy of the identity of the communicating parties, and the main objectives of the scheme are:

$$G1: \ A| \equiv A \overset{K_{A-B}}{\longleftrightarrow} B$$

$$G2: \ A| \equiv B| \equiv A \overset{K_{A-B}}{\longleftrightarrow} B$$

$$G3: \ B| \equiv A \overset{K_{A-B}}{\longleftrightarrow} B$$

$$G4: \ B| \equiv A| \equiv A \overset{K_{A-B}}{\longleftrightarrow} B$$

$$G5: \ A| \equiv ID_B$$

$$G6: \ A| \equiv B| \equiv ID_B$$
$$G7: \ B| \equiv ID_A$$
$$G8: \ B| \equiv A| \equiv ID_A$$
$$G9: \ A| \equiv SK_B$$
$$G10: \ A| \equiv B| \equiv SK_B$$

In the data transfer phase, BAN Logic is applied to generate the idealized form as follows:

$$M: A \rightarrow B\ (\{ID_A, k_{A2}, M_{Request}\}_{Puk_B}, < h(ID_A, k_{A2}, M_{Request}) >_{K_{A-B}})$$

$$M: B \rightarrow A\ (\left\{ID_B, k_{B3}, SK_B, M_{Reply}\right\}_{Puk_A}, < h(ID_B, k_{B3}, SK_B, M_{Reply}) >_{K_{A-B}})$$

The proposed scheme is analyzed and the following assumptions made:

$$A1: \ B| \equiv \#(k_{A2})$$
$$A2: \ A| \equiv \#(k_{B3})$$
$$A3: \ A| \equiv B| \Rightarrow A \xleftrightarrow{K_{A-B}} B$$
$$A4: \ B| \equiv A| \Rightarrow A \xleftrightarrow{K_{A-B}} B$$
$$A5: \ A| \equiv B| \Rightarrow ID_B$$
$$A6: \ B| \equiv A| \Rightarrow ID_A$$
$$A7: \ A| \equiv B| \Rightarrow SK_B$$
$$A8: \ A| \equiv \xrightarrow{Puk_B} B$$
$$A9: \ B| \equiv \xrightarrow{Puk_A} A$$

According to the assumptions and rules of BAN Logic, the main proofs of the data transfer phase are as follows:

(1) The administrator of Enterprise B (B) authenticates the administrator of Enterprise A (A). Through *M1* and the seeing rule, we derive:

$$B \triangleleft (\{ID_A, k_{A2}, M_{Request}\}_{Puk_B}, < h(ID_A, k_{A2}, M_{Request}) >_{K_{A-B}}) \tag{1}$$

Through *M1* and the seeing rule, we derive:

$$B| \equiv \#(\{ID_A, k_{A2}, M_{Request}\}_{Puk_B}, < h(ID_A, k_{A2}, M_{Request}) >_{K_{A-B}}) \tag{2}$$

Through Formula (1), A9, and the message meaning rule, we derive:

$$B| \equiv A| \sim (\{ID_A, k_{A2}, M_{Request}\}_{Puk_B}, < h(ID_A, k_{A2}, M_{Request}) >_{K_{A-B}}) \tag{3}$$

Through Formulas (2)–(3), and the nonce verification rule, we derive:

$$B| \equiv A| \equiv (\{ID_A, k_{A2}, M_{Request}\}_{Puk_B}, < h(ID_A, k_{A2}, M_{Request}) >_{K_{A-B}}) \tag{4}$$

Through Formula (4) and the belief rule, we derive (G4)–(G8):

$$B| \equiv A| \equiv A \xleftrightarrow{K_{A-B}} B \tag{5}$$

$$B| \equiv A| \equiv ID_A \tag{6}$$

Through Formula (5), A4, and the jurisdiction rule, we derive (G3):

$$B| \equiv A \overset{K_{A-B}}{\longleftrightarrow} B \tag{7}$$

Through Formula (5), A6, and the jurisdiction rule, we derive (G7):

$$B| \equiv ID_A \tag{8}$$

(2) The administrator of Enterprise A (A) authenticates the administrator of Enterprise B (B). Through $M2$ and the seeing rule, we derive:

$$A \lhd (\left\{ID_B, k_{B3}, SK_B, M_{Reply}\right\}_{Puk_A}, < h(ID_B, k_{B3}, SK_B, M_{Reply}) >_{K_{A-B}}) \tag{9}$$

Through A2 and the freshness rule, we derive

$$A| \equiv \#(\left\{ID_B, k_{B3}, SK_B, M_{Reply}\right\}_{Puk_A}, < h(ID_B, k_{B3}, SK_B, M_{Reply}) >_{K_{A-B}}) \tag{10}$$

Through Formula (9), A8, and the message meaning rule, we derive:

$$A| \equiv B| \sim (\left\{ID_B, k_{B3}, SK_B, M_{Reply}\right\}_{Puk_A}, < h(ID_B, k_{B3}, SK_B, M_{Reply}) >_{K_{A-B}}) \tag{11}$$

Through Formulas (10) and (11), and the nonce verification rule, we derive:

$$A| \equiv B| \equiv (\left\{ID_B, k_{B3}, SK_B, M_{Reply}\right\}_{Puk_A}, < h(ID_B, k_{B3}, SK_B, M_{Reply}) >_{K_{A-B}}) \tag{12}$$

Through Formula (12) and the belief rule, we derive (G2), (G6), and (G10):

$$A| \equiv B| \equiv A \overset{K_{A-B}}{\longleftrightarrow} B \tag{13}$$

$$A| \equiv B| \equiv ID_B \tag{14}$$

$$A| \equiv B| \equiv SK_B \tag{15}$$

Through Formula (13), A3, and the jurisdiction rule, we derive (G1):

$$A| \equiv A \overset{K_{A-B}}{\longleftrightarrow} B \tag{16}$$

Through Formula (14), A5, and the jurisdiction rule, we derive (G5):

$$A| \equiv ID_B \tag{17}$$

Through Formula (15), A7, and the jurisdiction rule, we derive (G9):

$$A| \equiv SK_B \tag{18}$$

Through Formulas (6), (8), (16), and (17), it can be proven that, in the proposed scheme, A and B authenticate each other. Moreover, it can also be proven that the proposed scheme can authenticate the private key of A and B.

In the proposed scheme, B authenticates A by verifying:

$$x_{A2}{}' = r_{A2} \bmod n \tag{19}$$

If it passes the verification, B authenticates the legality of A. A authenticates the B by verifying:

$$x_{B3}' = r_{B3} \; mod \; n \tag{20}$$

If it passes the verification, A authenticates the legality of B. The data transfer phase of the proposed scheme, thus, guarantees mutual authentication between A and B.

4.2. Data Integrity

In our scheme, the parties' transaction data will be permanently stored in the blockchain network while we use ECDSA and AES to sign and encrypt the transactions to ensure data integrity. For example, in the data storage phase, B will sign and add timestamps to the Keyword-index table, and then upload it to the blockchain network, which will verify the timestamp $T_{NOW} - T_2 \leq \tau$ and signature (r_{B2}, s_{B2}) upon receipt. If the data is tampered with, then $x_{B2}' \neq r_{B2} \; mod \; n$, M_2 does not match (r_{B2}, s_{B2}), and the attacker's attack failed. During the data transfer phase, both communicating parties also verify the signature upon receipt of the message to ensure the integrity of the data. The data uploaded to the blockchain is stored in the blocks in a chained data structure, and each block is linked to the previous block through a hash function. If an attacker wants to tamper with the data, he needs to modify the hash value of the whole chain, which is unrealistic in a decentralized network system.

4.3. Traceability

Every transaction data stored in the blockchain is signed and stored forever, and the data is transparent and can be publicly verified. For example, the message is uploaded to the blockchain with the signed hash $Submit_B$ of B in the data storage phase. In the data query phase, the signature hash $Query_A$ of A is uploaded to the blockchain M_{Query}. All members can trace the transaction process and determine whether the data in the blockchain is legitimate by verifying $Submit_B \stackrel{?}{=} h(r_{B2}, s_{B2})$ and $Query_A \stackrel{?}{=} h(r_{A1}, s_{A1})$.

4.4. Non-Repudiation

In the proposed scheme, ECDSA's private key signature is used to achieve non-repudiation. The messages sent by all members of the system use their private keys to sign the messages. The receiver will verify the signature after receiving the message. If the verification is successful, the sender cannot deny the content of the message sent. Table 3 shows the non-repudiation of each role in the proposed scheme.

Table 3. The non-repudiation description.

Phase \ Item	Signature Value	Sender	Receiver	Signature Verification
Phase 2	(r_{B2}, s_{B2})	B	HFB	$Verify(z_{B2}', r_{B2}, s_{B2})$
Phase 3	(r_{A1}, s_{A1})	A	HFB	$Verify(z_{A1}', r_{A1}, s_{A1})$
Phase 4	(r_{A2}, s_{A2})	A	B	$Verify(z_{A2}', r_{A2}, s_{A2})$
	(r_{B3}, s_{B3})	B	A	$Verify(z_{B3}', r_{B3}, s_{B3})$

4.5. Resist Known Attacks

In this phase, we analyzed possible attacks against the system, including man-in-the-middle attacks and replay attacks.

4.6. Man-in-the-Middle Attack

The attacker tries to intercept and tamper with the message content. In our scheme, both communicating parties do not have to send their public keys to each other, and both parties can query each other's public keys in the blockchain network, which can effectively prevent the attacker from intercepting the message and replacing the public key. For example, A uses B's public key to encrypt the message $C_{A-B} = E_{Puk_B}(M_{Request})$. B uses A's public key to encrypt the message $C_{B-A} = E_{Puk_A}(M_{Reply})$. The attacker does not know the private keys of the communicating parties, so he cannot decrypt the message.

4.7. Replay Attacks

The messages of the two communicating parties may be intercepted by the attacker, who pretends to be a legitimate sender and sends the same message to the recipient. In our scheme, a timestamp mechanism is added between two parties of arbitrary communication to prevent such attacks. For example, during the data transfer phase, B sends a timestamped message M_{B-A} to A, who checks that the timestamped message $T_{NOW} - T_{B-A} \leq \tau$ is valid. Even if the attacker tampers with the timestamp data, because B has added a timestamp $TB - A$ ($s_{B3} = k^{-1}(z_{B3} + r_{B3}d_B)$ $mod\ n$, $z_{B3} = h(M_{B-A})$) to the signature (r_{B3}, s_{B3}), A checks that the timestamp does not match the signature and the replay attack fails.

5. Performance Evaluation

5.1. Communication Cost

Table 4 shows the communication cost analysis of the proposed scheme. In the Gigabit Ethernet environment, the maximum transmission speed is 1 Gbps, and in the 10 Gigabit Ethernet environment, the maximum transmission speed is 10 Gbps. We assume that the ECDSA signature and key are 160 bits, the asymmetric encryption message is 1024 bits, the hash function operation requires 160 bits, and the length of other messages (such as ID and timestamp, etc.) is 80 bits. Taking the data transmission phase with the highest communication cost as an example, A needs to send two signatures, one hash, one asymmetric encrypted message, and one other message to B. The total size is 2×160 bits + 160 bits + 1024 bits + 80 bits = 1584 bits. B needs to send two signatures, one hash, one asymmetric encrypted message, and one other message to A. The total size is 2×160 bits + 160 bits + 1024 bits + 80 bits = 1584 bits. The total communication cost for the data transfer phase is 1584 bits + 1584 bits = 3168 bits, which takes 3.168 µs in a Gigabit Ethernet communication environment and 0.3168 µs in a 10 Gigabit Ethernet environment. These communication costs are very low, so the proposed scheme has good communication performance.

Table 4. Analysis of the communication cost.

Phase \ Item	Message Length	Rounds	Gigabit Ethernet (1 Gbps)	10 Gigabit Ethernet (10 Gbps)
Phase 1	560 bits	2	0.56µs	0.056µs
Phase 2	560 bits	1	0.56µs	0.056µs
Phase 3	560 bits	2	0.56µs	0.056µs
Phase 4	3168 bits	2	3.168µs	0.3168µs

5.2. Computation Cost

In Table 5, we analyze the computational cost of each phase of the scheme, and we use asymmetric encryption and decryption, hashing operations, and addition, subtraction, multiplication, and division operations as the basis for the computational cost analysis. Taking the data transfer phase (phase 4) with the highest computational cost as an example, A requires three encryption/decryption operations, two comparison operations, five modular operations, two hash operations, eight multiplication operations, and one signature operation. B requires two encryption/decryption operations, two comparison operations, five modular operations, two hash operations, eight multiplication operations, and one signature operation. Thus, in our scheme, the calculation cost is acceptable.

5.3. Blockchain Architecture Comparison

There are currently at least four types of blockchain networks: public blockchains, private blockchains, consortium blockchains, and hybrid blockchains [30]. Private blockchains are too centralized and not suitable for data sharing between enterprises but only for resource management within a specific individual or company. We summarize the comparison between two blockchain platforms, Hyperledger Fabric, a typical representative of consortium blockchains, and Ethereum, a typical representative of public blockchains, as shown in Table 6.

Table 5. Analysis of the communication cost.

Phase \ Party	*A*	*B*	*HFB*
Phase 2	N/A	$1T_{E/D}+4T_{Mod}+2T_H+8T_{Mul}+1T_{Sym}+2T_{Sig}$	$2T_{Cmp}+3T_{Mod}+4T_{Mul}+1T_H$
Phase 3	$1T_{Cmp}+2T_{Mod}+1T_H+4T_{Mul}+1T_{Sig}$	N/A	$2T_{Cmp}+3T_{Mod}+4T_{Mul}+1T_H$
Phase 4	$3T_{E/D}+2T_{Cmp}+5T_{Mod}+2T_H+8T_{Mul}+1T_{Sig}$	$2T_{E/D}+2T_{Cmp}+5T_{Mod}+8T_{Mul}+2T_H+1T_{Sig}$	N/A

Notes: $T_{E/D}$: Encryption/Decryption operation, T_H: Hash function operation, T_{Mul}: Multiplication operation, T_{Cmp}: Comparison of operation, T_{Mod}: Modular operation, T_{Sym}: Symmetric encryption operation, T_{Sig}: Signature operation.

Table 6. Comparison between Ethereum and Hyperledger Fabric.

	Hyperledger Fabric	**Ethereum**
Category	Consortium Blockchain	Public Blockchain
Description	Generic blockchain platform	Modular blockchain platform
Consensus algorithms	Practical Byzantine Fault Tolerance (PBFT)	Proof of Work (PoW)
Throughput	≥1000 TPS	≥25 TPS
Decentralization	Partial de-centralization	Completely decentralization
Fault tolerance rate	33%	50%
Success rate	Lower	Higher
Privacy	Yes	No
Authentication	Yes	No
Scalability	Yes	No
Pluggability	Yes	No

From the above table, we can see that although Ethereum has advantages in fault tolerance and the transaction success rate, Hyperledger Fabric outperforms Ethereum in terms of the average transaction latency, throughput, privacy, and scalability, and the modularity and channel design of Hyperledger Fabric is more suitable for data sharing among enterprises [31].

5.4. Function Comparison

Table 7 shows the comparison of the previous scheme with our proposed scheme. It can be seen from the table that this scheme overcomes the shortcomings of the previous scheme.

Table 7. Functionality comparison of previous schemes and the proposed scheme.

Authors	Year	Objective	1	2	3	4	5	6
Teslya et al. [13]	2017	Proposed a blockchain-based IIOT trust information sharing platform	Y	N	Y	N	N	N
Wang et al. [14]	2018	To use blockchain double-link structure combined with proxy re-encryption for data sharing	Y	N	N	N	Y	N
Zhang et al. [15]	2018	To realize data sharing in the electronic medical system through alliance chain	Y	N	N	N	Y	Y
Ra Lee et al. [16]	2019	To use blockchain registry and FHIR to share healthcare data	Y	Y	N	N	Y	Y
Kumar et al. [17]	2020	To provide controlled access and secure transmission of patient health information	Y	N	Y	N	Y	N
Ours	2021	Propose a solution for corporate privacy-preserved and data sharing based on Fabric blockchain	Y	Y	Y	Y	Y	Y

Notes: 1: Blockchain architecture, 2: Data integrity, 3: Mutual Authentication, 4: No-repudiation, 5: Scalability, 6: Off-chain storage; (Y) Yes; (N) No.

We compare with previous studies, which, as mentioned before, have some flaws, we improve on the flaws based on the previous work. Teslya et al. [13] proposed a blockchain-based IIOT trust information sharing platform. Tis paper describes a possible way of integrating IoT and blockchain technology to solve these problems. To this end, an architecture combining the Smart-M3 information sharing platform and the blockchain platform was developed. However, it only proposes an architecture without detailed deployment and experiments. Furthermore, this paper does not discuss the security of the architecture and lacks a theoretical basis. This paper has detailed instructions on system security and experimental testing. Wang et al. [14] proposed a new data-sharing scheme based on blockchain technology, which combines the blockchain with a double-chain structure and proxy re-encryption to achieve safe and reliable data sharing. This scheme only discusses the security and complexity of the system and does not have actual experimental tests. In addition, this scheme cannot detect the source of data leakage, and the segmentation of data blocks lacks theoretical support. We experimentally test the proposed scheme, and we employ signature technology to ensure data traceability. Zhang et al. [15] proposed a blockchain-based security and privacy-preserving PHI sharing (BSPP) scheme for improving diagnosis in e-health systems. However, the scheme uploads all PHI data to the blockchain network, which undoubtedly increases the overhead of the blockchain client, and the scheme does not provide discussion on the authentication between the nodes of the Consortium chain. Our solution uses off-chain storage of data to reduce the overhead of the blockchain network, and we use ban logic proof to prove the identity security among the nodes. Ra Lee et al. [16] proposed a standards-based sharing framework SHAREChain that combines two properties to deal with reliability and interoperability issues and Kumar et al. [17] proposed a healthcare application based on a blockchain network with a Hyperledger fabric structure, but these two schemes do not discuss the security and efficiency of the system. We illustrate the safety of the proposed scheme, and the experimental results show the good efficiency of our scheme.

We propose a complete system framework focusing on the security issues of enterprise data transmission among blockchain networks. Therefore, we focus on the security issues of the system in the analysis phase. Compared to previous studies, our solution has advantages in data privacy, data protection, and data traceability, which are lacking in previous solutions, while we adopt off-chain storage of data to increase the scalability of the blockchain network and use digital signature technology to ensure the authenticity of data. Finally, the experimental results show that our scheme has good efficiency and practical prospects.

6. Deployment and Testing

In this section, we experimentally evaluate the proposed scheme. The HyperLedger Fabric uses Docker container technology to run the Chaincode containing the system application logic. The Fabric framework includes a certificate authority (CA), order nodes, and peer nodes. Each peer node maintains a full copy of the blockchain data, and in our scenario, the Enterprise Administrator is the peer node. Each peer node uses CouchDB to maintain the state of its ledger. All nodes are run in their own Docker containers. We deployed 6 peer nodes, 1 order node and 2 CA on a server with Intel Core i7-8700 @3.2GHz CPU and 8 GB RAM. The operating system of the physical machine is Ubuntu 18.04.2 LTS. The version of Fabric we used is v1.4.

6.1. *Performance of File Transmission in Traditional and IPFS Network*

In this experiment, we compared the file upload performance of different file sizes in traditional TCP/IP networks and IPFS networks. Because the number of IoT devices is huge in industrial IoT networks and each device can only generate a small amount of data, we chose files of sizes 1, 5, 10, 50, and 100 MB, respectively. As can be seen from Figure 7, The latency of the IPFS network was 0.11, 0.26, 0.95, 10.55, and 25.34 s, while the latency of the TCP/IP network was 0.25, 0.88, 1.55, 10.71, and 25.65 s, respectively. In terms of transmission speed, the transmission speed of IPFS is 9.09, 19.23, 10.52, 6.73, and 4.94 MB/s while the transmission speed of TCP/IP is 4.05, 5.68, 6.45, 4.88, and 3.89 MB/s, respectively.

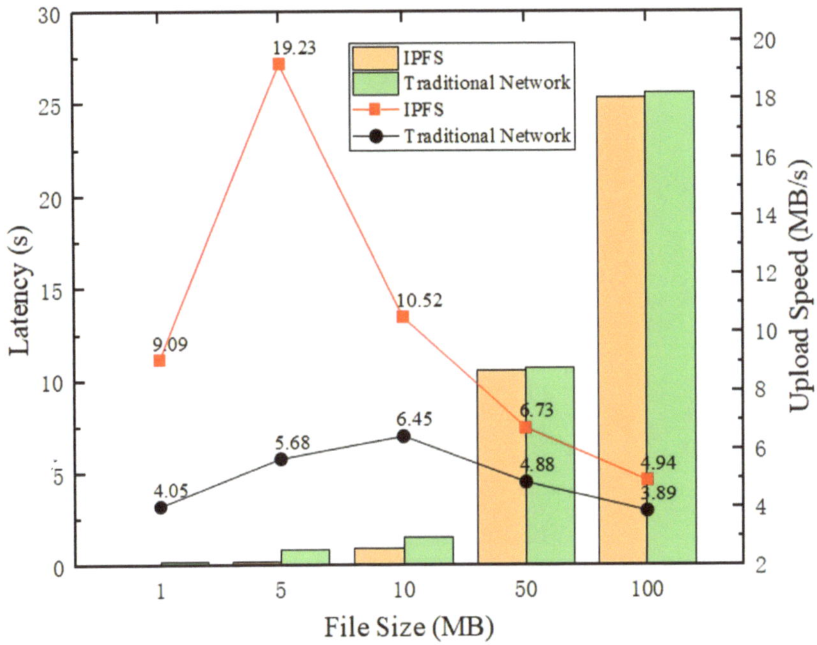

Figure 7. Performance comparison of file transfers in traditional and IPFS networks using different file sizes.

From the experimental results, almost all the transfer rates in the IPFS network are faster than in the TCP/IP network, and the IPFS networks are almost 4 times larger than TCP/IP networks when transferring data of 5 MB file size. Moreover, IPFS networks take less time than TCP/IP networks, which is more evident when transferring small files (File Size $\leq$ 10 MB), IPFS networks take one-half the time of TCP/IP networks when transferring 1 MB files, and one-quarter the time of TCP/IP networks when transferring 5 MB files. The data transfer performance in IPFS networks is generally better than that in traditional networks.

6.2. Throughput and Latency of Smart Contract Calling

We designed two smart contracts for the blockchain network and used throughput and transaction latency as the main performance metrics in our benchmarking. Throughput is the rate at which transactions are committed to the ledger, measured in terms of how many transactions are executed per second (tps). Latency is the time it takes from the time the application sends a transaction proposal to the time the transaction is committed to the ledger. As can be seen from Figure 8, when the block size and send rate is fixed, the TPS remains essentially constant as the number of transactions increases. "Querfile" fluctuates around 110 tps, with a minimum of 101.3 tps and a maximum of 115.6 tps; and "Subfile" fluctuates around 50 tps, with a minimum of 44. 3 tps and a maximum of 53.2 tps. In addition, as shown in Figure 9, the latency increases with the increase in the number of transactions.

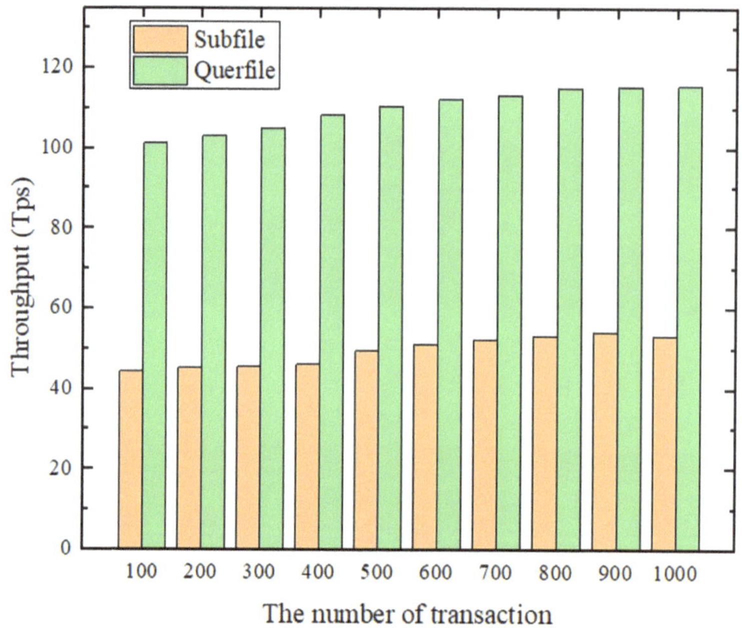

Figure 8. System throughput at different transaction volumes.

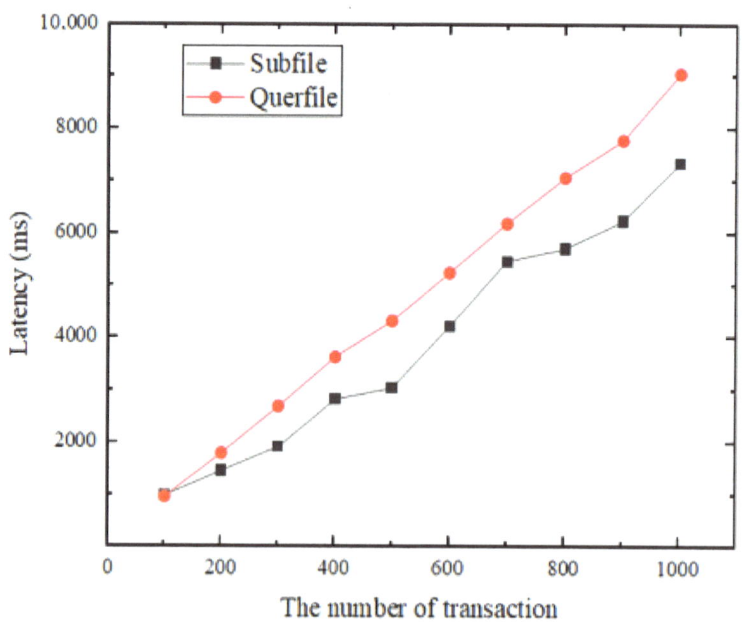

Figure 9. System latency at different transaction volumes.

6.3. Performance Comparison of Different Systems

To demonstrate the good performance of our proposed scheme, we compare it with other blockchain systems mainly used today: Bitcoin, Ethereum, Litecoin, BitcoinCash, and Primecoin in terms of the system transaction average latency and average throughput [32].

The sending rate, block size, and some transactions are set to 200 tps, 2 MB, and 400. Figure 10 gives the comparison results.

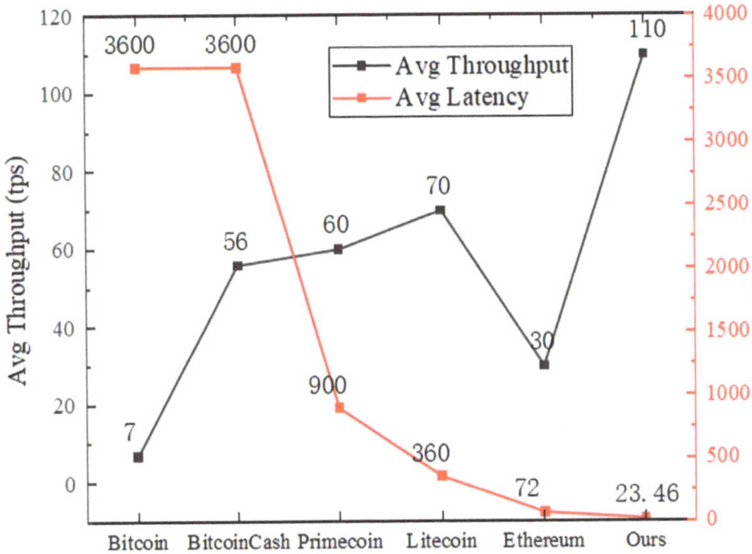

Figure 10. Comparison with current major blockchain systems.

From the comparison, it is clear that our scheme has better performance than existing blockchain systems in terms of the average transaction latency and average throughput. In terms of throughput, the block size limits the throughput of Bitcoin to only seven transactions per second. In total, 70, 60, and 56 transactions per second are achieved for Primecoin, Litecoin, and Bitcoin Cash, respectively, while Ethereum processes about 30 transactions per second. The average throughput of our solution can reach 110 tps, and the minimum throughput in experimental tests can reach 101 tps. In terms of system overhead, since the blockchain platform used in this system is Hyperledger Fabric, it does not need to consume a lot of computational resources for mining; therefore, the overhead of our solution is extremely low.

7. Conclusions

To solve the data sharing and privacy protection problems brought by the rapid growth of data in industrial IoT, we proposed an enterprise privacy protection and data sharing scheme based on the Hyperledger Fabric blockchain. We focused on the security and privacy of data transmitted by all parties in industrial systems. We utilized the Hyperldeger Fabric channel mechanism to enable enterprises to share data while keeping sensitive data private, isolating data between different channels, and all transaction data will carry time stamps and be permanently stored in the blockchain ledger, and be open, transparent, and traceable. Moreover, we achieved a high degree of automation in data recall through the designed Chaincode. The under-chain storage approach can effectively increase the scalability of the system. In addition, our scheme achieves mutual authentication of all parties in the system and data integrity protection. Finally, the analysis results show that our scheme has good traceability, non-repudiation, and resistance against known cyber attacks, and good performances.

In the future, a potential research direction is how to optimize the consensus algorithm of Hyperldeger Fabric, in which the backing nodes are responsible for endorsing the legitimacy of all transaction contents and carry a large amount of sensitive transaction data. How to protect the backing nodes from attacks and enhance the processing power of backing nodes to improve the transaction speed of the whole blockchain network is one of the valuable research directions.

Author Contributions: Conceptualization, C.-L.C. and J.Y.; methodology, C.-L.C., J.Y. and W.W.; validation, W.-J.T., C.-M.W. and X.W.; investigation, C.-L.C. and J.Y.; data analysis, C.-L.C., J.Y., W.W., C.-M.W. and X.W.; writing—original draft preparation, C.-L.C. and J.Y.; writing—review and editing, W.-J.T., W.W., C.-M.W. and X.W.; supervision, C.-L.C. and W.W. All authors have read and agreed to the published version of the manuscript.

Funding: This work was supported in part by the National Natural Science Foundation of China (No. 51808474), the Ministry of Science and Technology in Taiwan (No. MOST 110-2218-E-305-001-MBK), the Education and Teaching Reform Project of the Xiamen University of Technology (No. JG2021007), and the Education Research Project for Yong and Middle-aged Teachers of Fujian Province (No. JAT190679).

Institutional Review Board Statement: This study is based entirely on theoretical basic research. It does not involve humans.

Informed Consent Statement: This study is based entirely on theoretical basic research. It does not involve humans.

Data Availability Statement: The data used to support the findings of this study are available from the corresponding author upon request.

Conflicts of Interest: The authors declare no conflict of interest.

Notations

$Cert_X$	Certificate of X issued by CA
ID_X	X's identity
(r_X, s_X)	Elliptic curve signature value of X
T_X	Timestamp message of X
C_X	Ciphertext sent by X
DT_X	Data sent by X
SK_X	The AES key of party X
$E_{SK_X}(M)$	Use X's AES key to encrypt M
$D_{SK_X}(M)$	Use X's AES key to decrypt M
Puk_X	The public key of party X
Prk_X	The private key of party X
$E_{Puk_X}(M)$	Use X's public key to encrypt M
$E_{Prk_X}(M)$	Use X's private key to decrypt M
$h(.)$	The one-way hash function
τ	Valid timestamp interval
M_{SUBMIT}	The submitted registration information
$A \stackrel{?}{=} B$	Verify whether A is equal to B

References

1. Qi, S.; Lu, Y.; Zheng, Y.; Li, Y.; Chen, X. Cpds: Enabling Compressed and Private Data Sharing for Industrial Internet of Things Over Blockchain. *IEEE Trans. Ind. Inform.* **2021**, *17*, 2376–2387. [CrossRef]
2. TechCrunch is Part of the Yahoo Family of Brands. Available online: https://techcrunch.com/2021/01/11/ubiquiti-says-customer-data-may-have-been-accessed-in-data-breach/ (accessed on 20 December 2021).
3. SITA Data Breach Affects Millions of Airline Passengers. Available online: https://www.techradar.com/news/sita-data-breach-affects-millions-of-airline-passengers (accessed on 20 December 2021).
4. Journal, H. CaptureRx Ransomware Attack Affects Multiple Healthcare Provider Clients and 1,919,938 Individuals. Available online: https://www.hipaajournal.com/capturerx-ransomware-attack-affects-multiple-healthcare-provider-clients (accessed on 20 December 2021).
5. Volkswagen, Audi Notify 3.3 Million of Data Breach. Available online: https://www.bankinfosecurity.com/volkswagen-audi-notify-33-million-people-data-breach-a-16875 (accessed on 20 December 2021).
6. Smits, M.; Hulstijn, J. Blockchain applications and institutional trust. *Front. Blockchain* **2020**, *3*, 5. [CrossRef]

7. Tomescu, A.; Devadas, S. Catena: Efficient non-equivocation via bitcoin. In Proceedings of the 2017 IEEE Symposium on Security and Privacy (SP), San Jose, CA, USA, 22–26 May 2017; pp. 393–409.
8. Nakamoto, S. Bitcoin: A peer-to-peer electronic cash system. *Decent. Bus. Rev.* **2008**, 21260.
9. Hyperledger—Open Source Blockchain Technologies. Available online: https://www.hyperledger.org/ (accessed on 20 December 2021).
10. Heron, S. Advanced Encryption Standard (AES). *Netw. Secur.* **2009**, *2009*, 8–12. [CrossRef]
11. Jianjun, S.; Ming, L.; Jingang, M. Research and application of data sharing platform integrating Ethereum and IPFs Technology. In Proceedings of the 2020 19th International Symposium on Distributed Computing and Applications for Business Engineering and Science (DCABES), Xuzhou, China, 16–19 October 2020; pp. 279–282.
12. Johnson, D.; Menezes, A.; Vanstone, S. The Elliptic Curve Digital Signature Algorithm (ECDSA). *Int. J. Inf. Secur.* **2001**, *1*, 36–63. [CrossRef]
13. Teslya, N.; Ryabchikov, I. Blockchain-based platform architecture for industrial IoT. In Proceedings of the 2017 21st Conference of Open Innovations Association (FRUCT), Helsinki, Finland, 6–10 November 2017; pp. 321–329.
14. Wang, Z.; Tian, Y.; Zhu, J. Data sharing and tracing scheme based on blockchain. In Proceedings of the 2018 8th International Conference on Logistics, Informatics and Service Sciences (LISS), Toronto, ON, Canada, 3–6 August 2018; pp. 1–6.
15. Zhang, A.; Lin, X. Towards secure and privacy-preserving data sharing in e-health systems via consortium blockchain. *J. Med. Syst.* **2018**, *42*, 140. [CrossRef] [PubMed]
16. Lee, A.R.; Kim, M.G.; Kim, I.K. SHAREChain: Healthcare data sharing framework using Blockchain-registry and FHIR. In Proceedings of the 2019 IEEE International Conference on Bioinformatics and Biomedicine (BIBM), San Diego, CA, USA, 18–21 November 2019; pp. 1087–1090.
17. Kumar, N.; Dakshayini, M. Secure Sharing of Health Data Using Hyperledger Fabric Based on Blockchain Technology. In Proceedings of the 2020 International Conference on Mainstreaming BlockChain Implementation (ICOMBI), Bengaluru, India, 21–22 February 2020; pp. 1–5.
18. Kang, B.; Shao, D.; Wang, J. A fair electronic payment system for digital content using elliptic curve cryptography. *J. Algorithms Comput. Technol.* **2017**, *12*, 13–19. [CrossRef]
19. Available online: https://hyperledgerfabric.readthedocs.io/en/release-2.2 (accessed on 20 December 2021).
20. Transaction Flow—Hyperledger-Fabricdocs Master Documentation. Available online: https://hyperledger-fabric.readthedocs.io/en/release-2.2/txflow.html (accessed on 20 December 2021).
21. Foschini, L.; Gavagna, A.; Martuscelli, G.; Montanari, R. Hyperledger Fabric Blockchain: Chaincode Performance Analysis. In Proceedings of the ICC 2020–2020 IEEE International Conference on Communications (ICC), Online, 7–11 June 2020; pp. 1–6.
22. Uddin, M. Blockchain Medledger: Hyperledger fabric enabled drug traceability system for counterfeit drugs in pharmaceutical industry. *Int. J. Pharm.* **2021**, *597*, 120235. [CrossRef] [PubMed]
23. Nizamuddin, N.; Salah, K.; Ajmal Azad, M.; Arshad, J.; Rehman, M. Decentralized document version control using ethereum blockchain and IPFS. *Comput. Electr. Eng.* **2019**, *76*, 183–197. [CrossRef]
24. Burrows, M.; Abadi, M.; Needham, R. A logic of authentication. *ACM Trans. Comput. Syst.* **1990**, *8*, 18–36. [CrossRef]
25. Chen, C.; Deng, Y.; Weng, W.; Sun, H.; Zhou, M. A Blockchain-Based Secure Inter-Hospital EMR Sharing System. *Appl. Sci.* **2020**, *10*, 4958. [CrossRef]
26. Chen, C.; Deng, Y.; Li, C.; Zhu, S.; Chiu, Y.; Chen, P. An IoT-Based Traceable Drug Anti-Counterfeiting Management System. *IEEE Access* **2020**, *8*, 224532–224548. [CrossRef]
27. Kiayias, A.; Tsiounis, Y.; Yung, M. Traceable Signatures. *Adv. Cryptol. Eurocrypt* **2004**, *2004*, 571–589.
28. Zhou, J.; Gollman, D. A fair non-repudiation protocol. In Proceedings of the 1996 IEEE Symposium on Security and Privacy, Oakland, CA, USA, 6–8 May 1996; IEEE: Piscataway, NJ, USA; pp. 55–61.
29. Denning, P.; Denning, D. Discussing cyber attack. *Commun. ACM* **2010**, *53*, 29–31. [CrossRef]
30. Blockchain—Wikipedia. Available online: https://en.wikipedia.org/wiki/Blockchain (accessed on 20 December 2021).
31. Nasir, Q.; Qasse, I.; Abu Talib, M.; Nassif, A. Performance Analysis of Hyperledger Fabric Platforms. *Secur. Commun. Netw.* **2018**, *2018*, 3976093. [CrossRef]
32. Lin, H.; Garg, S.; Hu, J.; Kaddoum, G.; Peng, M.; Hossain, M. A Blockchain-Based Secure Data Aggregation Strategy Using Sixth Generation Enabled Network-In-Box For Industrial Applications. *IEEE Trans. Ind. Inform.* **2021**, *17*, 7204–7212. [CrossRef]

sensors

MDPI

Article

A Blockchain-Based Privacy Information Security Sharing Scheme in Industrial Internet of Things

Yue Wang [1,2], Tingyu Che [1,2], Xiaohu Zhao [1,*], Tao Zhou [1,2], Kai Zhang [1,2] and Xiaofei Hu [3]

1 National and Local Joint Engineering Laboratory of Internet Applied Technology on Mines, China University of Mining and Technology, Xuzhou 221008, China; wyxz@cumt.edu.cn (Y.W.); tingyu@cumt.edu.cn (T.C.); kdstutaozhou@cumt.edu.cn (T.Z.); kaizhang@cumt.edu.cn (K.Z.)
2 School of Information and Control Engineering, China University of Mining and Technology, Xuzhou 221008, China
3 School of Information and Business Management, Chengdu Neusoft University, Chengdu 611844, China; huxiaofei@nsu.edu.cn
* Correspondence: zhaoxiaohu@cumt.edu.cn

Abstract: Due to the competitive relationship among different smart factories, equipment manufacturers cannot integrate the private information of all smart factories to train the intelligent manufacturing equipment fault prediction model and improve the accuracy of intelligent manufacturing equipment fault detection. The use of a low fault recognition rate model for smart factories will cause additional losses for them. In this work, we propose a blockchain-based privacy information security sharing scheme in Industrial Internet of Things (IIoT) to solve the sharing problem of private information in smart factories. Firstly, we abstract smart factories as edge nodes and build decentralized, distributed trusted blockchain networks based on Ethereum clients on simulated edge devices and propose an Intelligent Elliptic Curve Digital Signature Algorithm (IECDSA) to guarantee the ownership of shared information by edge nodes. Secondly, we propose the Reputation-based Delegated Proof of Stake (RDPoS) consensus algorithm to improve the security and reliability of the Delegated Proof of Stake (DPoS) consensus algorithm. Furthermore, we design and implement an incentive mechanism based on information attributes to increase the motivation of edge nodes to share information. Finally, the proposed solution is simulated. Through theoretical and simulation experiments, it is proved that the blockchain-based privacy information security sharing scheme in IIoT can improve the enthusiasm of edge nodes to share information on the premise of ensuring the security of information sharing.

Keywords: privacy information security sharing; Industrial Internet of Things; blockchain; consensus algorithm; incentive mechanism

1. Introduction

With the continuous advancement of Industry 4.0 on a global scale, more and more intelligent manufacturing equipment is widely used in smart factories. Usually, the smart factory will push the computing resources of the cloud server to the edge of the network to build a cloud-side collaboration architecture to meet the requirements of intelligent manufacturing equipment for high computing power and low latency and ensure the stable production of the smart factory [1,2]. However, the failure of intelligent manufacturing equipment will bring unpredictable losses to smart factories. Intelligent manufacturing equipment manufacturers minimize the loss of smart factories by using predictive maintenance methods to issue failure warnings before products fail [3]. Intelligent equipment manufacturers use the data fusion theory of multi-intelligent manufacturing equipment data fusion to improve the accuracy of intelligent manufacturing equipment failure prediction models and reduce the false early warning information caused by the waste of maintenance personnel manpower [4]. However, for smart factories, when their products are manufactured, a large amount of private information about the product is stored in the intelligent manufacturing equipment. Smart factories will instinctively protect private information and will not easily share information, eventually forming an IIoT information island phenomenon [5,6].

Information sharing is one of the most effective ways to solve the phenomenon of information islands, and it is also very suitable for IIoT scenarios [7–9]. The security of the information sharing process is a thorny issue for private information. Traditional information sharing is mostly achieved through cloud data sharing. However, traditional cloud storage is faced with the problems of single point attack, transmission delay, and resource waste [10]. Blockchain technology, as a new type of distributed architecture technology, has been widely used, providing a new solution for information security sharing, and it can effectively solve the problems faced by traditional information sharing [11–13]. Privacy information is stored in the form of a Merkel tree after being multi-hashed in the blockchain [14]. When the privacy information stored in the blockchain is tampered with by the attacker, the hash value of the Merkle tree root will also be changed due to the private information, and other nodes in the blockchain environment will immediately detect the tampered data and ensure the consistency of the global data through the unique consensus mechanism of the blockchain [15]. The blockchain consensus mechanism is also one of the important factors affecting the secure sharing of privacy information. In addition, the speed of the consensus mechanism will also directly affect the speed of privacy information sharing [16].

As the earliest consensus algorithm, the Proof of Work (PoW) consensus algorithm was not only used to maintain the smooth operation of the Bitcoin blockchain network but also applied to the Ethereum client [17]. This algorithm not only wastes node resources when competing for bookkeeping rights but also affects the throughput of the blockchain network system. The Proof of Stake (PoS) consensus algorithm uses proof-of-stake, which not only reduces the waste of blockchain node resources caused by the PoW consensus algorithm to a certain extent but also improves the throughput of the blockchain network system [18]. However, the PoS consensus algorithm is prone to the risk of forking blockchains. The Delegated Proof of Stake (DPoS) consensus algorithm uses voting campaigns instead of mining in PoS consensus algorithms and PoW consensus algorithms, with low consumption, high throughput, and low latency [19]. However, the DPoS consensus algorithm not only lacks reward and punishment measures for blockchain nodes in the process of carrying out voting campaign operations but also risks malicious nodes being selected as proxy nodes to participate in the campaign. The Practical Byzantine Fault Tolerance (PBFT) consensus algorithm can achieve the same transaction processing speed as the DPoS consensus algorithm but is not suitable for blockchain networks with a large number of nodes [20]. At present, the consensus algorithm still has the problems of node scale, performance, and fault tolerance, which are difficult to balance.

In order to solve the above problems existing in the consensus algorithm and ensure the speed and reliability of private data sharing among smart factories, this work aims to promote the sharing of private information about intelligent manufacturing equipment in smart factories and provide privacy information for the training of intelligent manufacturing equipment failure prediction models. We propose a blockchain-based privacy information security sharing scheme in IIoT to ensure the secure sharing of privacy information among different smart factories. The main contributions of this work are as follows:

- The cloud-edge collaboration architecture of the smart factory is analyzed, and the edge-end network architecture based on edge servers is established. Then, the Intelligent Elliptic Curve Digital Signature Algorithm (IECDSA) is proposed to determine the ownership of the smart factory's private information. In contrast to the traditional method, trusted storage and distribution of keys was implemented by the Key Distribution Smart Contract (KDSC), which reduces the risk of keys being tampered with and more securely guarantees the ownership of the shared private information by smart factories.

- The working principle of the DPoS consensus algorithm is analyzed, and in view of the situation that the malicious node is selected as a proxy node due to "hoarding" in the election process, the Reputation-based Delegated Proof-of-Stake consensus algorithm (RDPoS) is proposed. The algorithm performs a weighted operation on the number of node votes and reputation values and selects proxy nodes to participate in the consensus process according to the weighted operation results. Compared with the existing DPoS consensus algorithm, the probability of malicious nodes being selected as proxy nodes is reduced, and the security and reliability of the consensus reached between blockchain nodes are effectively improved.
- In view of the phenomenon that smart factories protect their own private information and refuse to participate in information sharing, a trusted incentive smart contract based on information attributes is constructed. Furthermore, a trusted network incentive environment without third party involvement is implemented, sending reward points to smart factories that provide private information sharing and ensuring the enthusiasm of smart factories in sharing information. Compared with the traditional incentive mechanism, the incentive mechanism realized by smart contracts is not interfered with by external factors, ensuring the fairness, impartiality, and openness of the incentive mechanism.

The rest of this work is organized as follows: Section 2 is the related work. Section 3 provides a detailed description of the proposed solution, including an overview of the overall solution, a network architecture, and a security analysis. Section 4 focuses on theories related to smart factory data ownership, blockchain data storage, the RDPoS consensus algorithm, and the incentive mechanisms based on information property. Section 5 provides an analytical discussion of the experimental results. Section 6 summarizes the full text.

2. Related Work

In [21], the authors propose a privacy-preserving data sharing framework for the Industrial Internet of Things that provides privacy protection for data contributors by interfering with the data provided by them. However, the framework does not take into account the enthusiasm of data contributors in contributing data. In [22], an asynchronous federated learning scheme is proposed that uses deep reinforcement learning (DRL) for node selection to improve efficiency, integrates machine learning models into the blockchain, and performs two-stage verification to ensure the reliability of shared data. The scheme also ignores the incentives of information providers to share data and malicious decisions in the process of the blockchain consensus mechanism [23].

The DPoS consensus algorithm is widely used due to its low energy consumption, high throughput, and dynamic scalability. However, the DPoS consensus algorithm suffers from the problems of malicious nodes being easily selected as proxy nodes and the low motivation of participating nodes to vote during the working process. In [24], in view of the low voting motivation of nodes and the lack of reference basis for nodes in the voting process, the concepts of token investment and side chains were introduced into the DPoS consensus algorithm, which effectively improved the voting motivation of nodes. However, the algorithm's excessive reliance on tokens can easily lead to the emergence of malicious nodes. In [25], an improved ring-based coordinator election algorithm is proposed to optimize the election process in the DPoS consensus algorithm, which further improves the decentralization and fairness of the DPoS consensus algorithm, but the algorithm suffers from the risk of not reaching consensus among nodes. In [26], the introduction of node behavior monitoring and Borda count voting to select proxy nodes in the DPoS consensus algorithm reduces the probability of a malicious node being elected as a proxy node and improves the fairness of the election, although the scheme only considers the detection of the behavior of witness nodes generating blocks.

The incentive mechanism can improve the enthusiasm of smart factories to share private information, and the incentive mechanism can be roughly divided into two types:

game-based incentives and external incentive-based incentives [27]. Kang et al. introduced reputation as an indicator of the reliability and trustworthiness of mobile devices in federated learning and proposed an effective incentive mechanism that combines reputation with contract theory to incentivize high-reputation mobile devices with high-quality data to participate in model learning, but the accuracy of reputation calculation in this scheme needs to be improved [28]. Zhan et al. studied the incentive mechanism of federated learning and designed an incentive mechanism based on deep reinforcement learning (DRL) to determine the optimal pricing strategy of the parameter server and the optimal training strategy of the edge node to motivate the edge node to contribute to the model training, but the method relies on the computing resources of the edge node [29]. In [30], an incentive mechanism for rational miners to purchase computing resources in the blockchain network environment of edge computing establishes a two-stage Stackelberg game model between miners and edge service providers (ESP) to maximize profits under two mining schemes, but this mechanism is not suitable for multiple ESP.

This work analyzes the above literature, and in view of the shortcomings of existing methods, proposes a blockchain-based privacy information security sharing scheme in IIoT to ensure the secure, fast, and active sharing of information among edge nodes (e.g., smart factories).

3. Scheme in Detail

This section provides a detailed description of the blockchain-based privacy information security sharing scheme in IIoT in terms of general scheme overview, system model, and security analysis.

3.1. The Overall Scheme

In this work, we abstract smart factories as edge nodes and build decentralized, distributed trusted blockchain networks based on Ethereum clients on simulated edge devices. Based on this architecture, a blockchain-based privacy information security sharing scheme in IIoT is proposed. In this scheme, the following steps are performed by the edge nodes of the information sender for information sharing.

Step 1. The information sender edge node uses the Intelligent Elliptic Curve Digital Signature Algorithm (IECDSA) to sign the information to be shared and stores the signed shared information in the blockchain.

Step 2. Information sharing among edge nodes is via the Reputation-based Delegated Proof of Stake (RDPoS) consensus algorithm.

Step 3. The received information is verified by the IECDSA at the edge node of the information receiver.

Step 4. The information receiver edge node provides rewards to the edge node of the information sender based on the incentive mechanism of information attribute to ensure the enthusiasm of the edge node in the network to share information.

The workflow of the blockchain-based privacy information security sharing scheme in IIoT is shown in Figure 1.

Sended information: The information sender edge node signs the information to be shared using IECDSA, declaring its ownership of the shared information.

Information is uploaded to the block: The information sender edge node uploads the signed message to the block in preparation for information sharing.

Broadcast information: The blockchain uses the RDPoS consensus algorithm to achieve consensus on the information stored in the block, enabling information sharing among edge nodes.

Received information: The information receiver edge node uses IECDSA to verify the identity of the information sender edge node.

Trigger incentive mechanism: After the information receiver edge node determines the identity of the information sender edge node, the incentive mechanism is triggered.

Send the reward: The triggered incentive mechanism provides a reward to the information sender edge node according to the pre-defined reward rules in the trusted incentive smart contract.

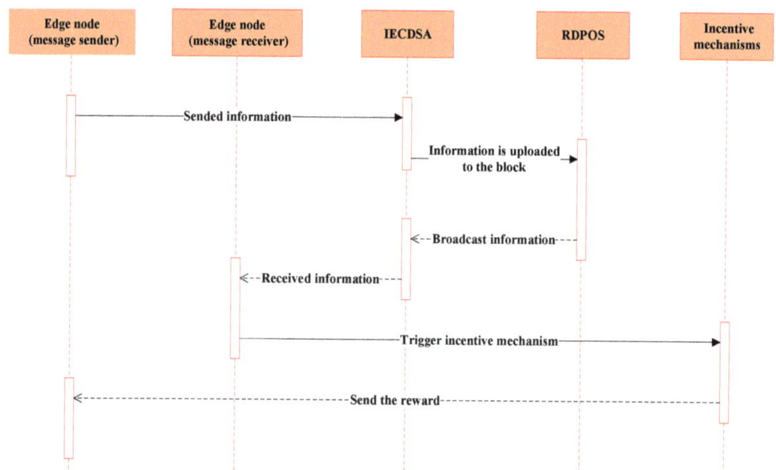

Figure 1. Scheme workflow.

3.2. The Network Architecture

As shown in Figure 2, the network architecture of the blockchain-based privacy information security sharing scheme in IIoT can be divided into three layers: the terminal layer, the edge layer, and the blockchain layer from the bottom up.

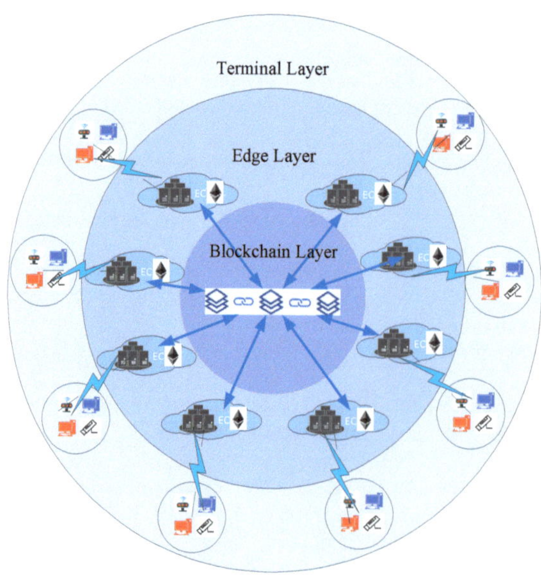

Figure 2. Network architecture.

Terminal Layer: A group of terminal devices $TD = \{td_1, td_2, td_3, \cdots, td_n\}$ are connected to a high-performance edge device using the terminal device access technology. The edge device stores and processes the data of its subordinate terminal devices and provides services for the terminal devices.

Edge Layer: Divide the physical environment into different regions $R = \{r_1, r_2, r_3, \cdots, r_n\}$ by region and deploy a set of edge devices in different regions based on the load capacity of the edge devices $ED = \{ed_1, ed_2, ed_3, \cdots, ed_n\}$. These edge devices (edge nodes) are made up of edge layers $EL = \{r1_{ed_{1 \cdots i}}, r2_{ed_{i+1 \cdots k}}, r3_{ed_{k+1 \cdots l}}, \cdots, rn_{ed_{m \cdots n}}\}$.

Blockchain Layer: At the edge layer, a blockchain network (blockchain layer) is formed by deploying an Ethereum client with a RDPoS consensus algorithm on high-performance edge devices. The consensus mechanism of the blockchain ensures the consistency of data among blockchain nodes (edge nodes).

3.3. Security Analysis

This section provides a security analysis of the blockchain-based privacy information security sharing scheme in IIoT.

3.3.1. The Security of Information Storage

In our proposed scheme, we still use the Merkle tree of data storage structure used by the blockchain to store information. If the shared information stored in a block is tampered with, then the hash value of the Merkle tree root is changed, at which point the consensus mechanism of the blockchain will calculate the proportion of the shared information stored in the current block to the shared information stored in blocks of the same block height in all blockchain nodes (edge nodes). If the percentage is less than 51%, the information has been tampered with. In this case, the RDPoS consensus algorithm will overwrite the tampered information with the correct information to ensure that the data are consistent among the edge nodes. If the attacker wants to make the ratio exceed 51%, he needs to control 51% of the edge nodes in the network to achieve this, and it is not realistic for the attacker to control 51% of the edge nodes at the same time in a short period of time.

3.3.2. The Security of Information Sharing

We have improved the problems with the widely used Delegated Proof of Stake consensus algorithm, which suffers from the problems of malicious nodes being easily selected as proxy nodes and the low motivation of nodes involved in voting and propose a Reputation-based Delegated Proof of Stake (RDPoS) consensus algorithm. The RDPoS consensus algorithm first supervises the behavior of nodes through a reputation model and assigns corresponding behavior scores according to the normality of nodes' historical behavior, calculates the reputation value and trustworthiness status of nodes, and finally selects proxy nodes to participate in the consensus process. In addition, the algorithm also designs a hybrid mechanism model to ensure the motivation of nodes to participate in voting. The RDPoS consensus algorithm not only guarantees the security of the proxy node election process but also ensures the motivation of participating voting nodes, and it improves the security of the consensus process.

3.3.3. The Fairness of Information Sharing

In our proposed scheme, an incentive mechanism based on information properties is designed, which is implemented by a trusted incentive smart contract. The trusted incentive smart contract can be executed automatically without the involvement of a third party. In a blockchain environment, smart contracts can only be changed through version replacement, and if an edge node wants to modify the sharing rules inside a trusted incentive smart contract to give itself additional revenue, it needs to redeploy the contract to do so. However, this process is open and transparent and monitored by all edge nodes. In addition, we use an incentive mechanism based on information attributes, where the information receiver edge nodes can provide rewards to the information sender edge nodes based on the value of the shared information for their own use, reducing the waste of edge node assets to a certain extent. Therefore, our scheme is therefore fair and frugal.

In summary, we analyzed the security of our proposed scheme from three aspects: information storage security, information sharing security, and information sharing fairness, and the results proved that our proposed blockchain-based privacy information security sharing scheme in IIoT is safe and reliable.

4. Methods

In this section, the Intelligent Elliptic Curve Digital Signature Algorithm (IECDSA), the Reputation-based Delegated Proof-of-Stake consensus algorithm (RDPoS), and the incentive mechanism based on information attributes are implemented separately.

4.1. Intelligent Elliptic Curve Digital Signature Algorithm (IECDSA)

Our proposed scheme uses the Intelligent Elliptic Curve Digital Signature Algorithm to ensure ownership of shared information by edge nodes, as follows.

Generator $G(x,y)$ is used as a public parameter on the elliptic curve $E_p(a,b)$. We chose PrK as the private key for the digital signature in $E_p(a,b)$. The public key can then be expressed as:

$$PuK = PrK * G \tag{1}$$

The message sender edge node signs the message m to be shared (the signature consists of two parts s_b and s_b).

The pseudo-random numbers are generated using the linear congruence algorithm, as shown in Equation (2) [31].

$$RandSeed = (a * RandSeed + c)\%m \tag{2}$$

where a, c, m are the constants set by the generator.

Point multiplication of $RandSeed$ with the generator $G(x,y)$ to the point P.

$$P{=}RandSeedG(x,y) = (x_1, y_1) \tag{3}$$

The s_a part of the signature is generated by performing the operation according to Equation (4) using the horizontal coordinates of the point $P(x_1, y_1)$ and the prime number n.

$$s\_a = x_1 \bmod n \tag{4}$$

Calculate the hash value of the shared information by Equation (5).

$$h = Hash(m) \tag{5}$$

The signature information is obtained by Equation (6).

$$s\_b = RandSeed^{-1}(h + \mathrm{P}rKs\_a) \bmod n \tag{6}$$

where PrK is the private key, h is the hash of the shared information, s_a is the signature information, $RandSeed$ is a random number, and n is a prime number.

The message sender edge node's signature information for shared messages is (s_a, s_b).

We designed the key distribution smart contract as shown in Algorithm 1 to enable intelligent, supervised, and secure distribution of public keys among edge nodes without the involvement of third parties. When the information sender edge node sends the shared information, the public key is stored in the blockchain through the key distribution smart contract, and the storage structure and procedure of the block in the blockchain are described in Appendix A.

After receiving the shared information, the information receiver edge node uses the public key to verify the signature information of the message sender. The information receiver edge node verifies that s_a and s_a are integers in $[1, n-1]$; then, it computes the hash h of the shared information according to Equation (5), followed by the value of w, u_1, u_2.

$$w = s\_b^{-1} \bmod n \tag{7}$$

$$u_1 = hw \bmod n \tag{8}$$

$$u_2 = s\_a^{-1} w \bmod n \tag{9}$$

Algorithm 1 Key Distribution Smart Contract (KDSC)

Input: *PuK, EList* // *PuK* is public key. *EList* is information set of edge nodes in the network.
Output: State of public key distribution.

$KdList$ // This list is used to store the public key distribution status.
for i to size($EList$) **do**
  $PuK \rightarrow EList(i)$ // Assigns public keys to edge nodes in the network.
  $KdList.Add(Stae(EList(i))$ // $Stae(EList(i))$ is the status in which the current edge node distributes the public key.
end for
return $KdList$

Bringing the parameters $u_1, u_2, G(x,y)$ and the public key PuK into Equation (10) yields the point X.

$$X{=}u_1 G + u_2 PuK = (x_1, y_1) \tag{10}$$

Take the horizontal coordinate x_1 of point X and prime number n for modular arithmetic according to Equation (11). If the equation is true, the signature is valid; otherwise, the signature is invalid.

$$x_1 \bmod n = s\_a \tag{11}$$

The IECDSA is not only resistant to plaintext attacks but also to ciphertext attacks, so that even if an attacker intercepts the signature message, he cannot forge a valid signature message. The IECDSA is shown in Algorithm 2.

Algorithm 2 Intelligent Elliptic Curve Digital Signature Algorithm (IECDSA)

Input: *MList, EList* // *MList* is the information shared set by the edge nodes of the message sender. *EList* is information set of edge nodes in the network.
Output: Status of signatures and verification of signatures.

for i to size($MList$) **do**
  Selecting the data signature private key.
  $PuK = PrK * G$ // Calculating a digitally signed public key. distribution the public key.
  $RandSeed = (a * RandSeed + c)\%m$ // Generate random numbers.
  $P{=}RandSeedG(x,y) = (x_1, y_1)$ // Calculation of the parameter P.
  Generate data signature s_a by Equation (4).
  $h(MList(i)) = Hash(MList(i))$ //Calculating hash values of shared information.
  Generate data signature s_b by Equation (6).
  KDSC(PuK,EList) // Key Distribution Smart Contracts enables intelligent distribution of public keys.
  if $s\_a \&\& s\_b \notin [1, n-1]$ **then** // Information receiver edge nodes verify signatures.
    Signature verification failure.
  else
    Calculating hash values of shared information by Equation (5).
    Calculate the parameters w, u_1, u_2 according to Equations (7)–(9).
    $X{=}u_1 G + u_2 PuK = (x_1, y_1)$ // Calculate the parameter X.
    if $x_1 \bmod n = s\_a$ **then**
      Successful signature verification.
    else
      Signature verification failure.
    end if
  end if
end for
return Status of signatures and verification of signatures.

4.2. Reputation-Based Delegated Proof of Stake (RDPoS)

The process of the Reputation-based Delegated Proof of Stake consensus algorithm is as follows. Firstly, the node's behavior is monitored by the reputation model and assigned a corresponding behavior score; then, the node's reputation value is calculated and the node's trustworthiness status is defined by the reputation value, and finally the proxy node is selected to participate in the consensus process based on the reputation value and the number of votes. In the RDPoS consensus algorithm, blockchain nodes are divided into three categories: normal nodes, candidate nodes (voting nodes), and proxy nodes.

In the reputation model, a node is evaluated for trustworthiness based on its performance throughout the period and it is assigned a reputation value (R) to indicate the trustworthiness of the node. Assuming that R is a real number between 0 and 1, the larger the value of R, the higher the trustworthiness of the node. When new nodes join the blockchain network, the reputation value defaults to 0.5.

All acts *behavior_j* of the i-th node *node_i* in the period T are denoted as:

$$behavior\_j = \{B_1, B_2, B_j, \ldots, B_n\} \tag{12}$$

where B_j is the score of the j-th act of *behavior_j* and n is the number of acts.

Assuming that $B_{i(j)}$ denotes the behavior value of the j-th behavior of node *node_i* during the period T, the $(j+1)$-th behavior value $B_{i(j+1)}$ of node *node_i* is determined according to the type of node *node_i* (agent node and voting node). The rules for calculating the behavior value of a node are shown in Table 1.

Table 1. The rules for calculating the behavior value of a node.

Value of Behavior	Voting Node	Agent Node
$min\left(1, (1+y)B_{i(j)}\right)$ [1]	Voting active	Generate blocks and upload them to the blockchain
$xB_{i(j)}$ [2]	Voting inactivity	Block not generated on time
0	Vote invalid	Generate invalid blocks

[1] Where $0 < y < 0.03$; [2] Where $0 < x < 1$.

The time interval for scoring the behavior of a node is the period T, and when the period ends the behavior value of node *node_i* is calculated according to Equation (13).

$$B_i = \sum_{j=1}^{n} \left(B_{i(j)}\right) \tag{13}$$

Calculate the reputation value of node *node_i* using the behavior value of node *node_i*.

$$R_{\mathrm{T}}^{node\_i} = \frac{1}{1 + \mathrm{e}^{-\varphi B_i}} \tag{14}$$

where T is the current period and φ is an adjustable parameter.

The Reputation Model Algorithm is shown in Algorithm 3.

To reduce the probability of a malicious node being selected as a proxy node, we classify the nodes into trusted status (e.g., Good, Normal, Abnormal and Error) by their reputation values. The node status corresponding to the reputation value, the weight of the reputation value and the weight of the number of votes are shown in Table 2.

When an edge node first joins the blockchain network, its default status is Normal, its reputation value is initialized to 0.5, and it is in the preferred position for subsequent participation in the campaign process. When an edge node has a good record, a block is generated and validated according to expectations, and as its reputation value gradually rises and will exceed threshold a, the node status switches to Good and the node in Good has a greater chance of being selected as a proxy node. Furthermore, in the second category, it will be selected in subsequent participation in the campaign process. If an edge node has generated incorrect records, validated invalid blocks and other irregularities, when its reputation value gradually drops below 0.5, the blockchain network converts the status of this edge node to Abnormal, and it is in the third category to be selected in the subsequent participation in the campaign process. If an edge node consistently generates invalid blocks or has persistent irregularities, when the reputation value of the edge node drops below *b*, the node's status will switch to Error and it will be in the last category of the selected positions in the subsequent participation in the campaign process.

Table 2. Parameters corresponding to the status of the node.

Trusted Status	Reputation Value (R)	Weight of the R (w_1)	Weight of the Number of Votes (w_2)
Good	[a, 1] [1]	[0.3, 0.5)	(0.5, 0.7]
Normal	[0.5, a)	0.5	0.5
Abnormal	[b, 0.5) [2]	(0.5, 0.7]	[0.3, 0.5)
Error	$R < b$	0	0

[1] Where 0.5 < a < 1; [2] Where 0 < b < 0.5. a, b represent thresholds, respectively.

Algorithm 3 Reputation Model Algorithm (RMA)

Input: node *node_i*, penalty coefficient x, incentive increase factor y.
Output: R_{node_i}. // The reputation value of node *node_i* is R_{node_i}

R_{node_i}=0.5 // Initialize the node reputation value.
B_i // The sum of the historical behavior of the node *node_i*.
$B_{i(j)} = 0$ // Initialize the behavioral score of node *node_i*.
if *node_i* is a non-proxy node **then**
  if *node_i* active participation in voting **then**
    $B_{i(j+1)} = min\left(1, (1+y)B_{i(j)}\right)$
    $B_i = B_i + B_{i(j+1)}$
  end if
  if *node_i* inactive participation in voting **then**
    $B_{i(j+1)} = xB_{i(j)}$
    $B_i = B_i + B_{i(j+1)}$
  end if
  if *node_i* cast an invalid vote **then**
    $B_{i(j+1)} = 0$
    $B_i = B_i + B_{i(j+1)}$
  end if
end if
if *node_i* is a proxy node **then**
  for $t = 0$ to T **do** // t is the time in the cycle and T is the whole cycle time.
    if *node_i* generates blocks and uploads them to the blockchain **then**
      $B_{i(j+1)} = min\left(1, (1+y)B_{i(j)}\right)$
    end if
    if *node_i* did not generate the block on time **then**
      $B_{i(j+1)} = xB_{i(j)}$
      $B_i = B_i + B_{i(j+1)}$
    end if
    if *node_i* generates invalid blocks **then**
      $B_{i(j+1)} = 0$
      $B_i = B_i + B_{i(j+1)}$
    end if
  end for
end if
$R_{node\_i} = \frac{1}{1+e^{-\varphi B_i}}$ // Calculate the reputation value of node *node_i*.
return R_{node_i}.

The weighting of reputation values and the weighting of the number of votes in Table 2 are used by Equation (15) to calculate the node scores.

$$score_{\mathrm{i}} = W_1 R_i + W_2 V_i \tag{15}$$

where R_i is the reputation value of node *node_i* and V_i is the total number of votes received by node *node_i*.

Ultimately, the edge nodes are ranked according to their scores, and the fixed number of nodes with the highest ranking are selected as proxy nodes to participate in block generation and verification.

The Reputation-based Delegated Proof of Stake Algorithm is shown in Algorithm 4.

Algorithm 4 Reputation-Based Delegated Proof of Stake Algorithm (RDPoS)

Input: Hash value of the current block.
Output: Status of the block on the chain.

```
DList. // The list of proxy nodes.
NList. // The list of blockchain network nodes.
RList. // The list of node reputation.
VList. // The list of total node votes.
Initialize the number of agent nodes to N.
Flag = False. // Block out status default failure.
for i = 0 to len(NList) do
    Use RMA(NList[i]) to calculate the node reputation value R.
    RList.Add(R)
    Calculate the total number of node votes V.
    VList.Add(V)
end for
Select a proxy node list DList.
for t = 0 to T do // t is the time in the cycle and T is the whole cycle time.
    Proxy nodes take turns generating blocks.
    if Other nodes validated successfully then
        Proxy nodes to upload blocks to the chain.
        Flag = True
    end if
end for
return Flag
```

In the DPoS consensus algorithm, there is the problem that only the proxy node that generates the block is rewarded and no other proxy nodes are rewarded. In addition, there are cases where both malicious and normal nodes receive the same reward when generating blocks, which undermines the original fairness of the blockchain. Our scheme combines a transaction fee incentive with a reputation incentive to propose a hybrid incentive mechanism to ensure the fairness of the RDPoS consensus algorithm. The transaction fee reward obtained by the node successfully generating the block is calculated according to Equation (16).

$$\Delta F = \begin{cases} F + \frac{R_{node\_i}}{\sum\limits_{j=1}^{j=N} R_j} * F, \; Node\ status\ is\ Good. \\ F, \; Node\ status\ is\ Normal. \\ F - \frac{R_{node\_i}}{\sum\limits_{j=1}^{j=N} R_j} * F, \; Node\ status\ is\ Abnormal. \\ 0, \; Node\ status\ is\ Error. \end{cases} \tag{16}$$

where ΔF is the transaction reward for successfully generated block, R_{node_i} is the current reputation of the node *node_i*, N is the total number of proxy nodes in the blockchain network, and F is the transaction fee reward allocated for a successfully generated block in the blockchain network.

4.3. Incentive Mechanism Based on Information Attributes

When the edge nodes share information, the negative situation of information sharing may appear due to some subjective factors, which affects the enthusiasm of information sharing in the whole edge network. In order to address the above phenomenon, we propose an incentive mechanism based on information attributed in the scheme to ensure the motivation of information sharing among the edge nodes in the network.

One must add the properties for shared information before sending the message by the information sender edge node.

$$m \rightarrow \{m_{type}, m_{quantity}, m_{limitation}, m_{expected}, m_{real}\} \tag{17}$$

where m_{type} is the type of shared data, $m_{quantity}$ is the number of shared messages, $m_{limitation}$ is the timeliness of the shared messages, $m_{expected}$ is the reward expected by the information sender edge node, and m_{real} is the real reward provided by the information receiver edge node for the information sender edge node.

The information sender edge nodes expect the following rewards.

$$m_{expected} = \sum_{i=1}^{n} m_{expected_i} \tag{18}$$

where i is the number of information receiver edge nodes.

The demand for shared information is different for different types of information receiver edge nodes. We assign different weights to the attributes of the information based on the demand for shared information by information receiver edge nodes.

$$\begin{aligned} m_{real} &= \sum_{i=1}^{n} w_{type_i} m_{type} + w_{quantity_i} m_{quantity} \\ &+ \sum_{i=1}^{n} w_{limitation_i} m_{limitation} \end{aligned} \tag{19}$$

where $w_{type_i} + w_{quantity_i} + w_{limitation_i} = 1$, w_{type_i} is the weight of the i-th information receiver edge node on the type of shared information, $w_{quantity_i}$ is the weight of the i-th information receiver edge node on the amount of shared information, and $w_{limitation_i}$ is the weight of the i-th information receiver edge node on the timeliness of shared information.

If the expected values of the information sender edge node and the information receiver edge node satisfy Equation (20), it means that the information sharing reward is provided successfully; otherwise, the information receiver edge node dynamically adjusts the proportion among the weights of the shared information attributes so that they satisfy Equation (20).

$$m_{expected} - m_{real} < 0.3 \tag{20}$$

We design trustworthy incentive smart contracts to process real and trustworthy incentive transaction information in the blockchain to achieve a trustworthy incentive network environment without third party participation, ensuring fair, open, and transparent incentive distribution among edge nodes.

5. Simulation Experiments

We simulated 15 edge nodes using VMware Workstation running 15 Ubuntu 19.04 virtual machines on 5 Windows 10 machines. All edge nodes have the same configuration: Intel(R) Core (TM) i5-8250U processor at 2.13 GHZ and 2G of RAM. At the same time, a private blockchain network based on both the DPoS consensus algorithm for the Ethereum client and the RDPoS consensus algorithm for the Ethereum client on each virtual machine.

5.1. The Experiments of IECDSA

We tested the Digital Signature Algorithm (DSA), the RSA digital signature algorithm, and the IECDSA separately using the Bot-IoT Dataset collected by Koroniotis et al. [32].

As shown in Figure 3, the DSA, RSA, and IECDSA for digital signature time consumption all show an increasing trend as the traffic information increases. The RSA is used to sign traffic information, consuming 130 s when the number of traffic information points reaches 100 and up to 2295 s when the number of traffic information points reaches 2000. The DSA is used to sign traffic information, consuming 211 s when the number of traffic information points reaches 100, and up to 3831 s when the number traffic information points reaches 2000. The IECDSA is used to sign traffic information, consuming 0 s when the number of traffic information points is 100 and only 5 s when the number of traffic information points reaches 2000. In general, the DSA consumes the most time signing traffic information and the IECDSA consumes the least time signing traffic information.

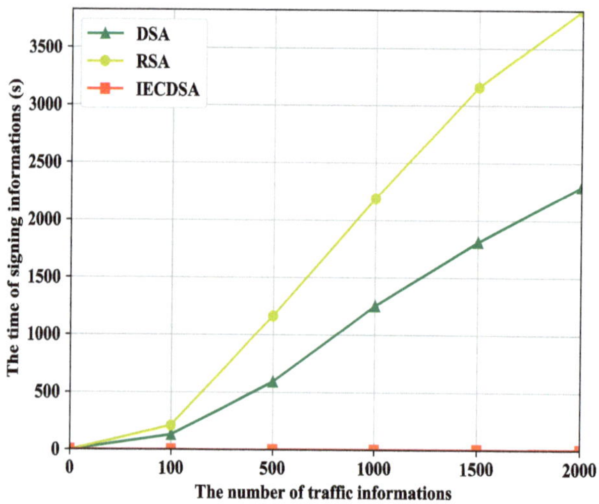

Figure 3. The time to sign information.

As shown in Figure 4, the DSA, RSA, and IECDSA algorithms all show an increasing trend in time spent on digital signature verification as the traffic information increases. The RSA is used to verify the traffic information signature, consuming 0 s when the number of traffic information points is 100 and 2 s when the number of traffic information points reaches 2000. The DSA is used to verify the traffic information signature, consuming 0 s when the number of traffic information points is 100 and 3 s when the number of traffic information points reaches 2000. The IECDSA is used to verify the traffic information signature, consuming 0 s when the number of traffic information points is 100 and 9 s when the number of traffic information points reaches 2000.

In summary, the time taken by the three digital signature algorithms to sign and verify traffic information shows that the IECDSA algorithm has a huge advantage when it comes to digital signatures. Although the IECDSA algorithm takes relatively more time to verify the signature, the difference is within a few seconds. Hence, we proposed an intelligent elliptic curve digital signature algorithm which is more advantageous when processing information shared by edge nodes.

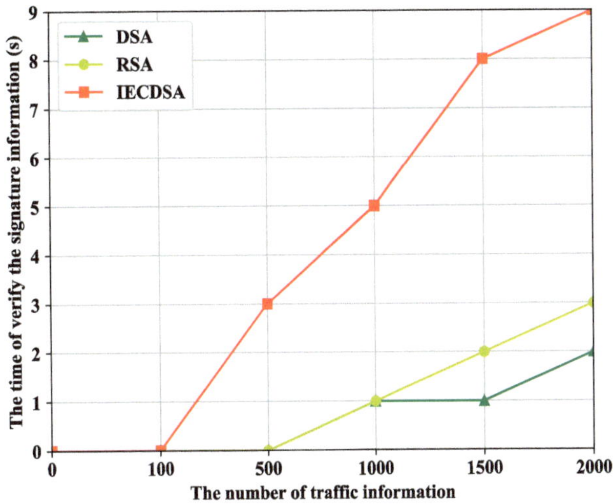

Figure 4. The time to verify the signature information.

5.2. The Experiments of RDPoS

This section analyses the rationality of proxy node selection and the RDPoS consensus algorithm.

Rationalization of proxy node selection: We chose four edge nodes with different reputation values for experimental validation: *Edge Node 1* (reputation value 0.8, trusted status Good), *Edge Node 2* (reputation value 0.6, trusted status Normal), *Edge Node 3* (reputation value 0.3, trusted status Abnormal), and *Edge Node 4* (reputation value 0.1, trusted status Error). The overall score corresponding to edge nodes receiving 10, 20, 30, 40, 50, and 60 votes was analyzed. The node status change is shown in Figure 5.

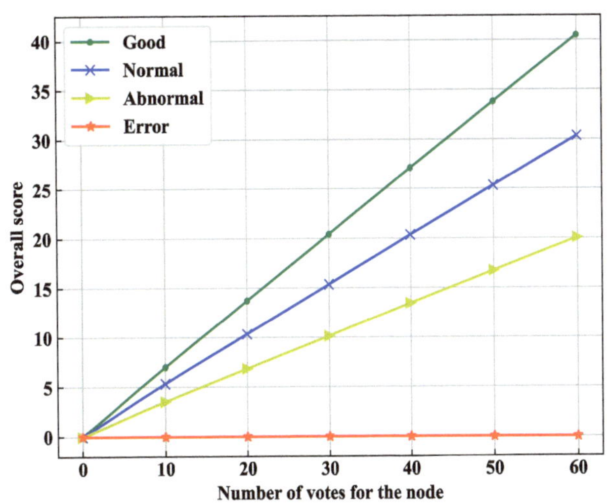

Figure 5. Changes in nodes with different reputation states.

The nodes with *Good* status have an increasing rating as the number of votes increases, and the nodes with *Good* status are always ahead of the nodes with *Normal*, *Abnormal*, and *Error* status. The nodes with the status *Normal* have an increasing rating as the number of votes increases, and nodes with the status *Normal* are always ahead of those with the status *Abnormal* and those with the status *Error*. The nodes with *Abnormal* status have an increasing rating as the number of votes increases, and the nodes with *Abnormal* status have a higher rating than the node with *Error* status. The nodes with the status *Error* have a constant score of 0 as the number of votes increases. As the number of votes continues to increase, nodes with high reputation values consistently score ahead of other nodes, and in general, nodes with high reputation values are more likely to be selected as proxy nodes.

RDPoS consensus algorithm: We select four edge nodes as proxy nodes out of the 15 simulated edge nodes and use a random generator to vote against the others to ensure that the voting is closer to the real voting scenario. The number of votes received by each edge node after the 1st round of voting is shown in Table 3. The four edge nodes with the highest number of votes were selected as proxy nodes for consensus based on the voting results.

Table 3. Results of the 1st round of voting.

Account	Edge Node	Reputation Value	Node Statu	Number of Vote
$0x511 \ldots 6202c7$	A	0.5	Normal	13
$0xbc3 \ldots 5a7abd$	B	0.5	Normal	4
$0x61f \ldots f6b828$	C	0.5	Normal	22
$0x59c \ldots 976260$	D	0.5	Normal	21
$0xd88 \ldots b33a21$	E	0.5	Normal	23
$0xf7a \ldots 0c927e$	F	0.5	Normal	0
$0xefa \ldots d2eb37$	G	0.5	Normal	3
$0x23d \ldots 0cb24f$	H	0.5	Normal	0
$0x929 \ldots 7dacef$	I	0.5	Normal	6
$0x640 \ldots e0576b$	J	0.5	Normal	2
$0x973 \ldots 9d023$	K	0.5	Normal	0
$0xdd9 \ldots b1364$	L	0.5	Normal	6
$0x72c \ldots a8c5b$	M	0.5	Normal	0
$0xfff \ldots 771332$	N	0.5	Normal	4
$0x5aa \ldots e645cc$	O	0.5	Normal	0

Round 2 voting sets *edge node C* to *abnormal* status and the other edge nodes to *normal* status. The 2nd round of voting is repeated for 30 rounds by voting on the basis of the end of the 1st round of consensus. Figure 6 shows the ranking of *edge node C* among the candidate nodes after voting by the DPoS consensus algorithm and the RDPoS consensus algorithm, respectively.

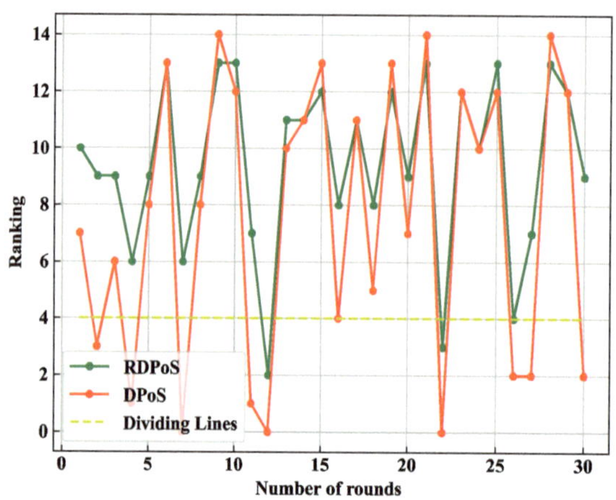

Figure 6. Ranking of anomalous *edge node* C per round.

From the ranking of *edge node* C in the 30 rounds of voting results, in the DPoS consensus algorithm, *edge node* C was selected as a proxy node for consensus 10 times, and the probability of an anomalous node being selected as a proxy node was 33.33%; in the RDPoS consensus algorithm, *edge node* C was selected as a proxy node for consensus only three times, and the probability of an anomalous node being selected as a proxy node was 10.00%.

The RDPoS consensus algorithm effectively reduces the probability of malicious nodes being selected in the process of selecting proxy nodes to ensure the security of the consensus.

5.3. The Experiments of Incentive Mechanism

In this section, we analyze trusted incentive smart contracts and incentive mechanisms based on information properties.

Trusted incentive smart contracts: Gas is finite for the users who need to consume it to send transactions and to deploy smart contracts and execute them. As shown in Figure 7, the deployment and execution costs of smart contracts vary linearly with the increasing number of set rules in the smart contract. The cost of deploying a smart contract increases as the number of rules in the contract increases, and the cost of executing a smart contract increases as the number of rules in the contract increases. Because smart contracts only need to be deployed once (paying for gas once on deployment) before they can be used, and they need to pay for gas every time they are executed, we see in the experimental results that their deployment cost is much higher than their execution cost. The red line in Figure 7 is the block of maximum gas, which is set when the Genesis block is initialized. It represents the maximum gas that a user is willing to pay to perform an operation or confirm a transaction, and if the block maximum gas is exceeded, the block will be rejected by the network.

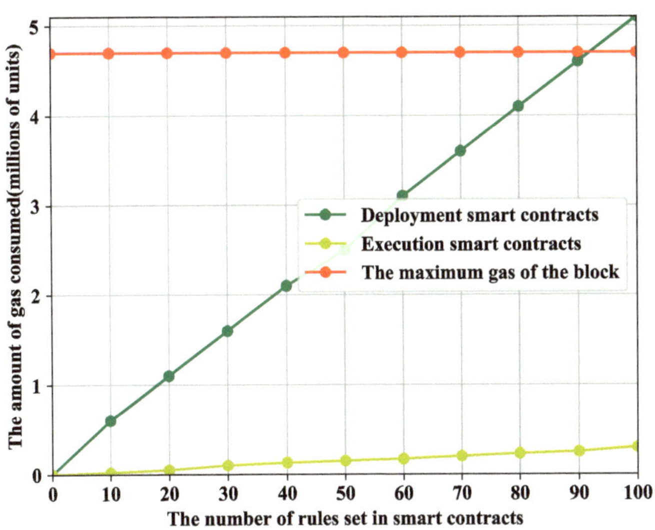

Figure 7. Smart contract gas consumption.

Incentive mechanisms based on information property: We tested the proposed incentive mechanism based on information properties by deploying trusted incentive smart contracts in a simulated private blockchain network based on the RDPoS consensus algorithm for the Ethereum client, and the test results are shown in Table 4.

Table 4. Edge node information sharing results.

Edge Nodes	Amount of Information Shared	
	Incentive Mechanisms	No Incentive Mechanisms
A	453	200
B	502	321
C	433	365
D	625	432
E	425	430
F	335	332
G	249	230
H	587	438
I	442	445
J	443	246
K	332	296
L	629	516
M	587	540
N	368	352
O	321	332

The experimental results show that 12 of the 15 simulated edge nodes with an incentive mechanism based on information attributes share more information than that shared in a normal case. The amount of information shared by *edge node E*, *edge node I*, and *edge node O* is slightly lower than the amount of information shared in a normal case. On the whole, the number of edge nodes sharing information under the incentive mechanism based on information attributes is significantly higher than the amount of information shared without the incentive mechanism. Thus, our proposed incentive mechanism based on information attributes stimulated the edge nodes to share information.

Although the number of simulated edge nodes during the experiment is limited, the above experimental results show that the blockchain-based privacy information security sharing scheme in the IIoT proposed in this work ensures the enthusiasm of smart factories in sharing private information under the premise of ensuring the security of private data.

6. Conclusions

In this work, we propose a blockchain-based privacy information security sharing scheme in IIoT to improve the motivation of smart factories to share information while ensuring the security of information sharing. Firstly, we propose an Intelligent Elliptic Curve Digital Signature Algorithm to sign the information shared by the smart factory and determine the ownership of the shared information. The algorithm not only protects the security of the key but also outperforms similar signature algorithms in terms of speed. Then, we propose a Reputation-based Delegated Proof-of-Stake consensus algorithm, which reduces the probability of malicious nodes being selected as proxy nodes and improves the security of data consistency among smart factories. Finally, we propose an incentive mechanism based on information attributes, and the amount of information shared by smart factories is significantly improved under the condition of using this incentive mechanism.

The scheme presented in this article was tested by the VMware Workstation, which affects the experimental results to a certain extent. In future work, we should test the proposed solutions on a real local area network. Although the incentive mechanism based on information attributes promotes the sharing of private data between smart factories to a certain extent, it ignores the competitive relationship among smart factories, and the introduction of game theory could be considered to improve the incentive mechanism in the future.

Author Contributions: Conceptualization, Y.W., X.Z. and T.C.; methodology, Y.W., T.C. and T.Z.; validation, T.Z., K.Z. and X.H.; investigation, Y.W. and T.C.; safety analysis, T.Z. and K.Z.; writing—original draft preparation, Y.W., T.C. and T.Z.; writing—review and editing, Y.W., K.Z. and X.H.; supervision, X.Z. All authors have read and agreed to the published version of the manuscript.

Funding: This work was supported by the Fundamental Research Funds for the Central Universities under Grant 2020ZDPY0223.

Institutional Review Board Statement: Not applicable.

Informed Consent Statement: Not applicable.

Data Availability Statement: Not applicable.

Conflicts of Interest: The authors declare no conflict of interest.

Appendix A. The Data Storage of Block

The block storage structure is shown in Figure A1. The block is divided into two parts: the block header, which consists of three sets of metadata: index data, consensus data, and transaction data, and the block body, which stores information shared by edge nodes in the form of the Merkle tree.

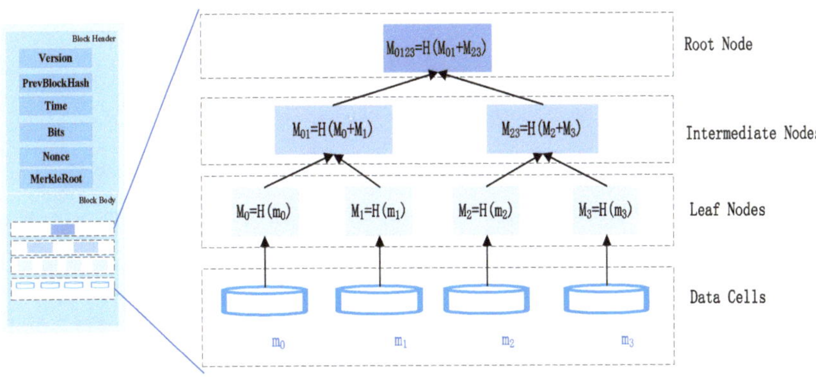

Figure A1. The storage structure of block.

The Merkle tree is generated by Equation (A1).

$$Y = Hash(x_i, x_j) \tag{A1}$$

where the length of Y is fixed. If one of the values of x_i and x_j changes, then the value of Y will also change.

Equation (A1) guarantees that any combination of two inputs will have a unique output value corresponding to it. It is impossible for an attacker to invert the values of x_i and x_j based on the value of Y.

The Merkle tree is divided from the bottom to the top into leaf nodes, intermediate nodes, and root nodes.

Leaf nodes: The hash value of the leaf node is obtained by hashing the data cell as a parameter. If the data block to be processed is an odd number, the last data cell needs to be copied so that the Merkle tree always remains a full Merkle tree.

$$M_i = H(m_i) \tag{A2}$$

Intermediate nodes: The hash of an intermediate node is the hash of the sum of its two child node hashes, calculated by Equation (A3).

$$\begin{aligned} M_{ij} &= H(m_i + m_j) \\ &= H(H(m_i) + H(m_j)) \end{aligned} \tag{A3}$$

Root node: The hash value in the root node is the hash value of the root of the Merkle tree, which is also stored in the block header.

References

1. Song, C.; Xu, W.; Han, G.; Zeng, P.; Wang, Z.; Yu, S. A cloud edge collaborative intelligence method of insulator string defect detection for power IIoT. *IEEE Internet Things J.* **2020**, *8*, 7510–7520. [CrossRef]
2. Ding, L.; Wang, Z.; Wang, X.; Wu, D. Security information transmission algorithms for IoT based on cloud computing. *Comput. Commun.* **2020**, *155*, 32–39. [CrossRef]
3. Zhang, W.; Yang, D.; Xu, Y.; Huang, X.; Zhang, J.; Gidlund, M. DeepHealth: A self-attention based method for instant intelligent predictive maintenance in industrial Internet of things. *IEEE Trans. Ind. Inform.* **2020**, *17*, 5461–5473. [CrossRef]
4. Meng, T.; Jing, X.; Yan, Z.; Pedrycz, W. A survey on machine learning for data fusion. *Inf. Fusion* **2020**, *57*, 115–129. [CrossRef]
5. Xiao, Y.; Jia, Y.; Liu, C.; Cheng, X.; Yu, J.; Lv, W. Edge computing security: State of the art and challenges. *Proc. IEEE* **2019**, *107*, 1608–1631. [CrossRef]
6. Jiang, D. The construction of smart city information system based on the Internet of Things and cloud computing. *Comput. Commun.* **2020**, *150*, 158–166. [CrossRef]
7. Nichols, K. Trust schemas and icn: Key to secure home iot. In Proceedings of the 8th ACM Conference on Information-Centric Networking, Online, 22–24 September 2021; pp. 95–106.
8. Yugha, R.; Chithra, S. A survey on technologies and security protocols: Reference for future generation IoT. *J. Netw. Comput. Appl.* **2020**, *169*, 102763. [CrossRef]
9. Shi, W.; Cao, J.; Zhang, Q.; Li, Y.; Xu, L. Edge computing: Vision and challenges. *IEEE Internet Things J.* **2016**, *3*, 637–646. [CrossRef]
10. Wang, Y.; Yu, J.; Yan, B.; Wang, G.; Shan, Z. BSV-PAGS: Blockchain-based special vehicles priority access guarantee scheme. *Comput. Commun.* **2020**, *161*, 28–40. [CrossRef]
11. Esposito, C.; Ficco, M.; Gupta, B.B. Blockchain-based authentication and authorization for smart city applications. *Inf. Process. Manag.* **2021**, *58*, 102468. [CrossRef]
12. Huang, H.; Zhu, P.; Xiao, F.; Sun, X.; Huang, Q. A blockchain-based scheme for privacy-preserving and secure sharing of medical data. *Comput. Secur.* **2020**, *99*, 102010. [CrossRef] [PubMed]
13. Torky, M.; Hassanein, A.E. Integrating blockchain and the internet of things in precision agriculture: Analysis, opportunities, and challenges. *Comput. Electron. Agric.* **2020**, *178*, 105476. [CrossRef]
14. Dhumwad, S.; Sukhadeve, M.; Naik, C.; Manjunath, K.; Prabhu, S. A peer to peer money transfer using SHA256 and Merkle tree. In Proceedings of the 2017 23RD Annual International Conference in Advanced Computing and Communications (ADCOM), Bangalore, India, 8–10 September 2017; pp. 40–43.
15. Sayeed, S.; Marco-Gisbert, H. Assessing blockchain consensus and security mechanisms against the 51% attack. *Appl. Sci.* **2019**, *9*, 1788. [CrossRef]
16. Xiao, Y.; Zhang, N.; Lou, W.; Hou, Y.T. A survey of distributed consensus protocols for blockchain networks. *IEEE Commun. Surv. Tutor.* **2020**, *22*, 1432–1465. [CrossRef]
17. Nakamoto, S. Bitcoin: A peer-to-peer electronic cash system. *Decentralized Bus. Rev.* **2008**, *2008*, 21260.
18. King, S.; Nadal, S. Ppcoin: Peer-to-Peer Crypto-Currency with Proof-of-Stake. 2012; Volume 19. Available online: https://www.semanticscholar.org/paper/PPCoin%3A-Peer-to-Peer-Crypto-Currency-with-King-Nadal/0db38d32069f3341d34c35085dc009a85ba13c13 (accessed on 1 April 2022).
19. Larimer, D. Delegated Proof-of-Stake (Dpos). Bitshare Whitepaper; 2014; Volume 81, p. 85. Available online: https://www.geeksforgeeks.org/delegated-proof-of-stake/ (accessed on 1 April 2022).
20. Castro, M.; Liskov, B.; et al. Practical Byzantine Fault Tolerance, 1999. In Proceedings of the Third Symposium on Operating Systems Design and Implementation, New Orleans, LA, USA, 22–25 February 2020.
21. Zheng, X.; Cai, Z. Privacy-preserved data sharing towards multiple parties in industrial IoTs. *IEEE J. Sel. Areas Commun.* **2020**, *38*, 968–979. [CrossRef]
22. Lu, Y.; Huang, X.; Zhang, K.; Maharjan, S.; Zhang, Y. Blockchain empowered asynchronous federated learning for secure data sharing in internet of vehicles. *IEEE Trans. Veh. Technol.* **2020**, *69*, 4298–4311. [CrossRef]
23. Liu, L.; Feng, J.; Pei, Q.; Chen, C.; Ming, Y.; Shang, B.; Dong, M. Blockchain-enabled secure data sharing scheme in mobile-edge computing: An asynchronous advantage actor–critic learning approach. *IEEE Internet Things J.* **2020**, *8*, 2342–2353. [CrossRef]
24. Zhang, W.; Ge, Y. Improvement of DPoS consensus based on block chain. In Proceedings of the 2019 4th International Conference on Intelligent Information Processing, Hong Kong, China, 16–17 November 2019; pp. 352–355.

25. Luo, Y.; Chen, Y.; Chen, Q.; Liang, Q. A new election algorithm for DPos consensus mechanism in blockchain. In Proceedings of the 2018 7th International Conference on Digital Home (ICDH), Guilin, China, 30 November–1 December 2018; pp. 116–120.
26. Tan, C.; Xiong, L. DPoSB: Delegated Proof of Stake with node's behavior and Borda Count. In Proceedings of the 2020 IEEE 5th Information Technology and Mechatronics Engineering Conference (ITOEC), Chongqing, China, 12–14 June 2020; pp. 1429–1434.
27. Ren, Y.; Liu, Y.; Ji, S.; Sangaiah, A.K.; Wang, J. Incentive mechanism of data storage based on blockchain for wireless sensor networks. *Mob. Inf. Syst.* **2018**, *2018*, 6874158. [CrossRef]
28. Kang, J.; Xiong, Z.; Niyato, D.; Xie, S.; Zhang, J. Incentive mechanism for reliable federated learning: A joint optimization approach to combining reputation and contract theory. *IEEE Internet Things J.* **2019**, *6*, 10700–10714. [CrossRef]
29. Zhan, Y.; Li, P.; Qu, Z.; Zeng, D.; Guo, S. A learning-based incentive mechanism for federated learning. *IEEE Internet Things J.* **2020**, *7*, 6360–6368. [CrossRef]
30. Chang, Z.; Guo, W.; Guo, X.; Zhou, Z.; Ristaniemi, T. Incentive mechanism for edge-computing-based blockchain. *IEEE Trans. Ind. Inform.* **2020**, *16*, 7105–7114. [CrossRef]
31. Tezuka, S. Linear congruential generators. In *Uniform Random Numbers*; Springer: Boston, MA, USA, 1995; pp. 57–82.
32. Koroniotis, N.; Moustafa, N.; Sitnikova, E.; Turnbull, B. Towards the development of realistic botnet dataset in the internet of things for network forensic analytics: Bot-iot dataset. *Future Gener. Comput. Syst.* **2019**, *100*, 779–796. [CrossRef]

sensors

MDPI

Review

Blockchain for Modern Applications: A Survey

Moez Krichen [1,2], Meryem Ammi [3], Alaeddine Mihoub [4] and Mutiq Almutiq [4,*]

1 Faculty of Computer Science and Information Technology, Albaha University, Alaqiq 65779, Saudi Arabia; mkreishan@bu.edu.sa or moez.krichen@redcad.org
2 ReDCAD Laboratory, National School of Engineers of Sfax, University of Sfax, Sfax 3038, Tunisia
3 Digital Forensics Department, Criminal Justice College, Naif Arab University for Security Sciences, Riyadh 14812, Saudi Arabia; mammi@nauss.edu.sa
4 Department of Management Information Systems and Production Management, College of Business and Economics, Qassim University, P.O. Box 6640, Buraidah 51452, Saudi Arabia; a.mihoub@qu.edu.sa
* Correspondence: mmatk@qu.edu.sa

Abstract: Blockchain is a modern technology that has revolutionized the way society interacts and trades. It could be defined as a chain of blocks that stores information with digital signatures in a distributed and decentralized network. This technique was first adopted for the creation of digital cryptocurrencies, such as Bitcoin and Ethereum. However, research and industrial studies have recently focused on the opportunities that blockchain provides in various other application domains to take advantage of the main features of this technology, such as: decentralization, persistency, anonymity, and auditability. This paper reviews the use of blockchain in several interesting fields, namely: finance, healthcare, information systems, wireless networks, Internet of Things, smart grids, governmental services, and military/defense. In addition, our paper identifies the challenges to overcome, to guarantee better use of this technology.

Keywords: blockchain; review; finance; healthcare; information systems; wireless networks; Internet of Things (IoT); smart grids; governmental services; military/defense

1. Introduction

Blockchain is a revolutionary paradigm that has introduced new concepts into securely sharing data and information. This modern technology consists of a chain of blocks that allows to securely store all committed transactions using shared and distributed networks [1,2]. To fulfill this goal, several basic technologies are adopted, such as the cryptographic hash function, distributed consensus algorithms, and digital signatures. All transactions are carried out in a decentralized way, removing the need for any mediators to confirm and verify them [3]. Blockchain has some key characteristics [4], such as:

- Decentralization: In the blockchain, a transaction can be performed between any two entities/actors without the need for central authentication. As a result, the use of blockchain can dramatically cut server expenses while also alleviating performance constraints at the central server.
- Persistency: it is nearly impossible to tamper with the system because each transaction must be validated and recorded in blocks dispersed across the whole network.
- Anonymity: With a created address, each user can communicate with the blockchain network. Furthermore, a user could generate a large number of addresses in order to protect his/her identity. It is (worth mentioning that just a few blockchain implementations offer anonymity. The majority of them are pseudonymous).
- Auditability: users can easily check and trace prior records by accessing any node in the distributed network because each transaction is confirmed and stored with a timestamp.

Blockchain was initially proposed for supporting the well-known cryptocurrency, Bitcoin [5]. However, during the last few years, blockchain was adopted in several new fields far beyond cryptocurrencies [6], including healthcare [7], intelligent transportation [8], and Internet of Things (IoT) [9]. Indeed, thanks to its ability to increase fairness and transparency and to help organizations save money and time, this technology is influencing a wide range of industries [10], ranging from basic individual entertainment activities to the management of critical and sensitive affairs of governments and states.

In this paper, we mainly focus on recent studies related to the incorporation of blockchain technology in modern applications, by comprehensively discussing the advantages and challenges related to the proposed solution. By doing so, we provide a survey on the use of blockchain in some modern applications (Figure 1 and Table 1):

- Financial Activities (Section 4);
- Healthcare (Section 5);
- Information systems (Section 6);
- Wireless networks (Section 7);
- Internet of Things (Section 8);
- Smart grids (Section 9);
- Governmental services (Section 10);
- Military and defense (Section 11).

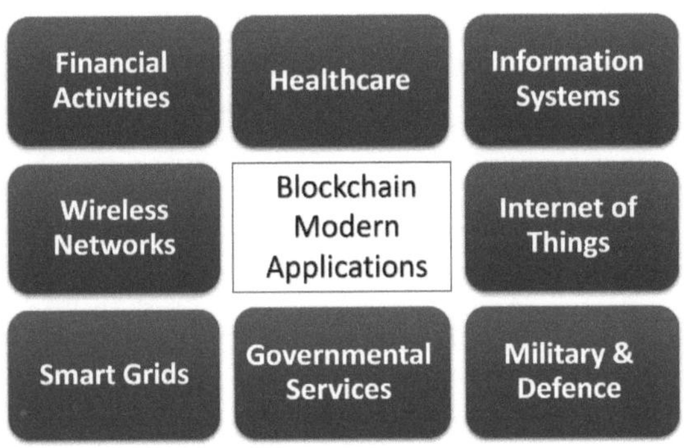

Figure 1. Blockchain application domains covered by this survey.

Table 1. Some examples of the use of blockchain technology in different fields.

<table>
<tr><th>Domain</th><th>Sub-Domains</th><th>Details</th></tr>
<tr><td rowspan="2">Finance [11–13]</td><td>Crowdfunding</td><td>Without the exorbitant fees charged by lawyers, creators obtain greater support for their initiatives with cheaper fees and overall costs.</td></tr>
<tr><td>Money transfer</td><td>Companies attempt to address a variety of concerns with this technology, including high transfer costs, limited money distribution methods, etc.</td></tr>
<tr><td rowspan="2">Healthcare [14–16]</td><td>Patient-centric health records</td><td>Developing a blockchain-based medical record system that can serve as a single, encompassing representation of a patient's data.</td></tr>
<tr><td>Staff credential verification</td><td>Blockchain technology can be used to track the experiences of medical experts, allowing trustworthy medical institutions to document the credentials of their employees.</td></tr>
</table>

Table 1. *Cont.*

Domain	Sub-Domains	Details
Information systems [17–19]	Preserving data integrity	The blockchain provides a secure, autonomous, and cost-effective proof-of-concept system that ensures that entries cannot be removed or changed.
	Cost efficiency and accuracy	Blockchain technology can reduce costs and increase accuracy while exchanging and storing vast amounts of data.
Wireless networks [20–22]	Security	Blockchain allows for secure communication with advanced wireless network technologies, such as edge computing, network slicing, open-source APIs, virtualization, etc.
	Access control	In wireless networks, blockchain technology provides a technique for anonymous access control.
Internet of Things [23–25]	Enhanced security	Blockchain offers a layer of security by encrypting data, eliminating single points of failure, and allowing users to rapidly discover the weakest link in a network.
	Reduced costs	The entire ecosystem may be made proactive at a lower cost by automating transaction validation and processing procedures on blockchain.
Smart Grids [26–28]	Renewable energy	To avoid double-counting, renewable energy certificates are recorded and awarded in real-time and automatically.
	Peer-to-peer trading	Automated smart contracts are used to sell excess renewable energy to other network participants.
Governmental services [29,30]	Registries	Using blockchain-based distributed ledgers to manage registries give the necessary transparencies to reduce fraud while also allowing for real-time modifications.
	Administration	Blockchain-based administration solutions allow for real-time collaboration across a wide range of stakeholders while also providing the necessary transparency.
Military and defense [31,32]	Marine aviation	Better tracking of aircraft replacement components, resulting in decreased operational costs.
	Logistics, procurement, and finance	The blockchain may be used to manage and register goods and services and it can be used to verify and register all financial transactions, improving efficiency.

For each domain aforementioned, we propose some related examples for the use of the blockchain technology while focusing on the corresponding benefits, limitations, and challenges. In Section 2, we present a quick summary of similar survey articles about blockchain technology that have been published between 2020 and 2022. In Section 3, an overview of the blockchain architecture is provided. Section 12 lists the main open challenges related to the use of blockchain technology. Section 13 presents a general conclusion of the paper.

2. Related Surveys

In this section, we provide a concise summary of nine related survey articles dealing with the use of blockchain technology in modern applications that have been published between 2020 and 2022.

An assessment of blockchain applications in smart grids with regard to cyber security perceptions and energy data protections was published in [33]. The authors discussed how big data and blockchain might help tackle major security challenges in smart grid scenarios. The researchers then identified a number of recent blockchain-based research papers that had been published in various journals, as well as examined security risks with smart grid technologies. They talked about a number of other recent practical designs, experiments, and products. Finally, they discussed some of the most pressing research issues as well as potential avenues for utilizing blockchain to address smart grid security challenges.

The writers of [34] conducted a survey and tutorial on blockchain applications in IoT systems. Based on the most important aspects, they suggested a blockchain taxonomy for IoT applications. They also looked at the most popular blockchain systems for IoT applications. They talked about how blockchain technology can be utilized to expand the range of IoT applications. Furthermore, they focused on new advancements and solutions for the IoT context. Finally, they discussed the obstacles and future research objectives for blockchain applications in IoT.

By assessing, arranging, and summarizing the literature, the authors of [35] offered a comprehensive overview of blockchain technology's role in tackling supply chain and logistics-related concerns. The proposed study demonstrated that blockchain technology may transform the supply chain and logistics services into secure, flexible, trustworthy, and transparent operations. The advantages of blockchain technology in giving provenance and traceability to crucial products are highlighted through an imagined application scenario.

The research [36] provides an overview of blockchains, including their construction, consensus techniques, and other topics. It compares algorithms based on their usefulness and drawbacks. The importance of blockchains in the sectors of smart healthcare, smart grids, and smart financial systems is also discussed in this study. Overall, this paper provides an overview of the blockchain domain's numerous protocols, algorithms, applications, difficulties, and potential.

The study provided in [37] focused on the potential applications of blockchain in future transportation systems that will be combined with connected and autonomous cars, in order to offer a general review of the current related literature and research activities on this subject. In addition, the writers focused on the problems, roadblocks, and future research areas associated with blockchain implementation in this context.

The writers of [38] provided an in-depth examination of blockchain technology's evolution, architecture, development frameworks, and security concerns. They also gave a comparison of frameworks, a categorization of consensus methods, and an examination of security threats and cryptographic primitives employed in the blockchain. Finally, they discussed critical future possible extensions and open research issues that researchers may investigate in order to make more progress in this field. The authors took a broad approach in this research and did not focus on the use of blockchain in any specific domains.

The paper [39] provides a comprehensive overview of blockchain technology's applications and use cases for securing and trusting smart systems. Readers of this article will gain a solid understanding of blockchain technology's applications and use cases.

The authors of [40] proposed a complete overview of blockchain applications, architectures, methodologies, and research issues in Industry 4.0. They presented a blockchain reference architecture for smart manufacturing, which drove their discussions on how to deploy blockchain technology to smart factory and smart supply chain applications. The authors covered only a limited number of limitations, namely, throughput and scalability; system integration, and privacy and security.

The authors of [41] proposed a taxonomy that incorporates both technical and application information and could help academics construct blockchain-based multimedia copyright protection systems. The study also explored several technical issues and suggested future research directions.

A summary of the previous studied survey articles is provided in Table 2. By studying these survey articles, we noticed that most concentrated on a few application domains in each article and did not provide enough details about the issues and challenges faced in the considered fields. For this reason, our paper aimed to cover a larger number of application fields and provide more insights into the problems and difficulties encountered in these domains.

Table 2. Summary of related surveys.

Ref.	Year	Domain	Goals	Limitations
[33]	2022	Blockchain for smart grid and energy trading	An assessment of blockchain applications in smart grids with regard to cyber security perceptions and energy data protections.	The authors concentrated only on security aspects and neglected other possible issues related to the use of blockchain technology.
[34]	2022	Blockchain for IoT systems	A survey and tutorial on blockchain applications, advancements, solutions, obstacles, and future research objectives for IoT systems.	The authors focused on a single application of blockchain technology (in the field of IoT systems).
[35]	2022	Blockchain for manufacturing supply chain and logistics	Comprehensive overview of blockchain technology's role in tackling supply chain and logistics-related concerns.	The authors focused on a single application of blockchain technology (in the field of manufacturing supply chain and logistics).
[36]	2021	Approaches toward blockchain innovation	Overview of blockchain and its importance in the sectors of smart healthcare, smart grid, and smart financial systems.	Only a few applications of blockchain technology were considered and a few challenges were covered.
[37]	2021	Blockchain for transportation systems	A survey on the use of blockchain technology for improving the operation and security of transportation systems.	Only one application of blockchain technology was considered and few challenges were covered.
[38]	2021	Blockchain evolution	In-depth examination of blockchain technology's evolution, architecture, development frameworks, and security concerns.	Adoption of a generic approach concerning the use of the blockchain; no specific application domains were covered.
[39]	2020	Blockchain-based smart systems	Comprehensive overview of blockchain technology's applications and use cases for securing and trusting smart systems.	Few details provided concerning the application fields and the corresponding challenges.
[40]	2020	Blockchain for Industry 4.0	A comprehensive review on blockchain in Industry 4.0 architectures, techniques, applications, and challenges.	A limited number of issues and challenges covered, such as throughput and scalability, system integration, and privacy and security.
[41]	2020	Blockchain-based protection of multimedia	Taxonomy incorporating technical and application information for constructing blockchain-based multimedia copyright protection.	Not enough details about possible challenges and eventual issues related to this topic have been provided.

3. Blockchain Architecture

A blockchain is a continuously expanding collection of data blocks linked together to form a long chain [42] as described in Figure 2. This network of connected data blocks represents a distributed ledger that is disseminated over a peer-to-peer network [43]. A distributed ledger contains a collection of digital data that are synced, replicated, distributed, and shared through a peer-to-peer network. Each device linked to the network maintains the latest version of the common ledger, i.e., each peer in the network has a copy of the ledger that is identical to the other. The ledger is mainly characterized by its safety, and the database can be expanded only by the addition of new blocks to the chain. Changes to records that have already been registered to the chain are computationally impossible. As a result, a primary benefit of the described distributed ledger is its decentralized nature. Indeed, there is no central authority that controls the ledger; however, each node updates its ledger when a new block is added to the blockchain, using a joint consensus mechanism [44]. Moreover, in the blockchain, and especially in the cryptocurrency networks, the authenticity of data is frequently verified by an asymmetric encryption technology known as public-key cryptography (PKC) [45]. In this technology, both the transmitter and receiver have a pair of keys consisting of a public key and a private one [46]. The private key is exclusively accessible to the nodes that created it, whereas the public key is spread rather freely throughout the network. The sender encrypts the data using the receiver's public key. Since data are encrypted using the receiver's public key, they can only be decrypted using the receiver's private key. Furthermore, in the case of sending transactions on a blockchain network, a transaction is deemed complete only after it is digitally signed. Following that, the transaction is signed by the sender using his private key. For the receiver, the transaction's authenticity i.e., the sender's' identity, can be checked using the associated public key (belonging to the sender). This way, all transactions are automatically checked and authenticated by nodes and the network rejects any unauthenticated transactions. Please note that on a blockchain network, an authentic, mined transaction is irreversible [47].

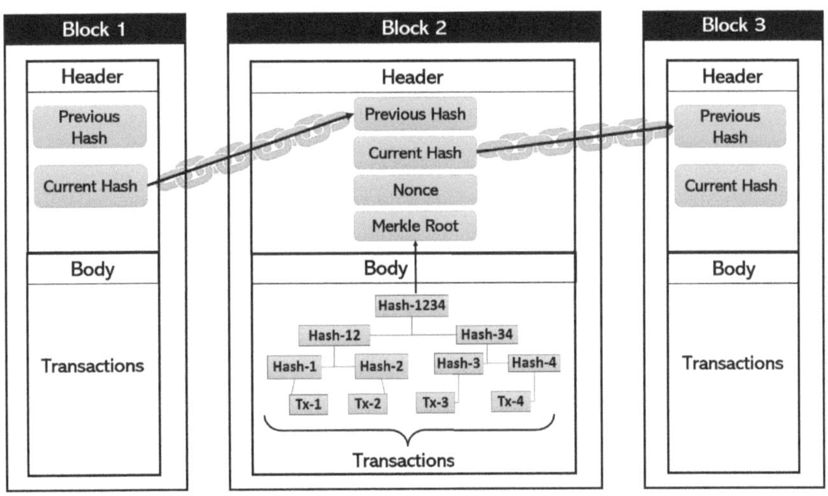

Figure 2. Blockchain general architecture.

Actually, it is difficult to alter the data contained in blocks thanks to the cryptographic qualities of the blockchain. Practically, the blocks are connected via a hash reference since each subsequent block carries the previous block's hash value in addition to the actual block's hash value (Figure 2). Generating a hash value is feasible through the use of a mathematical and sophisticated cryptographic hash algorithm, which accepts any input type and outputs a fixed-length number termed as the hash value. The primary characteristic of a hash function is that if a single fraction in the input is changed, the entire value in the output will be altered [48]. Consequently, if an attacker attempts to edit data in Block 1 (B1) for instance, the hash value of that block (B1) will be modified in the following block (B2), and so the intruder will have to modify the hash value of that block. Moreover, because B2 curries the hash of B1, any modification in the hash will alter the hash value of B2 in B3. As a result, if someone wants to modify a block, he or she must modify the data for all subsequent blocks on the blockchain. Additionally, even if the hash value of a block is known, calculating the hash function's input is difficult due to the hash function's non-invertible feature [46].

The next question is how to add new blocks to the network. Indeed, if we take the special case of the bitcoin cryptocurrency, there are particular types of nodes called "Miners" that are responsible for building new blocks in the chain [47]. The miner's job is to update (from prior transactions) the records of the blockchain public ledger. Any network node could be a miner. It takes miners hours to create a new block because they must resolve a mathematical puzzle called "Proof of Work" (PoW). Several miners can work in parallel to add a new block. Nevertheless, only one miner can add a novel block at a moment. The first miner to solve the PoW problem can mine that new block. To address the mining PoW problem, huge computing power is needed. We could break down the whole process into multiple steps:

- To begin mining a new block, a miner gathers transactions from the shared network and organizes them in a block.
- The miner will verify the blockchain's prior hash value and deposit it with the transactions in the intended new block.
- The miner will obtain and save in the same block a variable called "nonce" (Figure 2). This variable value can be altered at any time by the miner.
- The miner will now investigate the network's PoW puzzle. The problem consists of finding, for the whole new block, a special hash value starting with several zeros. This special hash value can be found by changing the nonce value which is the only parameter that the miner can modify. Once the miner discovers the same amount of beginning zeros for a given nonce value, he/she can broadcast the answer to the network and demonstrate that he/she succeeded in mining a new block. Note that the number of successive zeros indicates the mining difficulty level.

The nodes of type miners are also responsible for verifying all data contained within a block. To this end, the data of one block are saved with the shape of a Merkle tree, which represents a particular data structure in the form of a hash-based tree (Figure 2). Trees make data verification simple. Consider using the hash function of all transactions, not the structure of the Merkle tree. If a single transaction is altered, the entire hash result will be modified, making it impossible to detect the altered data. However, using the particular structure of the Merkle tree, we can see at any fraction of the tree which part delivers the erroneous hash value. Assume an attacker alters transaction Tx-3. As a result, we can easily detect that only the right side of the Merkle tree gives incorrect hash outputs. Because the hash values of Tx-3 and Tx-4 will be erroneous, we do not need to check Tx-1 and Tx-2. Consequently, the Merkle tree is extremely useful for data verification in peer-to-peer distributed systems [49].

4. Blockchain for Financial Activities

Blockchain technology has been massively used in the financial and economic sectors [12,13]. For instance, it has been used for the settlement of financial market transactions, trade finance, insurance, real-time money transfer, cross-border payments, etc. Bitcoin was the world's first decentralized cryptocurrency and a payment system not backed by a central bank. Without the need for an intermediary, transactions are performed directly between users through the P2P network [50] (Figure 3). Other cryptocurrencies, such as Bitcoin Cash, Ethereum, Ripple, and Dash are also available. The conventional cross-border payment system is based on the banking system, which has the disadvantages of being expensive, time-consuming, and less secure. However, by using blockchain to rebuild this payment system, all of these constraints may be efficiently solved [51]. Asset ownership (e.g., car, house, stocks, etc.) can be recorded, transferred, and verified using blockchain technology, as well as the integrity and validity of sensitive documents or data. The authors in [52] presented an extensive analysis of the differences between the main known cryptocurrencies in terms of release date, founder, the hash algorithm used, and the language used to develop it. Another interesting comparison between cryptocurrencies and the technology-based of blockchains and distributed ledgers behind them is found in [53].

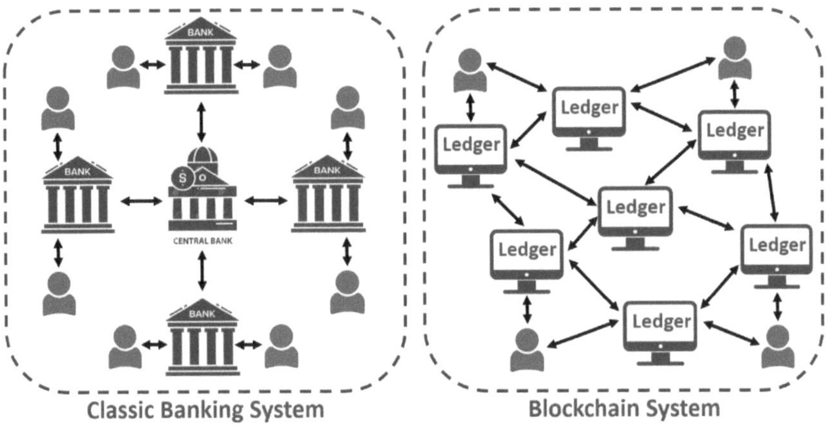

Figure 3. Illustration of the differences between the classic banking system (**left**) and the blockchain system (**right**).

Even though the use of blockchain technologies in the economic and financial fields appears to be highly promising, it still has a number of limitations [54,55]:

- Blockchain is too slow since it only allows for eight transactions per second. As a result, it has a significant disadvantage over the current third-party payment system Alipay [56], which can handle hundreds of transactions per second.
- If a private key or password is lost or disclosed, the blockchain system is impossible to recover, resulting in irreversible loss of consumer assets.
- Despite the fact that the blockchain is theoretically tough to crack violently, the risk of a data breach still exists.
- People still have a limited grasp and acceptance of blockchain technology, making it difficult to identify genuine and useful blockchain financial solutions.
- The lack of a centralized structure has made money laundering, fraud, and tax evasion more convenient, while also making supervision and control more complex.

5. Blockchain for Healthcare

Despite the significance of medical data sharing, health systems usually compel a patient to collect and exchange his/her medical information with medical staff, either in print form or electronically on some storage devices. This method of distributing medical records is inefficient since it is slow, insecure, and incomplete. Moreover, it is "provider-centric" instead of being "patient-centric". The inefficiency of this sharing method is mainly due to the lack of credibility between healthcare institutions and the lack of interoperability between the different IT platforms used by these institutions. According to [57], healthcare interoperability should cover three main levels, namely: foundational, structural, and semantic. This interoperability issue may be solved using blockchain technology [58,59]. Indeed, with blockchain implementation, patient medical information will be shared with necessary permissions using smart contracts for controlling operations, such as the change of viewership rights or the creation of new records. Next, we consider some examples of the use of blockchain technology in the healthcare field [60] (Figure 4):

- Patient identity: Patient identification [61] is a critical component of health information exchange. According to [61,62], medical errors cause 195,000 deaths every year in the USA, with identification problems accounting for 57% of the total number of errors. In such a situation, blockchain technology can impose a verifiable standardized identity for each patient through a universal patient index database, which may be shared between all healthcare facilities [63].

- Health records: Generally, the classical computerized centralized systems [64–66] do not address the root of the patient data sharing problem. However, thanks to blockchains [67–69], a patient may simply collect his/her medical history without asking for a copy from each provider he/she has visited. In this way, the blockchain technology allows for the creation of widely secure and accessible data distribution services that interface with different existing healthcare systems. Moreover, due to the use of a blockchain, data sharing between the patient and the doctor becomes easier and more secure [70].
- Telemedicine: Patients who are connected to the internet can avoid spending time in the healthcare center and receive fast treatment for small but critical problems. However, distant medical professionals may be unable to continuously access health data obtained during telemedicine treatment episodes, resulting in an incomplete medical history and putting the overall quality of care at risk. As a result, in this situation, the blockchain technology [71–74] can bridge the communication gap between different providers by eliminating the need for third-party authorities and empowering engaged participants to interact directly.

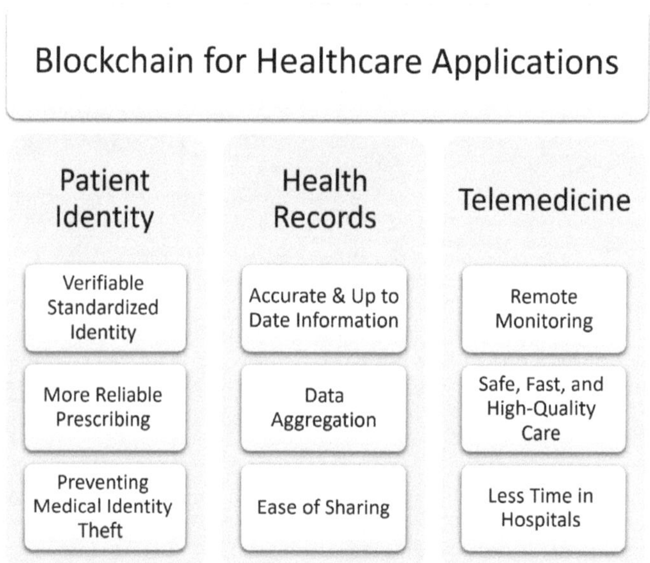

Figure 4. Blockchain for healthcare.

At this level, it is worth noting that the ability to store and handle large volumes of patient health data, ensure privacy and reduce operational costs are all requirements for implementing blockchain in healthcare [75–77].

6. Blockchain for Information Systems

An information system [78] is a collection of many different types of data that ensures the achievement of a business goal. Information systems are not really stand-alone IT business models. Integration with data and business processes, on the other hand, is a critical part of successful implementation. As a result, it is indeed easier to visualize the information system as a triangle. Processes, people, and computers are represented by the three elements of this triangle. To be successful, an information system must have all of these components working properly. The choice to integrate blockchain technology into information systems allows organizations to benefit from the vast array of applications and advantages that blockchain offers [79–81] (Figure 5).

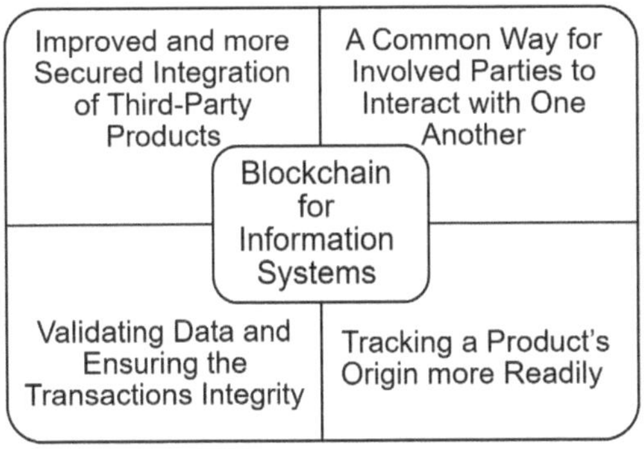

Figure 5. Blockchain for information systems.

Businesses, governments, and other organizations that maintain information systems sometimes rely on third-party agents or technologies to complete certain tasks. This necessitates the existence of a trust network among the partners involved, which is even more important when sensitive information is involved. Blockchain allows for improved and more secured integration of third-party products [82–85], while reducing the danger of revealing sensitive information to such parties. In addition, interoperability [86–88] fosters the promotion and acceptance of blockchain by providing a common way for involved agents to interact with one another via blockchain transaction ledgers and integrated networks, ensuring the validity of each engaged party.

Because blockchain is designed to be decentralized, it is an excellent contender for validating data and ensuring the transactions integrity. The adoption of the notion of "smart contract" [89–92] is one way to ensure transaction integrity. The purpose of smart contracts, as the name implies, is to allow the use of blockchains to ensure that two parties have an agreement being specifically composed into lines of code. This latter controls the execution, and exchanges are trackable and irreversible. If necessary, the blockchain could be utilized to resolve any disagreements that arise by confirming the authenticity of digital signatures in a safe, decentralized manner.

The fascinating utility of information systems, blockchain, and supply chain integration has been discovered for a range of businesses [93–95]. For example, many businesses consider product provenance to be critical. Blockchain can help track a product's origins more readily due to local regulations, preferences, tax reductions, and other incentives to identify provenance tracking. The entire supply chain, including logistical factors, can benefit from provenance. An item can be officially confirmed at any time, and transactions cannot be falsified or altered for the purpose of deceiving the final consumers of the products [96–98].

In conclusion, there are several considerable advantages to the use of the blockchain technology in the commercial world. However, there is a real significant risk that for many small- and medium-sized enterprises, the overhead costs of implementing integrated blockchain technology would be prohibitive and almost infeasible.

7. Blockchain for Wireless Networks

Wireless applications, such as broadband internet connections, mobile smartphones, and internet of vehicles [99,100] all require radio spectrum [101], which is precious and restricted resources. Wireless networks, such as cellular and Wi-Fi, are the most cost-effective ways to provide broadband internet access, particularly in low-income areas and emerging nations. As a result, diverse spectrum management regimes are needed to optimize advantages from the utilization of the available spectrum by mandating efficient spectrum usage while minimizing interference between consumers [102]. The traditional spectrum management regime has two major drawbacks. First, large portions of the licensed spectrum are underused. Second, this command-and-control spectrum management regime is slow to respond to market and technology changes [103]. Spectrum sensing [104], supporting secondary spectrum trading marketplaces [105], spectrum sharing [106], and policy enforcement [107] are all possible uses for the blockchain technology in spectrum management [108].

Blockchain technology may be used to create a secure spectrum sensing system as well as enable collaborative sensing, both of which improve the accuracy of spectrum sensing data. Mobile network operators can use spectrum sensing to combine available empty frequencies with their licensed frequencies to boost network capacity. Collaborative sensing, which includes fusing the sensing findings from a number of secondary sensors or users, can ensure the efficiency of spectrum sensing outcomes. The blockchain was first used as a peer-to-peer payment system. As a result, it naturally lends itself to the creation of a full-spectrum payment system based on digital currency that can be quickly converted to fiat currency. The blockchain technology can be used to accomplish the many functions of a geolocation database as well as the needs of spectrum management. The use of blockchain to actively store information about unoccupied spectrum bands and user geolocations is expected to increase spectrum access and utilization efficiency as well.

A secure spectrum sensing technique based on blockchain is presented in [109] to increase the energy efficiency and sensing accuracy of cognitive wireless networks at the same time. The mechanism can adapt to changes in the environment and adjust the number of nodes engaging in cooperative sensing in real-time, as well as evaluate the dependability of sensing nodes in real-time and calculate the node's trust value using an evaluation algorithm. Not only does the system record each node's energy consumption and sensing performance, but it also remembers the trust value of a single node. The trust value is recorded in the blockchain's reliability list, which is encrypted by the blockchain's management center to ensure that each node matches its own trust value. The suggested algorithm in this research may take into account both energy efficiency and sensing accuracy, extending the working life of cognitive wireless networks, according to experimental data.

Blockchain technology and reputation system were introduced into the spectrum sensing method in this research. A new secure spectrum sensing approach is presented. The user's direct reputation and referral reputation are both evaluated in this security sensing method. When a cooperative node asks for access to a certain frequency band, it must first determine whether the band is available. It will send a suggestion request to the fusion center if it is unresponsive. The sensing findings are more accurate in order to prevent collusion attacks and malicious node behaviors. The historical sensing records in the database and the distance of interaction history are regarded as a public ledger using blockchain technology, which can be shared by each neighbor node and no node in this situation can change the ledger information.

Spectrum management using blockchains is a new application with a lot of opportunities and challenges. Spectrum sensing and geo-location databases are the two main technologies used for providing dynamic spectrum access. Previously, these approaches were viewed as separate strategies in previous research. Because blockchain is a database technology, it may be used to create a unified method in which spectrum sensing techniques and geolocation database technology work in tandem. A more robust dynamic spectrum management framework will arise from combining these two spectrum access strategies. It is also necessary to investigate the integration of blockchains with the communication networks. The blockchain network could be set up as an overlay on top of the communication network, allowing communication network nodes to operate as complete nodes on the blockchain network. This network structure, however, is energy-intensive and necessitates a specialized control channel for transferring blocks and transactions over blockchain networks [110]. The possible applications of blockchain technology for wireless networks are illustrated in Figure 6.

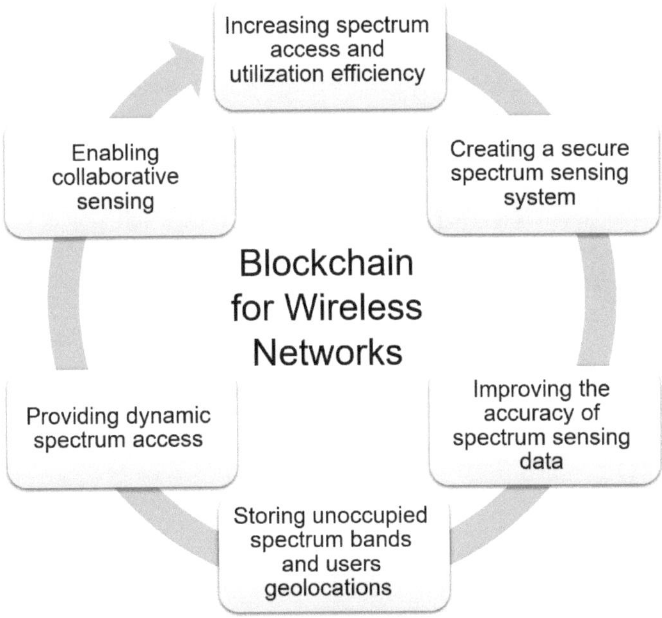

Figure 6. Blockchain for wireless networks.

8. Blockchain for Internet of Things

The Internet of Things (IoT) [111–113] is the linking of smart devices for data collection and intelligent decision-making. Yet, IoT is prone to privacy and security risks due to the absence of inherent security measures. The dispersed and centralized architecture of the Internet of Things is a significant challenge [114–116]. Every node in an redIoT infrastructure is typically a potential point of weakness that could be used to start cyber assaults. Data confidentiality and authentication are other continuous and serious threats. IoT data could be hacked and misused if data security is not established [117]. Data integrity is another issue for IoT. Decision support systems are one of the most important IoT applications. As a result, protecting the system from injection attacks, which attempt to insert bogus measures and, thus, impact decision-making, is critical. For automated systems, such as manufacturing sectors and vehicular networks [118], which handle real-time data, availability is crucial. The inclusion of a publicly verifiable audit trail that is not reliant on a trusted third-party is essential, as it addresses all of these issues. Blockchain may assist in solving major security concerns in IoT with its "security by construction" feature [119,120].

Blockchain is the final piece of the puzzle in resolving IoT privacy and dependability issues. The blockchain's inherent trustless, autonomous, and decentralized characteristics make it suited for use in a variety of scenarios. The blockchain technology, for example, may store a permanent record of smart gadgets [121,122]. Furthermore, the implementation of smart contracts may allow smart devices to perform autonomously, avoiding the need for human control or centralized authority. In addition, blockchain can establish a secure means for smart devices to communicate with one another [123,124].

The contribution in [125] can be viewed as a generic solution that can be used in any field of the IoT environment. Indeed, the authors of this paper developed a mechanism that would allow sensors to trade Bitcoin for data. Every node has a unique address that corresponds to the Bitcoin pub-key. When a user needs data from a sensor after locating it in a sensor repository, he sends a transaction directed to that sensor's public key. The sensor will reply by sending a transaction containing data to the client. This strategy is an extension of the solution provided in [126]. The Enigma framework [127] offers yet another intriguing solution. The latter makes use of a completely comparable concept—distributing data over multiple nodes while separating data from its references. Furthermore, in addition to making it difficult to reconstruct the original form of data, Enigma offers an extra layer of protection by encrypting such data chunks. As a result, Enigma is a P2P network that allows several participants to store and process data at the same time while maintaining privacy.

To summarize, the usage of blockchain for IoT applications provides excellent levels of security, which prevent unwanted data access (Figure 7). Yet, scalability [128] is still an open question since the blockchain can grow in size over time, making it difficult to acquire and save the ledger.

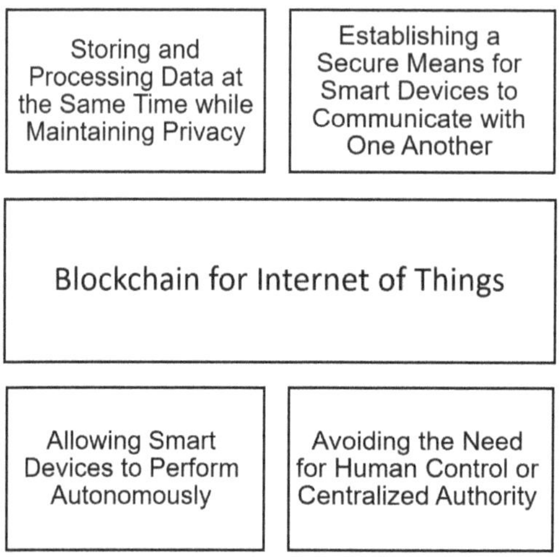

Figure 7. Blockchain for Internet of Things.

9. Blockchain for Smart Grids

A smart grid [129–132] is a digital communications-based electrical network that provides for the two-way flow of electricity and data, and also the identification, reaction, and avoidance of changes in usage and other difficulties. Current smart grids integrate communication and control techniques into power networks, allowing for considerable gains in energy efficiency and system safety. Traditional centralized techniques of managing smart grids pose significant hurdles. For instance, the centralized control method creates a dangerous single point of failure for the whole grid. In addition, many security issues have been growing and external security assaults could result in significant financial losses. To overcome these limitations, the use of blockchain technologies is considered a good choice in several research and industrial projects [133–135]. Indeed the use of blockchain for smart grids may have the following advantages (Figure 8):

- The blockchain has the potential to turn centralized grid administration into distributed intelligent administration.
- In terms of energy trading, a smart grid with blockchain technology can achieve optimum data flow and cash flow.
- Because of its decentralization and fault tolerance, blockchain can dramatically improve the privacy and security of power grids.

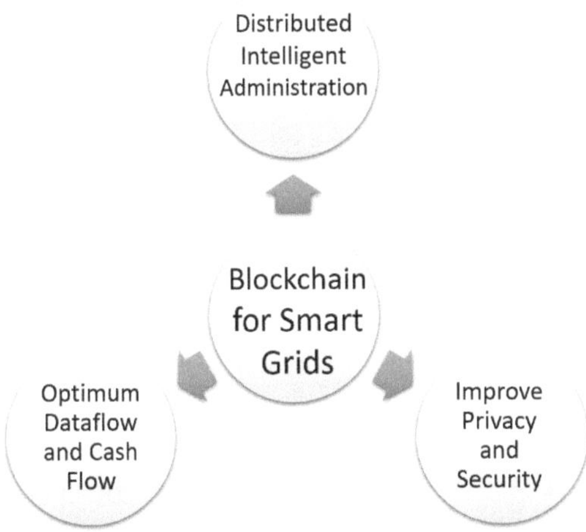

Figure 8. Blockchain for smart grids.

Incorporating cryptocurrencies for payment is one of the most important applications of blockchain for smart grids. BASNederland was the first company to use Bitcoin as payment for energy bills. This prompted numerous additional companies to develop blockchain-based billing and metering services, with several of them offering incentives to consumers who pay with cryptocurrency. For instance: Bankymoon in South Africa using Bitcoin, Spectral, and Alliander in the Netherlands using Jouliette, PowerLedger in Australia using Sparkz, LO3Energy, and ConsenSys in the USA using Ethereum, etc.

Electric vehicles [136,137] can be thought of as mobile power grid terminals that perform key services. This is known as V2G technology, and it has the potential to increase the power grid's reliability, efficiency, and stability. Electric vehicles, on the other hand, are not properly linked with smart grids, and there are a number of issues, such as energy shortages, security hazards, and data leakages. In this context, excessive charging loads and unsteady voltage in electric vehicles can be addressed with blockchain technology, as shown in [138,139]. In addition, using blockchain to connect smart grids and electric vehicles can lead to cost optimization through the use of smart contracts. Furthermore, using blockchain technology to connect smart grids and electric vehicles might reduce costs using smart contracts, as proposed in [140].

Although the use of blockchain technology for smart grids appears to be promising, as previously demonstrated, there are still hurdles in entirely converting to this new technology. For instance, re-architecting presents grid networks; implementing blockchain in the smart grid necessitates large infrastructural expenses, which will probably make grid operators hesitant to incorporate blockchains into their grid structures.

10. Blockchain for Governmental Services

Despite the fact that e-government initiatives have attempted to provide public services that are more straightforward, distributed, and adapted to the needs of inhabitants [141], they have never truly altered the functions of government agencies in record-keeping and management. One of the most important benefits of blockchain technology is the ability to promote direct interactions between government agencies, citizens, and businesses. As a result, blockchain technology has the potential to redefine how governments engage with individuals and each other, forcing public administrations to reconsider their roles in providing public services [142].

Governments might use this technology to take on supervisory functions over exchanges in a blockchain-based infrastructure. Blockchain has the potential to eliminate a considerable portion of the administrative functions that governments currently play in society, necessitating a shift in the governance of (public) service supply. This has the potential to change existing institutional frameworks, such as legal and public institutions [143].

Next, we provide a short overview of the adoption of blockchain technology by different governments in the world (Figure 9):

- China: the Chinese government declared that it would begin employing blockchains in invoice issuance and tax collection.
- Japan: The Japanese government announced that it will be experimenting with a blockchain-based system for handling government tenders. The technology consists of allowing users to obtain information electronically, such as tax payment documents.
- USA: the US government was looking for contractors to evaluate how blockchain technology may be incorporated into its contract bidding mechanism.
- Britain: The incorporation of blockchain technology into governmental operations in the United Kingdom was offered as an interesting case study. The main concept behind blockchain use is to automate the registration and payment of government grants and perks.
- Estonia: blockchain technology has been integrated by the Estonian government in official announcements, digital court files, property registries, succession registries, business registries, etc.
- Sweden: the Swedish government has begun to explore the use of blockchain technology to support real estate transactions.

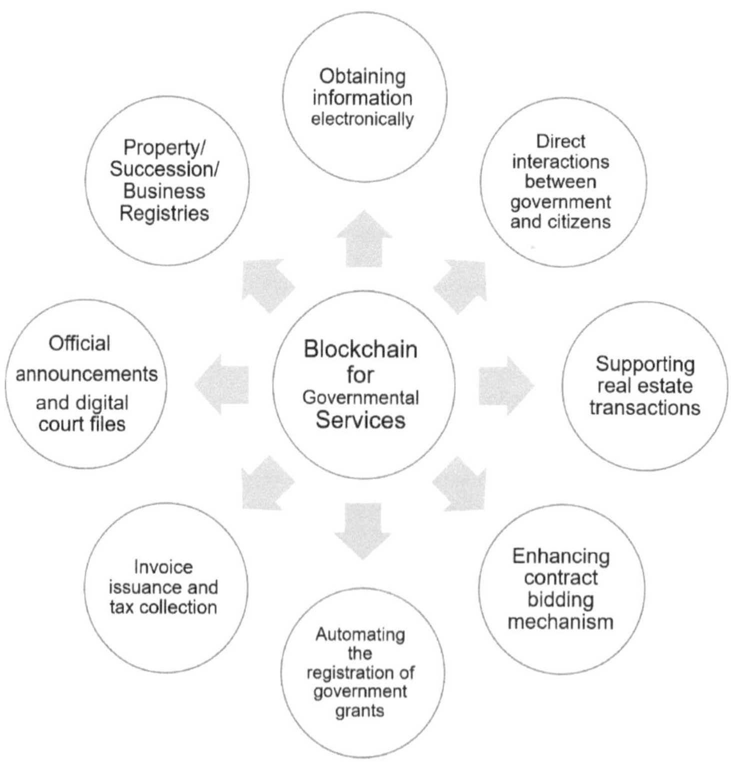

Figure 9. Blockchain for governmental services.

More research on the influences of these blockchain topologies on the technology–institution interface is required. Adopting blockchain technology for public services could result in not just a shift in the function of governments, but also a loss of jobs and a worsening of the digital divide. To minimize unforeseen repercussions when using this technology in the public sector, researchers should conduct research to compile a list of these effects. Finally, a study into public administration opinions toward blockchain technology could hasten its implementation.

11. Blockchain for Military and Defense

Military leaders who embraced cyber technology in the 1990s and early 2000s are now attempting to address the massive vulnerabilities that those same digital technologies produced [144,145]. Decades of hacking and exploiting cyber security systems have repeatedly proven how a determined cyber attacker may compromise military and civilian networks. The threat of sophisticated weapon systems being harmed or disabled by non-kinetic impacts have forced militaries to develop a long-term and ideally cost-effective defense for military systems [146]. Blockchain, and its as-yet-untested military uses, have the ability to shift the security vulnerabilities of some cyber systems from a single-point-of-failure vulnerability model, in which an attacker only needs to compromise one node to violate the system, to a majority-compromised vulnerability model, in which a malicious actor cannot exploit a single point of failure. The adoption of blockchain in the military field may cover the following aspects (Figure 10): (1) intrusion detection; (2) infrastructure monitoring; (3) battles management; (4) UAV management; (5) supply chain management; (6) encrypted communications.

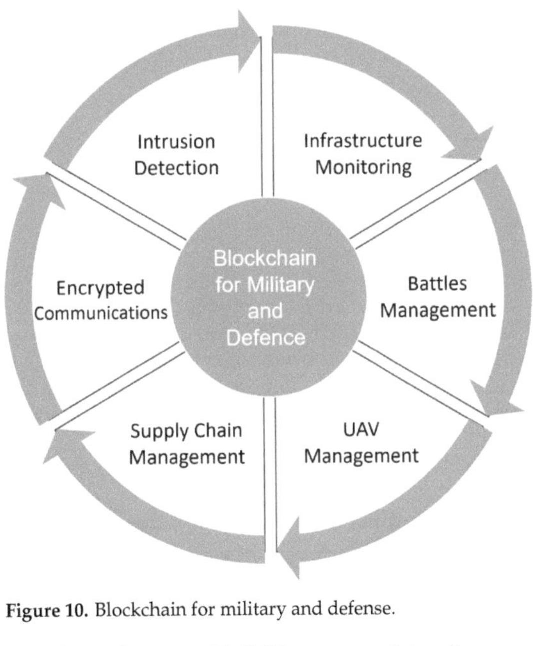

Figure 10. Blockchain for military and defense.

The work presented in [147] proposes an interesting comparison of the adoption of blockchain technology by three of the strongest armed forces in the world:

- USA: outside of the realm of cryptocurrencies, US military conversations have centered on improving data resiliency, with the premise that the US military could eliminate data compromise and corruption as threats to its data, and that the blockchain technology might act as a cyber security shield.
- Russia: the Russian Ministry of Defense announced the creation of a research laboratory tasked with establishing a blockchain system for detecting and mitigating cyber attacks [148] on crucial military digital infrastructure.

- China: the interest in military applications of the blockchain technology in China concentrated on equipment management, professional learning, logistics, and the conversion of commercial information technologies into defense programs.

Even though the military applications built on top of the blockchain, so far, do not seem to be completely ready for use. Defense logistics and data security are likely to be the applications that will be concretely implemented for the military blockchain in the near future. On the other hand, the adoption of blockchain by the world's strongest militaries is somewhat paradoxical. Indeed, while blockchain has the potential to share governance among citizens and guarantee more individual liberties, for the time being, the most centralized human organizations are committed to using this same technology to create a decentralized technology for military and defense applications.

12. Open Challenges

Many industrial unresolved problems need to be addressed and examined further in order to develop more usable and successful blockchain-based applications. In what follows, we discuss the main open problems.

- An in-depth study of the blockchain-based solution benefits: When applied to replace existing solutions [149], blockchain is a new technology that has the potential to destabilize the market, by introducing revolutionary ways that may transform society [150]. As a result, it is critical to establish whether a blockchain is truly required for a given application [151].
- Proper implementation: Blockchain is a general-purpose method of data manipulation that may be used in a variety of systems for various reasons, as long as its implementation has some degree of comprehension or maturity regarding its importance as well as the trade-offs. Indeed, the blockchain as a technology has various architectures, and different transaction processes; thus, its implementation is not a straightforward operation. Hence, its incorporation in different applications requires an in-depth and comprehensive study [152].
- Standard testing mechanism: another challenge faced when adopting a blockchain-based application is the need for a standard testing mechanism.
- Resilience to security risks: The resilience to security risks needs to be formally proved. With large-scale applications, the blockchain may face malfunctioning due to the system design or cyberattacks that intend to compromise its security.
- Scalability: This issue is raised basically from the fact that blockchain-based transactions are very slow to be processed and verified. Processing the transactions depend on the performance of the processing system. In [153], limitations of the proposed scaling methods are pointed out.
- Integration with other systems: This issue is a straightforward impact for organizations willing to adopt blockchain-enabled solutions. Indeed, the integration process will imply costs related to infrastructure change, trained staff, specialized developers, and management expectations [153].
- Energy challenges: the use of blockchain will undoubtedly require energy consumption much higher than the usual one. This challenge becomes an environmental issue when the energy used exceeds the load power and the equipment is fully utilized [154].
- Regulatory issues: regulations are of extreme importance to generalize and accept the use of blockchain-enabled solutions.
- Storage: The integration of blockchain with data-intensive applications, such as those based on the IoT, raises the problem of data storage. Indeed, blockchain stores data into blocks that cannot support large volumes of data. The authors in [7] proposed a hybrid architecture that combines blockchain with a decentralized database called IPFS. Another solution involves storing blocks in the cloud to benefit the extensible characteristic of the cloud, as proposed in [155].

In Table 3, we provide a summary of the main findings regarding the challenges associated with the use of the blockchain technology in the different considered fields. The abbreviation GDPR stands for "General Data Protection Regulation" and the abbreviation HIPAA for "Health Insurance Portability and Accountability Act", respectively.

Table 3. Summary of the main challenges associated with the use of blockchain technology.

Domain	Scalability	Regulations	Security	Resources and Architecture	Interoperability
Financial activities [12,156]	The huge gap with the current third-party fast payment systems	Difficulty in supervising and managing, especially internationally	Vulnerabilities related to hacking and other cyberattacks	The slowness of cryptocurrency transaction processing and the high costs	The integration of various payment systems
Healthcare [157–159]	The size of the blockchain database is growing continuously over time with the flowed medical records	Compliance with GDPR and HIPAA standards esp for privacy-preserving issues [160]	Healthcare data sharing and medical data access controls, authentication, non-repudiation of records [161]	IoT healthcare devices are computationally-limited while blockchain is energy-greedy with high bandwidth consumption	The integration of blockchain with existing health information technology (HIT)
Information systems [19,162]	The structure and maintainability of blockchain-based IS with large system companies	Legal and regulatory issues in a decentralized information systems and standards to transform the business process	Security vulnerabilities, such as the border gateway protocol (BGP) routing hijack attack in smart contracts and privacy issues [163]	Difficulty in implementing a distributed computing system for small or start-up businesses	Compatibility issues between implementations of existing platforms and cloud or edge computing architectures with blockchain
Wireless networks [164]	Different and increasing wireless networks, such as 5G [165], 6G [166], and in envisioned UAV networks [167]	Trust degrees among stakeholders and regulation requirements for different use cases in wireless networks	Data collection, filtering, and data sampling require security assurance and privacy protection [20]	Memory and resource consumption in large-scale networks are enormous	The heterogeneity demands of hyperconnected existence of 'everything' wireless networks
Internet of Things (IoT) [34,168]	The network size and transaction volume make scalable solutions in IoT challenging	Considerable regulatory uncertainties exist in many countries concerning blockchain	Security risks due to smart-contract bugs to defect prevention	Increasing computing power and energy for IoT devices validate the transactions	Cross platforms with various architectural designs and implementations
Smart grids [28,169–171]	Properly scale-up the platform to accommodate the requirements of the smart grid system	The current grid legal system does not support the trading of energy from consumers to consumers.	(1) Cybersecurity threats to energy data generated by grid members and processes. (2) Cyber-physical attacks [172]	The need of transaction rates as high as a few thousand per second	The integration of heterogeneous distributed energy resources at different voltage levels
Governmental services [143,173]	Large and complex networks with data management (digital identity, administration, voting, etc.)	The regulations of E-governmental blockchain services require intensive governmental efforts	Integrity verification, high availability requirements. Ensuring authentication and authorization	Energy-inefficient mechanisms in the governmental services when using blockchain	Different governmental systems require compatibility across various platforms for governmental services
Military and defense [174]	Increasing the military network that includes hundreds of sensors to collect and transfer data	(1) Standards and regulations for the military field. (2) Compliance with standards related to preserving privacy	Military operation requires high security mechanisms for data and privacy assurance	Minimum execution time for a transaction to meet the military objectives and minimize delays	Immense heterogeneous data in the aerospace and defense industry when dealing

13. Conclusions

In this paper, we shed light on recent studies related to the incorporation of blockchain technology in modern applications, namely: financial activities; healthcare; information systems; wireless networks; Internet of Things (IoT); smart grids; governmental services; and military and defense. For each field, we provided related examples for the use of blockchain technology, while focusing on corresponding benefits, limitations, and challenges. The reviewed solutions are summarized in Table 4.

Table 4. Summary of the main findings concerning the use of blockchain in different fields.

Domain	Papers	Main Applications	Limitations
Financial activities	[12,13,50,51,54,55]	(1) Settlement of financial market transactions; (2) trade finance; (3) insurance; (4) real-time money transfer; (5) cross-border payment.	(1) Too slow; (2) risk of irreversible loss of consumer assets; (3) risk of a data breach; (4) limited grasp and acceptance; (5) supervision is more complex.
Healthcare	[58–61,63–74]	(1) Verifiable standardized identity; (2) more reliable prescribing; (3) preventing medical identity theft; (4) accurate and up to date information; (5) data aggregation; (6) ease of sharing; (7) remote monitoring; (8) safe, fast, and high-quality care; (9) less time in hospitals.	(1) Storing large records may be inefficient and extremely expensive; (2) data in a is difficult to query, restricting clinical, statistical, and research applications.
Information systems	[79–98]	(1) Improved and more secured integration of third-party products; (2) a common way for involved parties to interact with one another; (3) validating data and ensuring the transaction integrity; (4) tracking a product's origin more readily.	Overhead costs of implementing integrated blockchain technology would be prohibitive and almost infeasible.
Wireless networks	[99,100,102–108,110]	(1) Increasing spectrum access and utilization efficiency; (2) creating a secure spectrum sensing system; (3) improving the accuracy of spectra sensing data; (4) storing unoccupied spectrum bands and user geolocations; (5) providing dynamic spectrum access; (6) enabling collaborative sensing.	(1) Energy-intensive; (2) necessitates a specialized control channel for transferring blocks and transactions over blockchain networks.
Internet of Things (IoT)	[117–127]	(1) Storing and processing data at the same time while maintaining privacy; (2) establishing a secure means for smart devices to communicate with one another; (3) allowing smart devices to perform autonomously; (4) avoiding the need for human control or centralized authority.	Scalability is still an open question since the blockchain can grow in size over time, making it difficult to acquire and save the ledger.
Smart grids	[133–140]	(1) Distributed intelligent administration; (2) improve privacy and security; (3) optimum dataflow and cash flow.	Large infrastructural expenses needed.
Governmental services	[142,143]	(1) Obtaining information electronically; (2) direct interactions between government and citizens; (3) supporting real estate transactions; (4) enhancing contract bidding mechanism; (5) automating the registration of government grants; (6) invoice issuance and tax collection; (7) official announcements and digital court files; (8) property/ succession/ business registries.	(1) Lack of legal and regulatory support; (2) issue of acceptability and the need of a new governance model.
Military and defense	[146,147]	(1) Infrastructure monitoring; (2) battles management; (3) UAV management; (4) supply chain management; (5) encrypted communications; (6) intrusion detection.	(1) Not completely ready for use; (2) somewhat paradoxical with the fact that military and defense applications need to be managed in a centralized fashion.

Blockchain is a revolutionary and exciting technology with enormous potential for usage in a wide range of modern applications. However, before the benefits of blockchain can be completely realized, a number of concerns and challenges must be addressed.

One approach to addressing blockchain's low throughput is to create new architectures and operational protocols for the system. The blockchain data, for example, may not be duplicated in every node in the network; instead, only the powerful nodes maintain a copy of the blockchain, while other light nodes simply save the block headers or do not save any data at all. To close the performance gap between a blockchain system and a typical database system, lightweight consensus techniques are also required.

While vertical and horizontal scaling of a blockchain system can help with scalability concerns, another research strategy is an interconnected multi-blockchain hierarchical structure with internal interconnections. Other approaches to reduce the amount of in-chain transactions could exist. Some transactions, for example, could be carried out directly between the parties without passing through the blockchain network; hence, enhancing blockchain scalability. Maintaining data security and privacy is difficult since all transactions committed to a blockchain are visible to all participants. Providing data auditability, on the other hand, may result in the loss of data and user anonymity. Manufacturing and enterprise solution data may have tremendous commercial value. As a result, in blockchain-based smart manufacturing systems, security and privacy are critical concerns. Before blockchain technology can be used on a broad basis, these and other security and privacy concerns must be addressed.

For more efficient, scalable, and secured blockchain industrial uses, additional work in the future is required. For instance, it will be interesting to investigate how machine learning (ML) techniques [175–177] may be used in the context of blockchain technology to increase security levels and the performances of blockchain-based systems. It will also be extremely useful to apply some formal testing techniques for blockchain-based solutions to improve their quality and increase their robustness [178–180].

Author Contributions: Conceptualization, M.K., M.A. (Meryem Ammi) and A.M.; methodology, M.K., M.A. (Meryem Ammi) and A.M.; investigation, M.K., M.A. (Meryem Ammi) and A.M.; resources, M.K., M.A. (Meryem Ammi) and A.M.; writing—original draft preparation, M.K., M.A. (Meryem Ammi) and A.M.; writing—review and editing, M.K., M.A. (Meryem Ammi), A.M. and M.A. (Mutiq Almutiq); visualization, M.K.; supervision, M.K.; project administration, M.K.; funding acquisition, M.A. (Mutiq Almutiq). All authors have read and agreed to the published version of the manuscript.

Funding: The APC was funded by the Deanship of Scientific Research, Qassim University.

Informed Consent Statement: Not applicable.

Data Availability Statement: Not applicable.

Acknowledgments: The researchers would like to thank the Deanship of Scientific Research, Qassim University, for funding the publication of this project.

Conflicts of Interest: The authors declare no conflict of interest.

References

1. Salah, K.; Rehman, M.H.U.; Nizamuddin, N.; Al-Fuqaha, A. Blockchain for AI: Review and open research challenges. *IEEE Access* **2019**, *7*, 10127–10149. [CrossRef]
2. Lahami, M.; Maâlej, A.J.; Krichen, M.; Hammami, M.A. A Comprehensive Review of Testing Blockchain Oriented Software. In Proceedings of the 17th International Conference on Evaluation of Novel Approaches to Software Engineering (ENASE 2022), Online, 25–26 April 2022; pp. 355–362.
3. Litke, A.; Anagnostopoulos, D.; Varvarigou, T. Blockchains for supply chain management: Architectural elements and challenges towards a global scale deployment. *Logistics* **2019**, *3*, 5. [CrossRef]
4. Kouhizadeh, M.; Sarkis, J. Blockchain practices, potentials, and perspectives in greening supply chains. *Sustainability* **2018**, *10*, 3652. [CrossRef]
5. Schilling, L.; Uhlig, H. Some simple bitcoin economics. *J. Monet. Econ.* **2019**, *106*, 16–26. [CrossRef]
6. Ravishankar, C.V.; Kavitha, K.S., Blockchain Applications that are Transforming the Society. In *Convergence of Internet of Things and Blockchain Technologies*; Gururaj, H.L., Ravi Kumar, V., Goundar, S., Elngar, A.A., Swathi, B.H., Eds.; Springer: Cham, Switzerland, 2022; pp. 23–39. [CrossRef]

7. Zaabar, B.; Cheikhrouhou, O.; Jamil, F.; Ammi, M.; Abid, M. HealthBlock: A secure blockchain-based healthcare data management system. *Comput. Netw.* **2021**, *200*, 108500. [CrossRef]
8. Jamil, F.; Cheikhrouhou, O.; Jamil, H.; Koubaa, A.; Derhab, A.; Ferrag, M.A. PetroBlock: A blockchain-based payment mechanism for fueling smart vehicles. *Appl. Sci.* **2021**, *11*, 3055. [CrossRef]
9. Frikha, T.; Chaabane, F.; Aouinti, N.; Cheikhrouhou, O.; Ben Amor, N.; Kerrouche, A. Implementation of Blockchain Consensus Algorithm on Embedded Architecture. *Secur. Commun. Netw.* **2021**, *2021*, 9918697. [CrossRef]
10. Al-Jaroodi, J.; Mohamed, N. Blockchain in industries: A survey. *IEEE Access* **2019**, *7*, 36500–36515. [CrossRef]
11. Pal, A.; Tiwari, C.K.; Haldar, N. Blockchain for business management: Applications, challenges and potentials. *J. High Technol. Manag. Res.* **2021**, *32*, 100414. [CrossRef]
12. Zhang, L.; Xie, Y.; Zheng, Y.; Xue, W.; Zheng, X.; Xu, X. The challenges and countermeasures of blockchain in finance and economics. *Syst. Res. Behav. Sci.* **2020**, *37*, 691–698. [CrossRef]
13. Tapscott, A.; Tapscott, D. How blockchain is changing finance. *Harv. Bus. Rev.* **2017**, *1*, 2–5.
14. Prybutok, V.R.; Sauser, B. Theoretical and practical applications of blockchain in healthcare information management. *Inf. Manag.* **2022**, *59*, 103649.
15. Adere, E.M. Blockchain in healthcare and IoT: A systematic literature review. *Array* **2022**, *14*, 100139. [CrossRef]
16. Abbas, A.; Alroobaea, R.; Krichen, M.; Rubaiee, S.; Vimal, S.; Almansour, F.M. Blockchain-assisted secured data management framework for health information analysis based on Internet of Medical Things. *Pers. Ubiquitous Comput.* **2021**, 1–14.
17. Morozova, M.; Stepanov, Y.G.; Burlov, D. Innovations in Tourism and Hospitality through Modern Information Systems and Blockchain Technologies. *Components Sci. Technol. Prog.* **2022**, *42*.
18. Cao, H.; He, H.; Tian, J. A Scientific Research Information System via Intelligent Blockchain Technology for the Applications in University Management. *Mob. Inf. Syst.* **2022**, *2022*, 7512692. [CrossRef]
19. Berdik, D.; Otoum, S.; Schmidt, N.; Porter, D.; Jararweh, Y. A survey on blockchain for information systems management and security. *Inf. Process. Manag.* **2021**, *58*, 102397. [CrossRef]
20. Rathod, T.; Jadav, N.K.; Alshehri, M.D.; Tanwar, S.; Sharma, R.; Felseghi, R.A.; Raboaca, M.S. Blockchain for Future Wireless Networks: A Decade Survey. *Sensors* **2022**, *22*, 4182. [CrossRef]
21. Roopa, V.; Pradhan, H.S. Blockchain Based Spectrum Sensing for Secured Cognitive Radio wireless networks. In Proceedings of the 2022 IEEE 11th International Conference on Communication Systems and Network Technologies (CSNT), Indore, India, 23–24 April 2022; pp. 553–559.
22. Choudhary, A.K.; Rahamatkar, S. Improving Trust Levels in Wireless Networks Using Blockchain Powered Dempster Shaffer Route Optimization. *ECS Trans.* **2022**, *107*, 2095–2115.
23. Huo, R.; Zeng, S.; Wang, Z.; Shang, J.; Chen, W.; Huang, T.; Wang, S.; Yu, F.R.; Liu, Y. A comprehensive survey on blockchain in industrial internet of things: Motivations, research progresses, and future challenges. *IEEE Commun. Surv. Tutor.* **2022**, *24*, 88–122. [CrossRef]
24. Wang, J.; Chen, J.; Ren, Y.; Sharma, P.K.; Alfarraj, O.; Tolba, A. Data security storage mechanism based on blockchain industrial Internet of Things. *Comput. Ind. Eng.* **2022**, *164*, 107903. [CrossRef]
25. Al Sadawi, A.; Hassan, M.S.; Ndiaye, M. A Survey on the Integration of Blockchain With IoT to Enhance Performance and Eliminate Challenges. *IEEE Access* **2021**, *9*, 54478–54497. [CrossRef]
26. Hua, W.; Chen, Y.; Qadrdan, M.; Jiang, J.; Sun, H.; Wu, J. Applications of blockchain and artificial intelligence technologies for enabling prosumers in smart grids: A review. *Renew. Sustain. Energy Rev.* **2022**, *161*, 112308. [CrossRef]
27. Pareek, A.; Singh, P.; Lather, J. Blockchain Technology in Smart Grids and Microgrids: A Critical Review of Challenges and Opportunities. *Power Electron. High Volt. Smart Grid* **2022**, 353–363.
28. Mollah, M.B.; Zhao, J.; Niyato, D.; Lam, K.Y.; Zhang, X.; Ghias, A.M.; Koh, L.H.; Yang, L. Blockchain for future smart grid: A comprehensive survey. *IEEE Internet Things J.* **2020**, *8*, 18–43. [CrossRef]
29. Verma, S.; Sheel, A. Blockchain for government organizations: Past, present and future. *J. Glob. Oper. Strateg. Sourc.* **2022**. [CrossRef]
30. Alexopoulos, C.; Charalabidis, Y.; Androutsopoulou, A.; Loutsaris, M.A.; Lachana, Z. Benefits and Obstacles of Blockchain Applications in E-Government. 2019. Available online: https://scholarspace.manoa.hawaii.edu/items/07e8c65a-7f32-4023-bde6-29a95dd425d7 (accessed on 15 February 2022).
31. Mohamed, R.; Abas, H.; Yusof, F.M. Blockchain resilient communication in military: A systematic literature review. *Open Int. J. Inform.* **2022**, *10*, 51–62.
32. Akter, R.; Golam, M.; Doan, V.S.; Lee, J.M.; Kim, D.S. IoMT-Net: Blockchain Integrated Unauthorized UAV Localization Using Lightweight Convolution Neural Network for Internet of Military Things. *IEEE Internet Things J.* **2022**. [CrossRef]
33. Hasan, M.K.; Alkhalifah, A.; Islam, S.; Babiker, N.; Habib, A.; Aman, A.H.M.; Hossain, M. Blockchain Technology on Smart Grid, Energy Trading, and Big Data: Security Issues, Challenges, and Recommendations. *Wirel. Commun. Mob. Comput.* **2022**, *2022*, 9065768. [CrossRef]
34. Abdelmaboud, A.; Ahmed, A.I.A.; Abaker, M.; Eisa, T.A.E.; Albasheer, H.; Ghorashi, S.A.; Karim, F.K. Blockchain for IoT Applications: Taxonomy, Platforms, Recent Advances, Challenges and Future Research Directions. *Electronics* **2022**, *11*, 630. [CrossRef]

35. Raja Santhi, A.; Muthuswamy, P. Influence of blockchain technology in manufacturing supply chain and logistics. *Logistics* **2022**, *6*, 15. [CrossRef]
36. Guru, D.; Perumal, S.; Varadarajan, V. Approaches towards blockchain innovation: A survey and future directions. *Electronics* **2021**, *10*, 1219. [CrossRef]
37. Khoshavi, N.; Tristani, G.; Sargolzaei, A. Blockchain Applications to Improve Operation and Security of Transportation Systems: A Survey. *Electronics* **2021**, *10*, 629.
38. Bhutta, M.N.M.; Khwaja, A.A.; Nadeem, A.; Ahmad, H.F.; Khan, M.K.; Hanif, M.A.; Song, H.; Alshamari, M.; Cao, Y. A survey on blockchain technology: Evolution, architecture and security. *IEEE Access* **2021**, *9*, 61048–61073. [CrossRef]
39. Rawat, D.B.; Chaudhary, V.; Doku, R. Blockchain technology: Emerging applications and use cases for secure and trustworthy smart systems. *J. Cybersecur. Priv.* **2020**, *1*, 4–18. [CrossRef]
40. Zuo, Y. Making smart manufacturing smarter—A survey on blockchain technology in Industry 4.0. *Enterp. Inf. Syst.* **2021**, *15*, 1323–1353.
41. Qureshi, A.; Megías Jiménez, D. Blockchain-based multimedia content protection: Review and open challenges. *Appl. Sci.* **2020**, *11*, 1.
42. Srivastava, G.; Dhar, S.; Dwivedi, A.D.; Crichigno, J. Blockchain education. In Proceedings of the 2019 IEEE Canadian Conference of Electrical and Computer Engineering (CCECE), Edmonton, AB, Canada, 5–8 May 2019; pp. 1–5.
43. Lemieux, V.L. Trusting records: Is Blockchain technology the answer? *Rec. Manag. J.* **2016**, *26*, 110–139. [CrossRef]
44. Johar, S.; Ahmad, N.; Asher, W.; Cruickshank, H.; Durrani, A. Research and applied perspective to blockchain technology: A comprehensive survey. *Appl. Sci.* **2021**, *11*, 6252. [CrossRef]
45. Li, X.; Jiang, P.; Chen, T.; Luo, X.; Wen, Q. A survey on the security of blockchain systems. *Future Gener. Comput. Syst.* **2020**, *107*, 841–853. [CrossRef]
46. Yan, B.; Yang, Z.; Ren, Y.; Tan, X.; Liu, E. Microblog sentiment classification using parallel SVM in apache spark. In Proceedings of the 2017 IEEE International Congress on Big Data (BigData Congress), Honolulu, HI, USA, 25–30 June 2017; pp. 282–288.
47. Nakamoto, S. Bitcoin: A peer-to-peer electronic cash system. *Decentralized Bus. Rev.* **2008**, 21260.
48. Zheng, Z.; Xie, S.; Dai, H.N.; Chen, X.; Wang, H. Blockchain challenges and opportunities: A survey. *Int. J. Web Grid Serv.* **2018**, *14*, 352–375.
49. Pilkington, M. Blockchain technology: Principles and applications. In *Research Handbook on Digital Transformations*; Edward Elgar Publishing: Cheltenham, UK, 2016.
50. Tschorsch, F.; Scheuermann, B. Bitcoin and beyond: A technical survey on decentralized digital currencies. *IEEE Commun. Surv. Tutor.* **2016**, *18*, 2084–2123. [CrossRef]
51. Yao, X.; Zhu, T. Blockchain is to create a new ecology of cross-border payment. *Financ. Expo* **2017**, *5*, 46–48.
52. Chhabra, V.; Bathla, S.; Maheshwari, H. An overview of blockchain technology and comparison between various cryptocurrencies. *J. Emerg. Technol. Innov. Res.* **2019**, *6*, 68–71.
53. Garriga, M.; Dalla Palma, S.; Arias, M.; De Renzis, A.; Pareschi, R.; Andrew Tamburri, D. Blockchain and cryptocurrencies: A classification and comparison of architecture drivers. *Concurr. Comput. Pract. Exp.* **2021**, *33*, e5992. [CrossRef]
54. Demirkan, S.; Demirkan, I.; McKee, A. Blockchain technology in the future of business cyber security and accounting. *J. Manag. Anal.* **2020**, *7*, 189–208. [CrossRef]
55. Hassani, H.; Huang, X.; Silva, E. Banking with blockchained big data. *J. Manag. Anal.* **2018**, *5*, 256–275.
56. Li, J.; Wang, J.; Wangh, S.; Zhou, Y. Mobile payment with alipay: An application of extended technology acceptance model. *IEEE Access* **2019**, *7*, 50380–50387.
57. Lumpkin, J.; Cohn, S.P.; Blair, J.S. Uniform data standards for patient medical record information. *Natl. Comm. Vital Health Stat.* **2003**, *53*.
58. Jabbar, R.; Fetais, N.; Krichen, M.; Barkaoui, K. Blockchain technology for healthcare: Enhancing shared electronic health record interoperability and integrity. In Proceedings of the 2020 IEEE International Conference on Informatics, IoT, and Enabling Technologies (ICIoT), Doha, Qatar, 2–5 February 2020; pp. 310–317.
59. Zhuang, Y.; Sheets, L.R.; Chen, Y.W.; Shae, Z.Y.; Tsai, J.J.; Shyu, C.R. A patient-centric health information exchange framework using blockchain technology. *IEEE J. Biomed. Health Inform.* **2020**, *24*, 2169–2176. [PubMed]
60. Zhang, P.; Schmidt, D.C.; White, J.; Lenz, G. Blockchain technology use cases in healthcare. In *Advances in Computers*; Elsevier: Amsterdam, The Netherlands, 2018; Volume 111, pp. 1–41.
61. Just, B.H.; Marc, D.; Munns, M.; Sandefer, R. Why patient matching is a challenge: Research on master patient index (MPI) data discrepancies in key identifying fields. *Perspect. Health Inf. Manag.* **2016**, *13*.
62. Alliance, S.C. Effective Healthcare Identity Management: A Necessary First Step for Improving US Healthcare Information Systems. 2014. Available online: https://www.securetechalliance.org/resources/pdf/Healthcare_Identity_Brief.pdf (accessed on 25 February 2022).
63. Krawiec, R.; Housman, D.; White, M.; Filipova, M.; Quarre, F.; Barr, D.; Nesbitt, A.; Fedosova, K.; Killmeyer, J.; Israel, A. Blockchain: Opportunities for health care. In Proceedings of the NIST Workshop Blockchain Healthcare, Gaithersburg, MD, USA, 26–27 September 2016; pp. 1–16.
64. Chen, D.; Chen, L.; Fan, X.; He, L.; Pan, S.; Hu, R. Securing patient-centric personal health records sharing system in cloud computing. *China Commun.* **2014**, *11*, 121–127. [CrossRef]

65. Barua, M.; Liang, X.; Lu, R.; Shen, X. PEACE: An efficient and secure patient-centric access control scheme for eHealth care system. In Proceedings of the 2011 IEEE Conference on Computer Communications Workshops (INFOCOM WKSHPS), Shanghai, China, 10–15 April 2011; pp. 970–975.
66. Li, M.; Yu, S.; Ren, K.; Lou, W. Securing personal health records in cloud computing: Patient-centric and fine-grained data access control in multi-owner settings. In *International Conference on Security and Privacy in Communication Systems*; Springer: Berlin/Heidelberg, Germany, 2010; pp. 89–106.
67. Zhang, P.; White, J.; Schmidt, D.C.; Lenz, G.; Rosenbloom, S.T. FHIRChain: Applying blockchain to securely and scalably share clinical data. *Comput. Struct. Biotechnol. J.* **2018**, *16*, 267–278. [CrossRef]
68. Dubovitskaya, A.; Baig, F.; Xu, Z.; Shukla, R.; Zambani, P.S.; Swaminathan, A.; Jahangir, M.M.; Chowdhry, K.; Lachhani, R.; Idnani, N. ACTION-EHR: Patient-centric blockchain-based electronic health record data management for cancer care. *J. Med. Internet Res.* **2020**, *22*, e13598. [CrossRef]
69. Jabbar, R.; Krichen, M.; Fetais, N.; Barkaoui, K. Adopting Formal Verification and Model-Based Testing Techniques for Validating a Blockchain-based Healthcare Records Sharing System. In Proceedings of the 22nd International Conference on Enterprise Information Systems, Prague, Czech Republic, 5–7 May 2020; pp. 261–268.
70. Panigrahi, A.; Nayak, A.K.; Paul, R. HealthCare EHR: A Blockchain-Based Decentralized Application. *Int. J. Inf. Syst. Supply Chain. Manag.* **2022**, *15*, 1–15. [CrossRef]
71. Ahmad, R.W.; Salah, K.; Jayaraman, R.; Yaqoob, I.; Ellahham, S.; Omar, M. The role of blockchain technology in telehealth and telemedicine. *Int. J. Med. Inform.* **2021**, *148*, 104399. [CrossRef]
72. Wang, W.; Wang, L.; Zhang, P.; Xu, S.; Fu, K.; Song, L.; Hu, S. A privacy protection scheme for telemedicine diagnosis based on double blockchain. *J. Inf. Secur. Appl.* **2021**, *61*, 102845. [CrossRef]
73. Kordestani, H.; Barkaoui, K.; Zahran, W. HapiChain: A blockchain-based framework for patient-centric telemedicine. In Proceedings of the 2020 IEEE 8th International Conference on Serious Games and Applications for Health (SeGAH), Vancouver, BC, Canada, 12–14 August 2020; pp. 1–6.
74. Parikh, D.P.; Dhanotiya, A.; Vetrivelan, P. Blockchain-Based Secure IoT Telemedicine System. In *Futuristic Communication and Network Technologies*; Springer: Berlin/Heidelberg, Germany, 2022; pp. 923–935.
75. Clohessy, T.; Acton, T. Investigating the influence of organizational factors on blockchain adoption: An innovation theory perspective. *Ind. Manag. Data Syst.* **2019**, *119*, 1457–1491. [CrossRef]
76. Park, Y.R.; Lee, E.; Na, W.; Park, S.; Lee, Y.; Lee, J.H. Is blockchain technology suitable for managing personal health records? Mixed-methods study to test feasibility. *J. Med. Internet Res.* **2019**, *21*, e12533.
77. Hawig, D.; Zhou, C.; Fuhrhop, S.; Fialho, A.S.; Ramachandran, N. Designing a distributed ledger technology system for interoperable and general data protection regulation–compliant health data exchange: A use case in blood glucose data. *J. Med. Internet Res.* **2019**, *21*, e13665. [CrossRef] [PubMed]
78. O'brien, J.A.; Marakas, G.M. *Introduction to Information Systems*; McGraw-Hill/Irwin: New York, NY, USA, 2005; Volume 13.
79. Fullana, O.; Ruiz, J. Accounting information systems in the blockchain era. *Int. J. Intellect. Prop. Manag.* **2021**, *11*, 63–80. [CrossRef]
80. Rossi, M.; Mueller-Bloch, C.; Thatcher, J.B.; Beck, R. Blockchain research in information systems: Current trends and an inclusive future research agenda. *J. Assoc. Inf. Syst.* **2019**, *20*, 14. [CrossRef]
81. Brandon, D. The blockchain: The future of business information systems. *Int. J. Acad. Bus. World* **2016**, *10*, 33–40.
82. Shaverdian, P. Start With Trust: Utilizing Blockchain to Resolve the Third-Party Data Breach Problem. *UCLA L. Rev.* **2019**, *66*, 1242.
83. Dos Santos, R.B.; Torrisi, N.M.; Pantoni, R.P. Third Party Certification of Agri-Food Supply Chain Using Smart Contracts and Blockchain Tokens. *Sensors* **2021**, *21*, 5307. [CrossRef]
84. Mut-Puigserver, M.; Cabot-Nadal, M.A.; Payeras-Capellà, M.M. Removing the trusted third party in a confidential multiparty registered eDelivery protocol using blockchain. *IEEE Access* **2020**, *8*, 106855–106871. [CrossRef]
85. Chiu, W.Y.; Meng, W.; Jensen, C.D. NoPKI-a Point-to-Point Trusted Third Party Service Based on Blockchain Consensus Algorithm. In *International Conference on Frontiers in Cyber Security*; Springer: Berlin/Heidelberg, Germany, 2020; pp. 197–214.
86. Schulte, S.; Sigwart, M.; Frauenthaler, P.; Borkowski, M. Towards blockchain interoperability. In *International Conference on Business Process Management*; Springer: Berlin/Heidelberg, Germany, 2019; pp. 3–10.
87. Lafourcade, P.; Lombard-Platet, M. About blockchain interoperability. *Inf. Process. Lett.* **2020**, *161*, 105976. [CrossRef]
88. Hardjono, T.; Lipton, A.; Pentland, A. Toward an interoperability architecture for blockchain autonomous systems. *IEEE Trans. Eng. Manag.* **2019**, *67*, 1298–1309. [CrossRef]
89. Singh, A.; Parizi, R.M.; Zhang, Q.; Choo, K.K.R.; Dehghantanha, A. Blockchain smart contracts formalization: Approaches and challenges to address vulnerabilities. *Comput. Secur.* **2020**, *88*, 101654. [CrossRef]
90. Cong, L.W.; He, Z. Blockchain disruption and smart contracts. *Rev. Financ. Stud.* **2019**, *32*, 1754–1797.
91. Kosba, A.; Miller, A.; Shi, E.; Wen, Z.; Papamanthou, C. Hawk: The blockchain model of cryptography and privacy-preserving smart contracts. In Proceedings of the 2016 IEEE Symposium on Security and Privacy (SP), San Jose, CA, USA, 22–26 May 2016; pp. 839–858.
92. Watanabe, H.; Fujimura, S.; Nakadaira, A.; Miyazaki, Y.; Akutsu, A.; Kishigami, J. Blockchain contract: Securing a blockchain applied to smart contracts. In Proceedings of the 2016 IEEE International Conference on Consumer Electronics (ICCE), Las Vegas, NV, USA, 7–11 January 2016; pp. 467–468.

93. Min, H. Blockchain technology for enhancing supply chain resilience. *Bus. Horiz.* **2019**, *62*, 35–45. [CrossRef]
94. Saberi, S.; Kouhizadeh, M.; Sarkis, J.; Shen, L. Blockchain technology and its relationships to sustainable supply chain management. *Int. J. Prod. Res.* **2019**, *57*, 2117–2135. [CrossRef]
95. Casado-Vara, R.; Prieto, J.; De la Prieta, F.; Corchado, J.M. How blockchain improves the supply chain: Case study alimentary supply chain. *Procedia Comput. Sci.* **2018**, *134*, 393–398.
96. Francisco, K.; Swanson, D. The supply chain has no clothes: Technology adoption of blockchain for supply chain transparency. *Logistics* **2018**, *2*, 2.
97. Dujak, D.; Sajter, D. Blockchain applications in supply chain. In *SMART Supply Network*; Springer: Berlin/Heidelberg, Germany, 2019; pp. 21–46.
98. Azzi, R.; Chamoun, R.K.; Sokhn, M. The power of a blockchain-based supply chain. *Comput. Ind. Eng.* **2019**, *135*, 582–592. [CrossRef]
99. Jabbar, R.; Dhib, E.; ben Said, A.; Krichen, M.; Fetais, N.; Zaidan, E.; Barkaoui, K. Blockchain Technology for Intelligent Transportation Systems: A Systematic Literature Review. *IEEE Access* **2022**, *10*, 20995–21031. [CrossRef]
100. Jabbar, R.; Fetais, N.; Kharbeche, M.; Krichen, M.; Barkaoui, K.; Shinoy, M. Blockchain for The Internet of Vehicles: How to use Blockchain to secure Vehicle-to-Everything (V2X) Communication and Payment? *IEEE Sens. J.* **2021**, *21*, 15807–15823. [CrossRef]
101. Leyton-Brown, K.; Milgrom, P.; Segal, I. Economics and computer science of a radio spectrum reallocation. *Proc. Natl. Acad. Sci. USA* **2017**, *114*, 7202–7209. [CrossRef]
102. Weiss, M.B.; Werbach, K.; Sicker, D.C.; Bastidas, C.E.C. On the application of blockchains to spectrum management. *IEEE Trans. Cogn. Commun. Netw.* **2019**, *5*, 193–205. [CrossRef]
103. Anker, P. From spectrum management to spectrum governance. *Telecommun. Policy* **2017**, *41*, 486–497. [CrossRef]
104. Ariyarathna, T.; Harankahadeniya, P.; Isthikar, S.; Pathirana, N.; Bandara, H.D.; Madanayake, A. Dynamic spectrum access via smart contracts on blockchain. In Proceedings of the 2019 IEEE Wireless Communications and Networking Conference (WCNC), Marrakesh, Morocco, 15–18 April 2019; pp. 1–6.
105. Qiu, J.; Grace, D.; Ding, G.; Yao, J.; Wu, Q. Blockchain-based secure spectrum trading for unmanned-aerial-vehicle-assisted cellular networks: An operator's perspective. *IEEE Internet Things J.* **2019**, *7*, 451–466. [CrossRef]
106. Han, S.; Zhu, X. Blockchain based spectrum sharing algorithm. In Proceedings of the 2019 IEEE 19th International Conference on Communication Technology (ICCT), Xi'an, China, 16–19 October 2019; pp. 936–940.
107. Careem, M.A.A.; Dutta, A. Sensechain: Blockchain based reputation system for distributed spectrum enforcement. In Proceedings of the 2019 IEEE International Symposium on Dynamic Spectrum Access Networks (DySPAN), Newark, NJ, USA, 11–14 November 2019; pp. 1–10.
108. Pei, Y.; Hu, S.; Zhong, F.; Niyato, D.; Liang, Y.C. Blockchain-enabled dynamic spectrum access: Cooperative spectrum sensing, access and mining. In Proceedings of the 2019 IEEE Global Communications Conference (GLOBECOM), Waikoloa, HI, USA, 9–13 December 2019; pp. 1–6.
109. Tangsen, H.; Li, X.; Ying, X. A Blockchain-Based Node Selection Algorithm in Cognitive Wireless Networks. *IEEE Access* **2020**, *8*, 207156–207166. [CrossRef]
110. Liang, Y.C. *Dynamic Spectrum Management: From Cognitive Radio to Blockchain and Artificial Intelligence*; Springer: Berlin/Heidelberg, Germany, 2020.
111. Mukhtar, H.; Rubaiee, S.; Krichen, M.; Alroobaea, R. An IoT framework for screening of COVID-19 using real-time data from wearable sensors. *Int. J. Environ. Res. Public Health* **2021**, *18*, 4022. [CrossRef]
112. Krichen, M.; Alroobaea, R. A New Model-based Framework for Testing Security of IoT Systems in Smart Cities using Attack Trees and Price Timed Automata. In Proceedings of the 14th International Conference on Evaluation of Novel Approaches to Software Engineering (ENASE 2019), Heraklion, Greece, 4–5 May 2019.
113. Jabbar, R.; Shinoy, M.; Kharbeche, M.; Al-Khalifa, K.; Krichen, M.; Barkaoui, K. Urban traffic monitoring and modeling system: An iot solution for enhancing road safety. In Proceedings of the 2019 International Conference on Internet of Things, Embedded Systems and Communications (IINTEC), Tunis, Tunisia, 20–22 December 2019; pp. 13–18.
114. Bhandari, K.S.; Ra, I.H.; Cho, G. Multi-topology based QoS-differentiation in RPL for internet of things applications. *IEEE Access* **2020**, *8*, 96686–96705. [CrossRef]
115. Bhandari, K.S.; Cho, G.H. An energy efficient routing approach for cloud-assisted green industrial IoT networks. *Sustainability* **2020**, *12*, 7358. [CrossRef]
116. Bhandari, K.S.; Cho, G.H. Resource oriented topology construction to ensure high reliability in IoT based smart city networks. *Int. J. Syst. Assur. Eng. Manag.* **2020**, *11*, 798–805. [CrossRef]
117. Krichen, M.; Lahami, M.; Cheikhrouhou, O.; Alroobaea, R.; Maâlej, A.J. Security testing of internet of things for smart city applications: A formal approach. In *Smart Infrastructure and Applications*; Springer: Cham, Switzerland, 2020; pp. 629–653.
118. Jabbar, R.; Kharbeche, M.; Al-Khalifa, K.; Krichen, M.; Barkaoui, K. Blockchain for the internet of vehicles: A decentralized IoT solution for vehicles communication using ethereum. *Sensors* **2020**, *20*, 3928.
119. Li, D.; Deng, L.; Cai, Z.; Souri, A. Blockchain as a service models in the Internet of Things management: Systematic review. *Trans. Emerg. Telecommun. Technol.* **2020**, *33*, e4139.
120. Pešić, S.; Radovanović, M.; Ivanović, M.; Tošić, M.; Iković, O.; Bošković, D. Hyperledger fabric blockchain as a service for the IoT: Proof of concept. In *International Conference on Model and Data Engineering*; Springer: Berlin/Heidelberg, Germany, 2019; pp. 172–183.

121. Kshetri, N. Can blockchain strengthen the internet of things? *IT Prof.* **2017**, *19*, 68–72. [CrossRef]
122. Dorri, A.; Kanhere, S.S.; Jurdak, R.; Gauravaram, P. Blockchain for IoT security and privacy: The case study of a smart home. In Proceedings of the 2017 IEEE International Conference on Pervasive Computing and Communications Workshops (PerCom Workshops), Kona, HI, USA, 13–17 March 2017; pp. 618–623.
123. Suliman, A.; Husain, Z.; Abououf, M.; Alblooshi, M.; Salah, K. Monetization of IoT data using smart contracts. *IET Netw.* **2019**, *8*, 32–37. [CrossRef]
124. Arumugam, S.S.; Umashankar, V.; Narendra, N.C.; Badrinath, R.; Mujumdar, A.P.; Holler, J.; Hernandez, A. IoT enabled smart logistics using smart contracts. In Proceedings of the 2018 8th International Conference on Logistics, Informatics and Service Sciences (LISS), Toronto, ON, Canada, 3–6 August 2018; pp. 1–6.
125. Wörner, D.; von Bomhard, T. When your sensor earns money: Exchanging data for cash with Bitcoin. In Proceedings of the 2014 ACM International Joint Conference on Pervasive and Ubiquitous Computing, Seattle, WA, USA, 13–17 September 2014; pp. 295–298.
126. Zhang, Y.; Wen, J. An IoT electric business model based on the protocol of bitcoin. In Proceedings of the 2015 18th International Conference on Intelligence in Next Generation Networks, Paris, France, 17–19 February 2015; pp. 184–191.
127. Shrobe, H.; Shrier, D.L.; Pentland, A. Enigma: Decentralized Computation Platform with Guaranteed Privacy. In *New Solutions for Cybersecurity*; MIT Press: Cambridge, MA, USA, 2018; Chapter 15, pp. 425–454.
128. Maâlej, A.J.; Krichen, M. A Model Based Approach to Combine Load and Functional Tests for Service Oriented Architectures. 2016. Available online: https://dblp.org/rec/conf/vecos/MaalejK16.html (accessed on 25 February 2022).
129. Zidi, S.; Mihoub, A.; Qaisar, S.M.; Krichen, M.; Al-Haija, Q.A. Theft detection dataset for benchmarking and machine learning based classification in a smart grid environment. *J. King Saud-Univ.-Comput. Inf. Sci.* **2022**. [CrossRef]
130. Fang, X.; Misra, S.; Xue, G.; Yang, D. Smart grid—The new and improved power grid: A survey. *IEEE Commun. Surv. Tutor.* **2011**, *14*, 944–980. [CrossRef]
131. Farhangi, H. The path of the smart grid. *IEEE Power Energy Mag.* **2009**, *8*, 18–28. [CrossRef]
132. Gungor, V.C.; Sahin, D.; Kocak, T.; Ergut, S.; Buccella, C.; Cecati, C.; Hancke, G.P. Smart grid technologies: Communication technologies and standards. *IEEE Trans. Ind. Inform.* **2011**, *7*, 529–539. [CrossRef]
133. Goranović, A.; Meisel, M.; Fotiadis, L.; Wilker, S.; Treytl, A.; Sauter, T. Blockchain applications in microgrids an overview of current projects and concepts. In Proceedings of the 43rd Annual Conference of the IEEE Industrial Electronics Society (IECON 2017), Beijing, China, 29 October–1 November 2017; pp. 6153–6158.
134. Xie, J.; Tang, H.; Huang, T.; Yu, F.R.; Xie, R.; Liu, J.; Liu, Y. A survey of blockchain technology applied to smart cities: Research issues and challenges. *IEEE Commun. Surv. Tutor.* **2019**, *21*, 2794–2830. [CrossRef]
135. Andoni, M.; Robu, V.; Flynn, D.; Abram, S.; Geach, D.; Jenkins, D.; McCallum, P.; Peacock, A. Blockchain technology in the energy sector: A systematic review of challenges and opportunities. *Renew. Sustain. Energy Rev.* **2019**, *100*, 143–174. [CrossRef]
136. Li, Z.; Khajepour, A.; Song, J. A comprehensive review of the key technologies for pure electric vehicles. *Energy* **2019**, *182*, 824–839. [CrossRef]
137. Das, H.; Rahman, M.; Li, S.; Tan, C. Electric vehicles standards, charging infrastructure, and impact on grid integration: A technological review. *Renew. Sustain. Energy Rev.* **2020**, *120*, 109618. [CrossRef]
138. Li, Y.; Hu, B. An iterative two-layer optimization charging and discharging trading scheme for electric vehicle using consortium blockchain. *IEEE Trans. Smart Grid* **2019**, *11*, 2627–2637. [CrossRef]
139. Li, Y.; Hu, B. A consortium blockchain-enabled secure and privacy-preserving optimized charging and discharging trading scheme for electric vehicles. *IEEE Trans. Ind. Inform.* **2020**, *17*, 1968–1977. [CrossRef]
140. Liu, C.; Chai, K.K.; Lau, E.T.; Chen, Y. Blockchain based energy trading model for electric vehicle charging schemes. In *International Conference on Smart Grid Inspired Future Technologies*; Springer: Cham, Switzerland, 2018; pp. 64–72.
141. Molnar, A.; Janssen, M.; Weerakkody, V. E-government theories and challenges: Findings from a plenary expert panel. In Proceedings of the 16th Annual International Conference on Digital Government Research, Phoenix, AZ, USA, 27–30 May 2015; pp. 160–166.
142. Atzori, M. Blockchain Technology and Decentralized Governance: Is the State Still Necessary? 2015. Available online: https://papers.ssrn.com/sol3/papers.cfm?abstract_id=2709713 (accessed on 15 February 2022).
143. Martinovic, I.; Kello, L.; Sluganovic, I. *Blockchains for Governmental Services: Design Principles, Applications, and Case Studies*; Centre for Technology and Global Affairs, University of Oxford: Oxford, UK, 2017.
144. Karaman, M.; Hayrettin, Ã.; Aybar, C. Institutional cybersecurity from military perspective. *Int. J. Inf. Secur. Sci.* **2016**, *5*, 1–7.
145. Armitage, W.D.; Gauvin, W.; Sheffield, A. Design and Launch of an Intensive Cybersecurity Program for Military Veterans. In Proceedings of the 17th Annual Conference on Information Technology Education, Boston, MA, USA, 28 September–1 October 2016; pp. 40–45.
146. Zhu, Y.; Zhang, X.; Ju, Z.Y.; Wang, C.C. A study of blockchain technology development and military application prospects. In *Journal of Physics: Conference Series*; IOP Publishing: Bristol, UK, 2020; Volume 1507, p. 052018.
147. Lilly, B.; Lilly, S. Weaponising Blockchain: Military Applications of Blockchain Technology in the US, China and Russia. *RUSI J.* **2021**, *166*, 46–56. [CrossRef]
148. Fredj, O.B.; Cheikhrouhou, O.; Krichen, M.; Hamam, H.; Derhab, A. An OWASP top ten driven survey on web application protection methods. In *International Conference on Risks and Security of Internet and Systems*; Springer: Cham, Switzerland, 2020; pp. 235–252.

149. Javed, A.R.; Shahzad, F.; ur Rehman, S.; Zikria, Y.B.; Razzak, I.; Jalil, Z.; Xu, G. Future smart cities requirements, emerging technologies, applications, challenges, and future aspects. *Cities* **2022**, *129*, 103794. [CrossRef]
150. Puri, N.; Garg, V.; Agrawal, R. Blockchain Technology Applications for Next Generation. In *Blockchain, Artificial Intelligence, and the Internet of Things*; Springer: Berlin/Heidelberg, Germany, 2022; pp. 53–73.
151. Wüst, K.; Gervais, A. Do you need a blockchain? In Proceedings of the 2018 Crypto Valley Conference on Blockchain Technology (CVCBT), Zug, Switzerland, 20–22 June 2018; pp. 45–54.
152. Monrat, A.A.; Schelén, O.; Andersson, K. A survey of blockchain from the perspectives of applications, challenges, and opportunities. *IEEE Access* **2019**, *7*, 117134–117151. [CrossRef]
153. Islam, M.R.; Rahman, M.M.; Mahmud, M.; Rahman, M.A.; Mohamad, M.H.S. A Review on Blockchain Security Issues and Challenges. In Proceedings of the 2021 IEEE 12th Control and System Graduate Research Colloquium (ICSGRC), Shah Alam, Malaysia, 7 August 2021; pp. 227–232.
154. Dwivedi, S.K.; Roy, P.; Karda, C.; Agrawal, S.; Amin, R. Blockchain-based internet of things and industrial IoT: A comprehensive survey. *Secur. Commun. Netw.* **2021**, *2021*, 7142048.
155. Zaabar, B.; Cheikhrouhou, O.; Ammi, M.; Awad, A.I.; Abid, M. Secure and Privacy-aware Blockchain-based Remote Patient Monitoring System for Internet of Healthcare Things. In Proceedings of the 2021 17th International Conference on Wireless and Mobile Computing, Networking and Communications (WiMob), Bologna, Italy, 11–13 October 2021; pp. 200–205.
156. Mosteanu, N.R.; Faccia, A. Digital systems and new challenges of financial management–FinTech, XBRL, blockchain and cryptocurrencies. *Qual. Access Success J.* **2020**, *21*, 159–166.
157. Singh, S.; Sharma, S.K.; Mehrotra, P.; Bhatt, P.; Kaurav, M. Blockchain technology for efficient data management in healthcare system: Opportunity, challenges and future perspectives. *Mater. Today Proc.* **2022**, *62*, 5042–5046. [CrossRef]
158. Durneva, P.; Cousins, K.; Chen, M. The current state of research, challenges, and future research directions of blockchain technology in patient care: Systematic review. *J. Med. Internet Res.* **2020**, *22*, e18619. [CrossRef]
159. Mazlan, A.A.; Daud, S.M.; Sam, S.M.; Abas, H.; Rasid, S.Z.A.; Yusof, M.F. Scalability challenges in healthcare blockchain system—A systematic review. *IEEE Access* **2020**, *8*, 23663–23673. [CrossRef]
160. Hasselgren, A.; Wan, P.K.; Horn, M.; Kralevska, K.; Gligoroski, D.; Faxvaag, A. GDPR Compliance for Blockchain Applications in Healthcare. *arXiv* **2020**, arXiv:2009.12913.
161. McGhin, T.; Choo, K.K.R.; Liu, C.Z.; He, D. Blockchain in healthcare applications: Research challenges and opportunities. *J. Netw. Comput. Appl.* **2019**, *135*, 62–75. [CrossRef]
162. Mendling, J.; Weber, I.; Aalst, W.V.D.; Brocke, J.V.; Cabanillas, C.; Daniel, F.; Debois, S.; Ciccio, C.D.; Dumas, M.; Dustdar, S. Blockchains for business process management-challenges and opportunities. *ACM Trans. Manag. Inf. Syst.* **2018**, *9*, 1–16. [CrossRef]
163. Li, X.; Zheng, Z.; Dai, H.N. When services computing meets blockchain: Challenges and opportunities. *J. Parallel Distrib. Comput.* **2021**, *150*, 1–14. [CrossRef]
164. Shen, X.S.; Huang, C.; Liu, D.; Xue, L.; Zhuang, W.; Sun, R.; Ying, B. Data management for future wireless networks: Architecture, privacy preservation, and regulation. *IEEE Netw.* **2021**, *35*, 8–15. [CrossRef]
165. Chaer, A.; Salah, K.; Lima, C.; Ray, P.P.; Sheltami, T. Blockchain for 5G: Opportunities and challenges. In Proceedings of the 2019 IEEE Globecom Workshops (GC Wkshps), Waikoloa, HI, USA, 9–13 December 2019; pp. 1–6.
166. Hewa, T.; Gür, G.; Kalla, A.; Ylianttila, M.; Bracken, A.; Liyanage, M. The role of blockchain in 6G: Challenges, opportunities and research directions. In Proceedings of the 2020 2nd 6G Wireless Summit (6G SUMMIT), Levi, Finland, 17–20 March 2020; pp. 1–5.
167. Mehta, P.; Gupta, R.; Tanwar, S. Blockchain envisioned UAV networks: Challenges, solutions, and comparisons. *Comput. Commun.* **2020**, *151*, 518–538.
168. Reyna, A.; Martín, C.; Chen, J.; Soler, E.; Díaz, M. On blockchain and its integration with IoT. Challenges and opportunities. *Future Gener. Comput. Syst.* **2018**, *88*, 173–190. [CrossRef]
169. Yapa, C.; de Alwis, C.; Liyanage, M.; Ekanayake, J. Survey on blockchain for future smart grids: Technical aspects, applications, integration challenges and future research. *Energy Rep.* **2021**, *7*, 6530–6564. [CrossRef]
170. Alladi, T.; Chamola, V.; Rodrigues, J.J.; Kozlov, S.A. Blockchain in smart grids: A review on different use cases. *Sensors* **2019**, *19*, 4862. [CrossRef]
171. Khan, F.A.; Asif, M.; Ahmad, A.; Alharbi, M.; Aljuaid, H. Blockchain technology, improvement suggestions, security challenges on smart grid and its application in healthcare for sustainable development. *Sustain. Cities Soc.* **2020**, *55*, 102018. [CrossRef]
172. Kim, S.M.; Lee, T.; Kim, S.; Park, L.W.; Park, S. Security issues on smart grid and blockchain-based secure smart energy management system. *MATEC Web Conf. EDP Sci.* **2019**, *260*, 01001.
173. Khayyat, M.; Alhemdi, F.; Alnunu, R. The Challenges and Benefits of Blockchain in E-government. *Int. J. Comput. Sci. Netw. Secur.* **2020**, *20*, 15–20.
174. Ahmad, R.W.; Hasan, H.; Yaqoob, I.; Salah, K.; Jayaraman, R.; Omar, M. Blockchain for aerospace and defense: Opportunities and open research challenges. *Comput. Ind. Eng.* **2021**, *151*, 106982. [CrossRef]
175. Abu Al-Haija, Q.; Krichen, M.; Abu Elhaija, W. Machine-Learning-Based Darknet Traffic Detection System for IoT Applications. *Electronics* **2022**, *11*, 556. [CrossRef]
176. Mihoub, A.; Fredj, O.B.; Cheikhrouhou, O.; Derhab, A.; Krichen, M. Denial of service attack detection and mitigation for internet of things using looking-back-enabled machine learning techniques. *Comput. Electr. Eng.* **2022**, *98*, 107716. [CrossRef]

177. Ben Fredj, O.; Mihoub, A.; Krichen, M.; Cheikhrouhou, O.; Derhab, A. CyberSecurity attack prediction: A deep learning approach. In Proceedings of the 13th International Conference on Security of Information and Networks, Merkez, Turkey, 4–7 November 2020; pp. 1–6.
178. Lahami, M.; Krichen, M. A survey on runtime testing of dynamically adaptable and distributed systems. *Softw. Qual. J.* **2021**, *29*, 555–593. [CrossRef]
179. Lahami, M.; Krichen, M.; Jmaïel, M. Runtime testing approach of structural adaptations for dynamic and distributed systems. *Int. J. Comput. Appl. Technol.* **2015**, *51*, 259–272. [CrossRef]
180. Lahami, M.; Krichen, M.; Barhoumi, H.; Jmaiel, M. Selective test generation approach for testing dynamic behavioral adaptations. In *IFIP International Conference on Testing Software and Systems*; Springer: Cham, Switzerland, 2015; pp. 224–239.

sensors

MDPI

Review

A Survey of Dummy-Based Location Privacy Protection Techniques for Location-Based Services

Shiwen Zhang [1], Mengling Li [1], Wei Liang [1], Voundi Koe Arthur Sandor [2] and Xiong Li [1,3,*]

[1] School of Computer Science and Engineering, Hunan University of Science and Technology, Xiangtan 411201, China
[2] School of Computer Science and Cyber Engineering, Guangzhou University, Guangzhou 510006, China
[3] School of Computer Science and Engineering, University of Electronic Science and Technology of China, Chengdu 611731, China
* Correspondence: lixiongzhq@163.com

Abstract: As smart devices and mobile positioning technologies improve, location-based services (LBS) have grown in popularity. The LBS environment provides considerable convenience to users, but it also poses a significant threat to their privacy. A large number of research works have emerged to protect users' privacy. Dummy-based location privacy protection solutions have been widely adopted for their simplicity and enhanced privacy protection results, but there are few reviews on dummy-based location privacy protection. Or, for existing works, some focus on aspects of cryptography, anonymity, or other comprehensive reviews that do not provide enough reviews on dummy-based privacy protection. In this paper, the authors provide a review of dummy-based location privacy protection techniques for location-based services. More specifically, the connection between the level of privacy protection, the quality of service, and the system overhead is summarized. The difference and connection between various location privacy protection techniques are also described. The dummy-based attack models are presented. Then, the algorithms for dummy location selection are analyzed and evaluated. Finally, we thoroughly evaluate different dummy location selection methods and arrive at a highly useful evaluation result. This result is valuable both to users and researchers who are studying this field.

Keywords: location privacy; privacy protection; dummy location

1. Introduction

In the United States, a large majority (90%) of smartphone owners used location-based services [1]. Locations are being used more frequently than ever before since the global pandemic. For example, the government should keep a record of every location ever visited, and track the whereabouts of people who have tested positive for COVID-19 to determine where the virus is likely to spread next [2]. Furthermore, location-based services will continue to gain attention and become more widely used in the future. According to Federica Laricchia, the annual worldwide blue-tooth location service device shipments reached 183 million units in 2021, with yearly shipments expected to reach 568 million units in 2026 [3].

While location-based services are widely used and provide significant convenience to users and society, they also pose a significant threat to privacy. According to risk-based security [4], the total amount of global data leakage in 2021 has reached 22 billion, which is about 14.5 billion less than in 2020. However, such an amount also quantifies the second highest year for confidential data leakage since 2005. As shown in Figure 1, a survey conducted by the China Consumers Association [5] in 2018 found that more than 80% of respondents had experienced personal information leakage. Moreover, it is common for mobile apps to collect excessive amounts of personal information, while location data have evolved into a type of profitable resource.

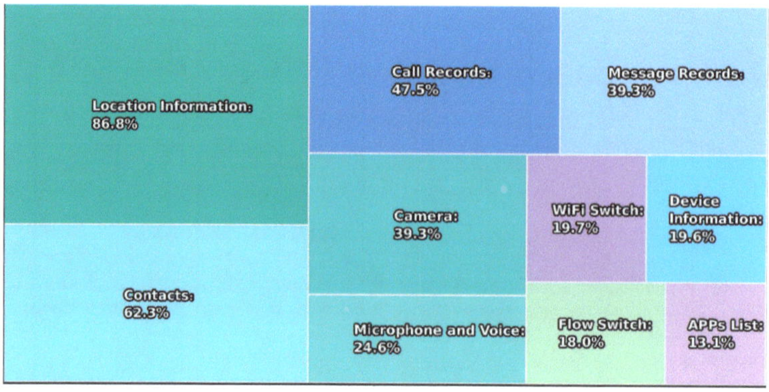

Figure 1. Permissions to install and use mobile apps.

In most cases, location data are linked to other sensitive attributes, such as health status, home address, behavioral habits, and other privacy concerns. As a result, protecting the location information of smartphone users, specially those who use location-based services, is critical and urgent.

Dummy refers to the method of adding multiple dummy locations and sending them to the LBS server along with the real query location to blur the real location. Domestic and foreign researchers have proposed a variety of location privacy protection schemes based on dummy. For example, Kido et al. proposed the first dummy-based location privacy protection techniques in the literature [6]. They generated dummy locations at random using the random walk model. Hara et al. [7] designed a method for selecting dummies that takes real-world constraints into account, such as excluding places where people are unlikely to exist. Niu B. [8] proposed a Cir-dummy- and Grid-dummy-based dummy location selection algorithm. Shu C. [9] proposed two new dummy selection algorithms, MaxMinDistDS and Simp-MaxMinDistDS, that take both the location semantic diversity and the physical dispersion into account.

There are numerous dummy-based schemes being made to deal with location privacy. However, reviews for dummy-based schemes are relatively rare, and some focus on aspects relating to cryptography [10], anonymity [11], or other comprehensive reviews [12], which focus on the whole picture, but there are not enough review on dummy-based privacy protection. In addition, these reviews fail to clarify the relationship between the level of privacy (LoP), the quality of service (QoS), and system overhead. These are struggle to explain the difference and relationship between dummy and other location privacy protection techniques, as well as analyze and summarize the dummy location attack model and how to choose dummies. Therefore, such studies cannot help readers understand the up-to-date challenges of dummy-based privacy protection brought on by attackers' expanding background knowledge and the intersection between LBS and other emerging technologies.

In this paper, we make a review of dummy-based location privacy protection techniques for location-based services. The main contributions are as follows.

- First, we distinguish the relationship between the LoP, the QoS, and the system overhead. Additionally, we make an overall comparison of several representative methods of location privacy protection techniques. Then, we describe the merits of dummy-based location privacy protection on LBS. Meanwhile, a summary of the major attacks on dummy-based location privacy protection techniques is also included.
- Second, we systematically and comprehensively analyze and summarize the ways of selecting dummies on three aspects, namely the query probability, the physical dispersion, and the semantic diversity of locations.
- Third, we provide an overview of the methods for achieving query probability, physical dispersion, and semantic diversity while choosing dummies. Furthermore, we

make comparative analysis to indicate the different privacy protection advantages of different selection rules when choosing dummies. Results of this comparative analysis can be of benefit both to users and researchers who are studying this field.

The remainder of this paper is organized as follows. Section 2 gives an overview of location privacy protection, and Section 3 provides a summary of the attack model of dummy-based location privacy protection techniques. Section 4 describes the system architecture and privacy protection methods, and also gives a detailed analysis and summary of how to choose a dummy location. Finally, in Section 5, we conclude our work.

2. Overview of Location Privacy Protection

In this section, we first introduce the key issues in location privacy protection and location privacy protection techniques. Then, we describe the difference and connection of various location privacy protection techniques, and finally make a comparison between them.

2.1. Location Privacy Protection

When users use LBS, their location privacy is compromised to some degree due to dishonest or semi-trusted LBS servers serving private interests. Nonetheless, because location privacy is closely related to explicit sensitive information, other implicit sensitive information about users is also leaked. Take, for example, John. He has been feeling uneasy lately, so he decides to go to the hospital to find out what's wrong with his body. However, because he does not want others to know about his medical condition, the hospital's location is important to him. In reality, "where are you staying" reveals the privacy of "what are you doing". Similarly, the user's historical location data expose the locations he frequently shows up at, and the routes he travels a lot by, which leads to his home address, behavioral habits, work nature, and other sensitive information he cares about potentially being leaked [13]. Therefore, there is no doubt that it is definitely vital to protect a user's location privacy.

2.2. Key Issues of Location Privacy Protection

When it comes to location privacy protection, it is naturally necessary to consider the connection between LoP, QoS, and the system overhead [14].

2.2.1. Issue on the Relationship between LoP and QoS

In location privacy protection, high LoP and QoS cannot be satisfied at the same time. To obtain location services, users must submit their location to the service provider in some way, which risks exposing their private information. Many techniques, such as using cloaking areas instead of the real location, adding noise to the real location, and so on, sacrifice some degree of location accuracy for higher LoP. However, if the location accuracy is too low to meet users' demands, availability will suffer, and privacy protection will be rendered ineffective. Furthermore, the requirements for location service quality vary depending on the user. Users who request to query a specific point of interest will be more concerned with QoS. Users seeking hospital location service, on the other hand, will be more concerned with their location information. As a result, they are willing to sacrifice some service quality in exchange for a higher LoP. Therefore, understanding the relationship between QoS and LoP is one of the most crucial matters in location privacy protection.

2.2.2. Issue on the Relationship between QoS and System Overhead

With the advent of the "fast" era, people are more concerned with speed, even when it comes to location privacy protection. People desire faster response times and lower latency. When a user initiates a query request, the user experience will suffer if the response time is too slow. However, the majority of existing studies improve LoP without taking system overhead into account, or at the expense of a significant increase in system overhead to achieve a minor improvement in LoP. Simultaneously, the costs of communication, storage, and computation, as well as the loss of precision, all have an impact on the user experience due to the limited resources on the user's device. For example, a large amount of computation cost slows down the processing speed of mobile devices, a large amount of communication cost raises the extra cost for users, and a large amount of electricity overhead affects outdoor use of mobile devices, ultimately hindering the development of location service [15]. Understanding the relationship between LoP and the system overhead is therefore another critical issue in location privacy protection.

2.3. Location Privacy Protection Techniques

Researchers proposed numerous approaches to protect location privacy, such as [16–19]. In general, location privacy protection techniques can be divided into four categories [20]: obfuscation [21], encryption [22–24], cache and collaboration [25], and anonymity mechanisms [26].

2.3.1. Location Privacy Protection Techniques Based on Obfuscation

Location privacy protection techniques based on obfuscation refer to the necessary disruption to the original location information in an LBS query in order to prevent the attacker from obtaining the user's true location while also ensuring that the user can acquire unrestricted services. Dummy [6], spatial cloaking [27,28], differential privacy [29], and other obfuscation techniques can reduce the accuracy of location information. The dummy method adds multiple dummies and sends them to the LBS server along with the real query location to blur the real location. To protect users' location privacy, Li et al. [30] proposed an attribute-aware privacy protection scheme (APS). The Voronoi dividing algorithm (VDA) and the dummy determining algorithm (DDA) are two algorithms included in APS. The VDA algorithm divides the local map into different Voronoi polygons to ensure that the selected dummy locations are scattered, whereas the DDA algorithm chooses dummy locations based on the four-color mapping theorem to ensure that dummy locations differ in attributes. The classical dummy method, which was later extended to trajectory, is frequently used to solve the single location problem. Ni et al. [31] proposed an R-constrained dummy trajectory-based privacy-preserving algorithm (RcDT). The generated dummy locations are in a specific range close to the real location because the generating range R of the dummy location is constrained. Furthermore, by constraining the exposure risk of each dummy location and trajectory, dummy trajectories with a higher similarity to the real trajectory are generated. Differential privacy protects location privacy by adding an appropriate number of noises to the returned value of the query function [29]. Several recent studies [32,33] have investigated the use of differential privacy in location protection. The concept of protecting user locations within a radius R, whose privacy level is dependent on R, is formally defined by the term of geographical indiscernibility [32]. To increase the user's LoP, controlled random noise is added to their location. In general, using the obfuscation strategy will result in a significant loss of precision in query results.

2.3.2. Location Privacy Protection Techniques Based on Encryption

To achieve the privacy goals, the cryptographic approach adopts encryption technology to make the user's query content and location information completely transparent to the LBS server. While ensuring QoS, this technique does not reveal any user's location information, ensuring stricter privacy protection. Private information retrieval (PIR) [34,35] is a popular encryption method. PIR prevents the server (the database owner) from determining the user's point of interest and drawing additional conclusions about the client's private information by ensuring that the server (the database owner) cannot determine the correct query object when the user requests the database. Paulet et al. [34] obtained and decrypted location data using a PIR-based protocol. The user's location information was kept private because the server was unable to determine it. The PIR method ensures the confidentiality of the entire communication process (user request, information retrieval, and result return process). However, the issue of over-collected storage and computation overhead in PIR needs to be investigated further. The primary challenge in using PIR is developing a good retrieval strategy and index structure. However, because the LBS server must store the entire map information of the local map, the server's limited storage space as well as retrieval efficiency make PIR only applicable to a small space range at the present time.

2.3.3. Location Privacy Protection Techniques Based on Collaboration and Cache

Collaboration and caching cut down the time spent communicating with the LBS server as much as possible in order to limit exposure to location-sensitive information. Domingo-ferrer et al. [36] proposed a cooperative method for disturbing users' location information by adding Gaussian noise. This method requests disturbed location information from other users and then forms a cloaking region according to that information. Rather than using the true location, the anonymous group's density center, formed by cooperative users, is used as the anchor point to replace it and launch query requests. Shokri et al. [37] proposed an effective collaborative location privacy protection approach. Zhang et al. [38] proposed a cache and spatial K-anonymity-based privacy enhancement technique.

This strategy employs a multi-level caching method to reduce the possibility of user location information being disclosed. Niu et al. [39] created a privacy protection algorithm using dummy locations and cache awareness. The research on privacy protection techniques based on caching and collaboration focuses on three main areas: reducing cache overhead, improving the cache hit ratio and location privacy, and quantifying the QoS level. Another consideration is how to reduce the expensive communication cost caused by such a collaborative technique architecture.

2.3.4. Location Privacy Protection Techniques Based on Anonymity

Methods based on anonymity to protect location privacy, such as k-anonymity and mix-zone, protect privacy by breaking the link between user identity and location data. The k-anonymity [40] technique ensures that the user's location information cannot be differentiated from that of other $k-1$ users through generalization. As a result, attackers have a $1/k$ chance of discovering users' true location. Stajano et al. [41] proposed the Mix-zone, which differs from the k-anonymity scheme. Attackers are unable to precisely pinpoint the user's real location by frequently changing the user's name or pseudonym in the anonymity area. In a variety of settings, anonymous approaches have been thoroughly researched and tested. However, this strategy raises concerns because maintaining the same level of anonymity in different scenarios is difficult.

The relationship between location privacy, location privacy protection techniques, obfuscation, and dummy generation is depicted in Figure 2. Table 1 compares existing location privacy protection techniques in terms of LoP, outlining their main advantages and disadvantages. The system overhead of the four location-based privacy protection techniques is compared in Table 2. Given that different privacy protection techniques provide different benefits, we must adopt location privacy protection methods that are appropriate for the given application in order to protect the user's location privacy.

Table 1. The comparison among four privacy protection techniques.

LPPT [1]	RM [2]	LoP [3]	TTP
Obfuscation	Dummy Spatial Cloaking Differential Privacy	low	yes
Encryption	PIR	high	no
Collaboration and Cache		medium	no
Anonymity	K-anonymity Mix-zone	medium	yes

[1] LPPT:location privacy protection techniques. [2] RM: representative method. [3] LoP: the level of protection privacy.

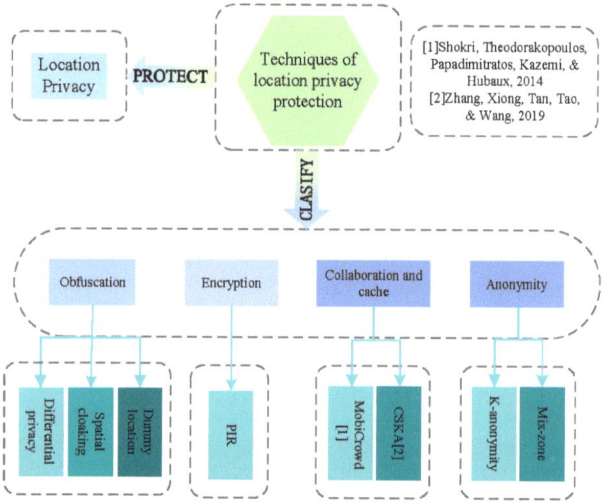

Figure 2. The relationship among location privacy, location privacy protection techniques, the obfuscation, and dummy location.

Table 2. The cost of four privacy protection techniques.

LPPT	Precision Loss	Communication Cost	Computation Cost	Storage Cost
Obfuscation	high	low	low	low
Encryption	low	low	high	medium
Collaboration and Cache	medium	high	low	high
Anonymity	medium	medium	high	medium

Dummy is an important obfuscation method that has stimulated the interest of researchers both at home and abroad. This is becayse it is simple to implement, does not require a trusted third party, and can protect location privacy while maintaining accuracy. Furthermore, we can see that dummy has other advantages over other privacy protections in Tables 1 and 2, such as low communication costs, low computation costs, and low storage costs.

3. Dummy-Based Attack Model

Malicious attackers aim to exploit various types of external information to find sensitive information about users, in addition to processing queries using various privacy protection mechanisms. However, the user's location contains inherent "side information", such as route information, human flow, and population distribution of the geographical region where the user is located [39,42]. Furthermore, attackers can obtain background knowledge in a variety of ways, including collaborative information systems, publicly available data aggregation, data brokers, data mining, and so on, in the age of Big Data and the Internet of Things.

Based on the attacker's prior knowledge in two dimensions, namely temporal information and context information, attacks can be classified into context dimension attacks and temporal dimension attacks [43]. In the former case, the attacker only has a single snapshot of a user's location, whereas in the latter case, the attacker has several locations collected over time or even a trajectory. We only consider the attack model on the context dimension in this paper because time is not taken into account. The most common threat to dummy-based location privacy protection techniques is background knowledge attacks in the context dimension. Such attacks can be classified into three types based on the attackers' prior knowledge: location-distributed attack, probability-distributed attack, and semantic similarity attack. This section will summarize the attack model of dummy-based location privacy protection techniques.

3.1. Location-Distributed Attack

The location distribution attack is a type of attack method in which the attacker explores the location distribution characteristics in the user-specified cloaking area. It is classified into three types. One is that the location distribution of the cloaking area is overly concentrated, resulting in a small hidden area. For example, all of the locations are in the same neighborhood. However, although it successfully blurs users' real locations, users' location privacy cannot be adequately protected. Regarding the second type, the user's true location is in the middle of the entire cloaking region, and the attacker can significantly reduce the user's range [44]. For instance, all of the dummy positions are centered on the real location. In the third type, the real cloaking area shrinks as a result of the uneven location distribution caused by the attacker's exclusion of some locations, which fails to meet the theoretical cloaking requirements. For example, if the majority of locations are distributed in a concentrated manner while one or two or a small portion of them are distributed in a relatively scattered manner, attackers can easily filter out those locations, reducing the original privacy protection intensity [45].

3.2. Probability-Distributed Attack

The probability distributed attack is defined as the attacker calculating historical query probability information by collecting historical service request records for all locations within a specific geographical region and over a specific time period [46]. When the probability distribution in the anonymous set generated by the user's query request is uneven, the attacker filters out the dummy locations with a large gap, resulting in a failure to achieve the true location privacy protection effect. If the chosen dummy locations set includes several dummy locations in the middle of the lake with zero query probability, the attacker can simply deduce that they are dummy locations and filter them out.

3.3. Semantic Similarity Attack

The semantic similarity attack refers to the attacker's speculation on the privacy information of users by parsing semantic information of locations in cloaking regions, such as behavior habits, health status, and professional attributes [47]. As long as all dummies' query probabilities and the real location of the user's query probability are equal or close, attackers can easily infer user behavior if all dummies in cloaking areas belong to the same kind of semantics.

4. Dummy-Based Location Privacy Protection Techniques

In this section, we outline the two system architectures of dummy-based location privacy protection techniques, then review the dummy-based location privacy protection techniques, and finally analyze and summarize how the dummy-based location privacy protection techniques choose dummies to handle background knowledge attacks.

4.1. System Architectures of Dummy-Based Location Privacy Protection

Dummy generation system architectures can be divided into two types: architecture with a third party and architecture without a third party, depending on whether a third party is deployed or not [48].

4.1.1. Architecture with a Third Party

This architecture consists of users, a third party, and an LBS server. One or more servers represent a third party [49,50], and these are the servers that generate the dummy location set for the query user in order to mask the true location. Figure 3 depicts a third-party architecture. The primary responsibility of the third party is to collect and process user query requests, protect sensitive location information using privacy protection techniques, and then forward the processed query requests to the LBS server. After receiving these requests from the third party, the LBS server retrieves the database and transmits the matching result sets to the third-party servers. Finally, the requesting users receive the result sets from the third-party servers. Third-party servers, for example, create a cloaking zone with multiple users, and all users in the zone submit the same query to LBS. In this case, the LBS server is unable to determine who initiated the query and, as a result, is unable to find out which location is the original requesting location.

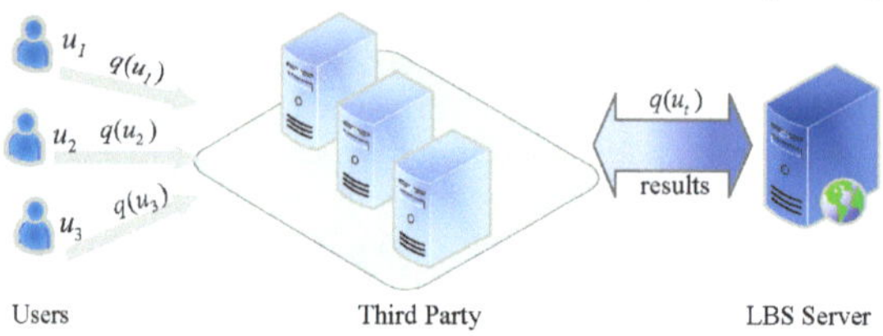

Figure 3. The architecture with a third party.

Obtaining a completely trustworthy third party, on the other hand, is difficult, and the "honest but curious" third party is vulnerable to a single point of attack and other vulnerabilities. As a result, the researchers have proposed an architecture that does not rely on a third party.

4.1.2. Architecture without a Third Party

Figure 4 depicts the architecture in the absence of a third party, which consists of users and an LBS server.

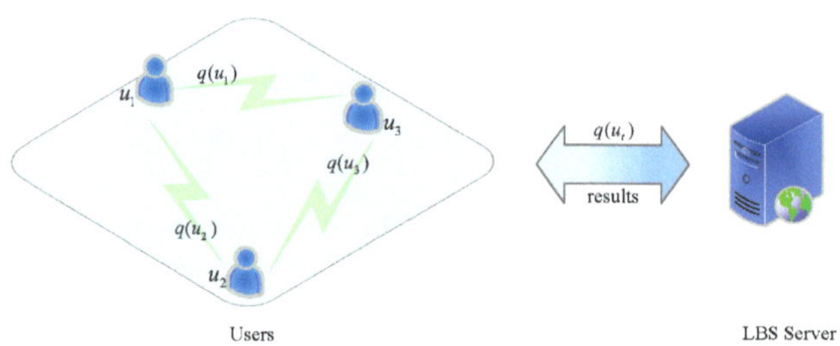

Figure 4. The architecture without a third party.

The architecture requires that mobile devices carried by users have certain computational and storage capabilities that can be used to select dummy locations, create cloaking areas, and save map data within a certain range. The non-third-party architecture can be divided into two types based on whether or not users collaborate. In the first type, users' location information is concealed in accordance with their privacy requirements [51]. For example, the Apple differential privacy team uses local differential privacy [52]. Users' personal data can be randomized on their devices before being uploaded to the server, which can improve the user experience without infringing on privacy. In the second type, users collaborate for the sake of secrecy [53]. Tor, for example, is a volunteer-run distributed relay network that enables users to conceal their location while providing a variety of services. When using this method of obscuring through user collaboration, it is important to consider the additional communication cost between users as well as the risk of collusion attack [54].

4.2. The Dummy-Based Location Privacy Protection Techniques

The dummy-based location privacy protection techniques select many dummy locations (assuming $k-1$ dummies) and send the same query request to the LBS server with the real location, making it difficult for the LBS server to distinguish the real ones from those k locations. However, if those dummies are chosen at random or without taking into account the attacker's background knowledge, some of the dummy locations will be too large for the attacker to filter out, and the theoretical LoP will be impossible to achieve. Figure 5 shows a cloaking zone with $k = 8$ users. The colorful one represents the user's true location, whereas the black ones represent the user's chosen dummy locations. The k locations cover the cloaking area.

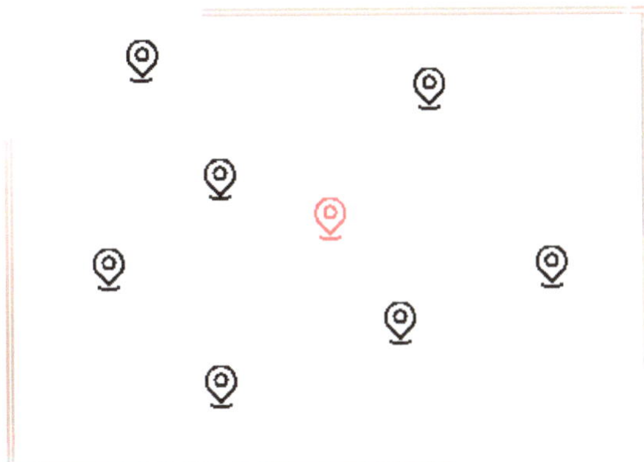

Figure 5. A cloaking area with $k = 8$ users.

In general, the higher the k value, the greater the privacy protection; otherwise, the lower the privacy security. When the value of k increases, the corresponding QoS decreases and the system overhead increases.

4.3. *Algorithms of Dummy Location Selection*

Researchers proposed a variety of approaches in the dummy-based locations' selection to withstand the background knowledge attack, such as [55]. The main work of these studies is to choose appropriate dummy locations to construct a candidate set that protects users' privacy effectively. The aim of dummy-based location privacy protection is to camouflage the user's real location in the dummy locations concentration; thus, the quality of these selected candidate dummies is crucial to attaining the desired level of location privacy in the overall system. As a result, it is critical to reduce the distinguishability of real and dummy locations in all aspects; that is, we must choose dummy locations that can satisfy user desires while also protecting user privacy. In this subsection, we summarize and discuss the rules on dummy selection for dummy-based location privacy protection techniques.

4.3.1. Take the Historical Query Probability of Locations into Consideration

The popularity of a location within a geographic location area over time is reflected by its historical query probability. The ratio of the number of times a location is queried to the total number of times all locations are requested in the global geographical area is used to calculate the historical query probability of a location in a certain period of time. For example, the following is the calculation formula for the historical query probability of location i inside a specific geographical area over time:

$$q_i = \frac{times \quad of \quad queries \quad in \quad location \quad i}{times \quad of \quad queries \quad in \quad all \quad locations}, \tag{1}$$

Because the LBS server has background information such as historical query probability of map locations, the server filters out dummy locations with obvious differences based on the probability distribution information of the candidate set, and thus the expected level of privacy protection cannot be achieved.

If the server filters out m dummy locations, the likelihood of identifying the user's dummy location increases from $\frac{1}{k}$ to $\frac{1}{k-m}$. In the entire map space, Figure 6 depicts the distribution of all locations and their historical query probability. Each little grid cell in the diagram represents a location. Varied shadow shapes portray different historical query probabilities, and the sum of the probabilities of all locations initiating query requests in the entire grid space is 1. Location A represents the user's real location, whereas B, C, and D are the dummy locations that have been chosen. Because their historical query probability is smaller than the real location's or even zero, the server can easily filter these dummy locations out.

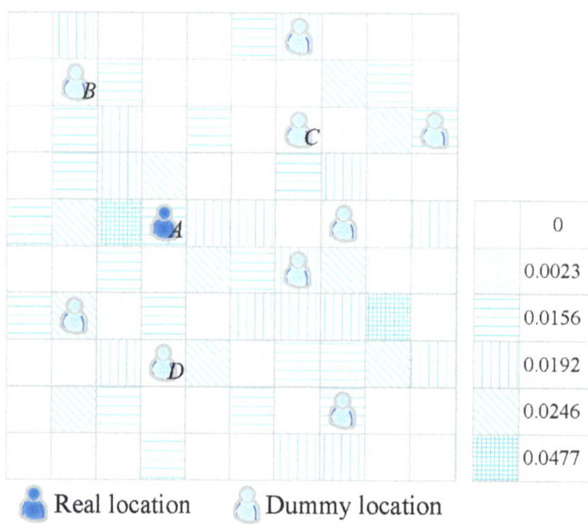

Figure 6. The historical query probability distribution of all locations.

Hara et al. [7] developed a dummy location selection algorithm that considers real-world environmental constraints and avoids dummy locations in inaccessible locations, such as the middle of a lake. However, this method only eliminates a small number of impossible locations, those where $q_i = 0$. As a result, the dummy quality is poor, as is the LoP. In order to improve the quality of dummies and the LoP, the DLS algorithm chooses dummy locations that have the same probability as the real ones. It not only keeps these $q = 0$ locations at bay, but it also reduces the difference in query probability between the real ones and dummies. In the literature [56], the greedy algorithm idea is used to select dummy locations so that the new location set composed of each new dummy location and the previously selected dummy locations have the best hiding effect. Other authors [57,58] have employed an information entropy-based method, with the historical query probability as a variable, to choose dummy locations. In [57,58], the set of dummy locations with the highest entropy value acts as the final set of candidate dummy locations. Because the historical query probability of each location over time is insufficient to convey the prevalence of each location, [59] introduced the concept of "current query probability", which was used to replace historical query probability as the criterion for selecting dummy locations. Users choose different geographical regions for different time periods, with each location's current query probability being different. As a result, the "historical query probability" is more diverse, posing a greater challenge to attackers.

4.3.2. Taking the Physical Dispersion of Locations into Consideration

The physical dispersion between locations describes the spatial distribution of locations. The obscuring of users' true locations will also perform poorly if an attacker learns this background knowledge in order to carry out location distribution attacks on them. As a result, selecting dummy locations solely on historical query probability is insufficient. In practical applications, physical dispersion between locations should be highlighted.

If the physical dispersion between locations is too small, the cloaking area will be too small. The cloaking area, as shown in Figure 7a, is small, allowing the attacker to quickly deduce that the real user is in a very small area. As a result, something like Figure 7b would be preferable because it provides a larger cloaking area for the real user. Simultaneously, the query probability of those chosen dummy locations is not too far off from the user's actual location. As a result, when selecting dummy locations, the spatial distribution of the $k-1$ dummy locations and the real ones should be guaranteed, while the historical query probability should be the same or similar.

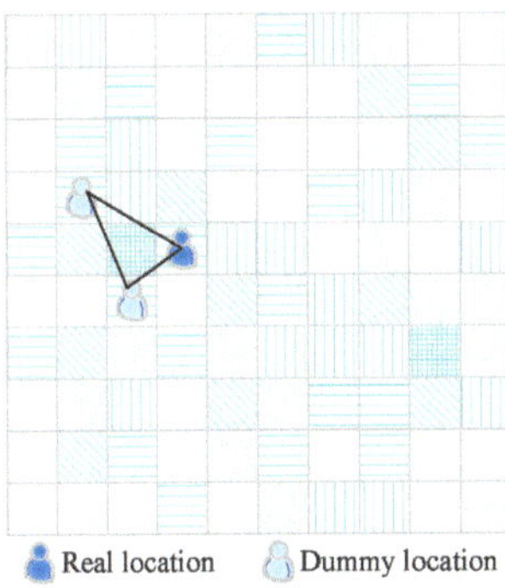

(**a**) small area

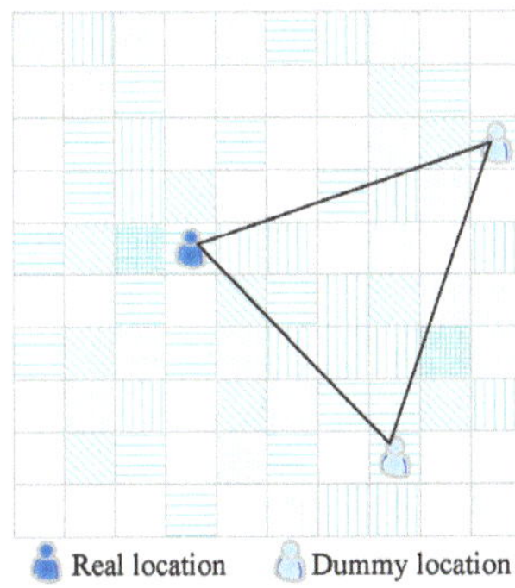

(**b**) large area

Figure 7. The physical dispersion situation between dummies and the real location.

To meet the requirements of physical dispersion of locations, Niu B. et al. [8] proposed a method for selecting candidate locations based on virtual circles and virtual grids. Because the user's true location is likely to be close to the center of the local map, the virtual circle algorithm may have performed poorly in terms of privacy. To provide a more obscured area, a virtual grid-based algorithm was introduced, which ensures that candidate locations are distributed fairly evenly around the true location and that the size of the cloaking area meets user needs.

In order to achieve physical dispersion between locations, Niu B et al. proposed the enhanced DLS algorithm in reference [42]. They argued that the product of locational distances was more accurate in depicting locational dispersion than the sum of locational distances. As Figure 8 shows, $CA+CB=DA+DB$ while $CA \cdot CB > DA \cdot DB$; as a result, we would prefer to choose C over D from the perspective of privacy. They used a multi-objective optimization model as well, where the probability and physical dispersion of locations are considered simultaneously to pick the best candidate set of dummy locations.

In addition to the previous research on location dispersion in physical space, there are numerous studies on how to portray the physical distance, such as the effective distance [60] or the road network distance [61,62], between two locations. The idea of effective distance was developed by Xu et al. [60] to characterize the distribution features of locations, and the effective distance between these two locations was defined as the shortest distance between the current location and any other location. Consider real user u_r and any other user u_i; their coordinates are (x_r, y_r) and (x_i, y_i), resulting in an effective distance of

$$d(u_r, u_i) = \min \mid u_r, u_i \mid = \min \sqrt{(x_r - x_i)^2 + (y_r - y_i)^2} \quad (2)$$

It is apparent from the effective distance calculation formula that the essence of the effective distance specified by them is the Euclidean distance. Despite the fact that the article is based on a real-world road network, Euclidean distance is nevertheless employed to measure location distribution features. Chen et al. [62] proposed a privacy protection method for the road network in response to the fact that the distance between any two points in real life is not a simple linear distance (Euclidean distance), and that users' activities are more restricted to the planned road. This approach requires that the number of road sections in the selected dummy sites satisfy the value given by the user in order to attain the purpose of physical dispersion between the selected dummy locations when picking dummy locations. This road network, however, is an undigraph road network model, which is insufficiently realistic for real-world road network simulation, as illustrated in Figure 9a. Zhou Changli et al. proposed a location privacy protection approach based on the digraph road network architecture (as shown in Figure 9b), in which an anchor point (dummy location) was used to replace users' real locations when initiating query requests. However, when choosing the anchor point, the historical query probability of the anchor and its geographical spatial distribution link with a real user were not taken into account.

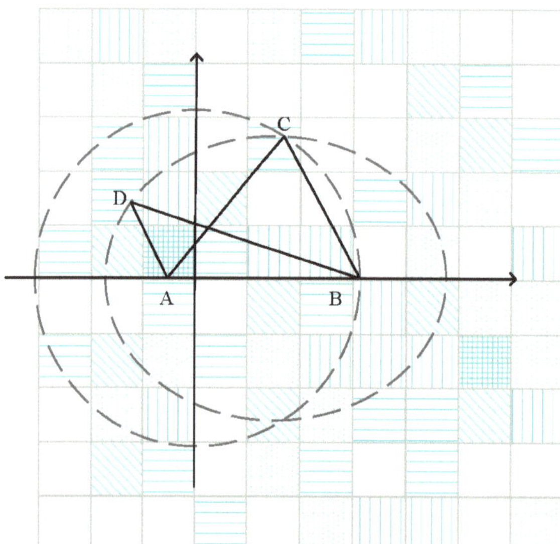

Figure 8. The enhanced DLS.

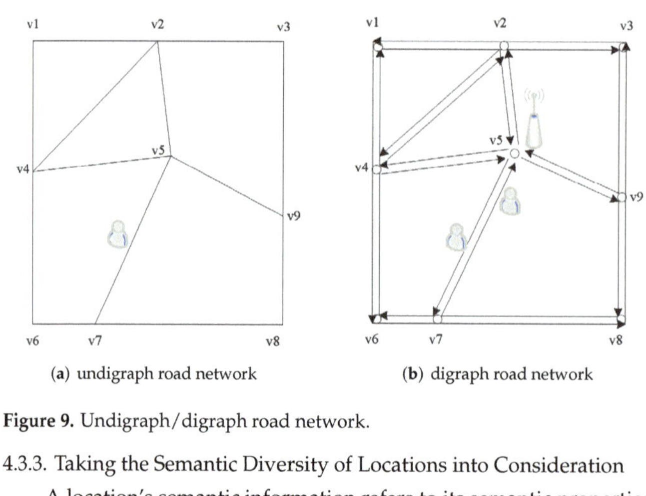

(**a**) undigraph road network (**b**) digraph road network

Figure 9. Undigraph/digraph road network.

4.3.3. Taking the Semantic Diversity of Locations into Consideration

A location's semantic information refers to its semantic properties, such as hospitals, restaurants, banks, schools, parks, and so on. Semantic features can be extracted using context information. The greater the number of location semantic features collected, the more accurate the semantic categorization, and the greater the ability to protect users' location privacy. Consider the semantics of a user's location for "hospital" which implies semantic information on the user's health, professional property, and so on. Because the user's state of health or professional attributes belong to the users of the important content of privacy, semantic information must be considered when selecting dummy locations.

Bostanipour [63] presented a method for combining obfuscation location information with semantic information to ensure that many semantically identical locations are cloaked, therefore preventing attackers from performing semantic inference attacks. The locations derived using this method, on the other hand, are semantically related to those of real users. For example, the real user's semantic tag is "Pizza Place", but the cloaking region includes venues such as "Noodle House" and "Hamburger Palace", all of which belong to the parent semantic tag "Restaurant". As a result, such a method is still vulnerable to semantic inference attacks.

In order to achieve semantic diversity, each location in the candidate dummy set should have a diverse set of semantic properties as much as possible. While representing semantic differences between locations is a challenge, Zeng et al. proposed the similarity of two semantic location types using Euclidian distance to calculate [64]. Tian et al. measure semantic distance based on the intersection and union of a location's semantic attributes:

$$sem_{dist}(A, B) = \frac{[sem_A \cup sem_B] - [sem_A \cap sem_B]}{sem_A \cup sem_B} \tag{3}$$

This then transforms the results to show semantic similarity [65]. Using Euclidean distance and the relationship between sets to quantify semantic difference not only consumes a lot of effort but also weakens the algorithm's efficiency. Another author [9] created a location semantic tree (LST) to arrange all locations, as shown in Figure 10, in order to achieve semantic similarity control that can serve the tailored needs of users and increase the efficiency of algorithm execution. The most fundamental semantic information is stored in the leaf nodes of the location semantic tree, and the hop number between the leaf nodes is used to calculate the semantic distance, which is then used to calculate the semantic difference degree. This approach can rapidly find and categorize the semantic associations of all locations in a specified geographical area.

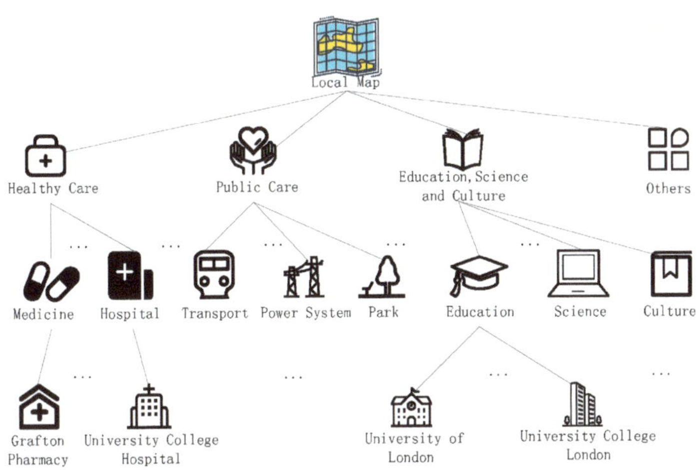

Figure 10. The location semantic tree.

When there are many different semantic varieties in a given geographical area and there are crossover circumstances, the depth and breadth of the semantic tree grow quite large, decreasing search efficiency. As a result, the location semantic tree is not ideal for such scenarios.

4.4. Summary

In general, we classify the existing dummy location selection methods into three categories according to the types of attacks they can defend against, as shown in Table 3. The first category of selection method can successfully defend against the probability similarity attack. The second category of the selection method can effectively prevent physical distribution attacks launched by attackers on the distribution pattern of locations. The third category of selection method can make it difficult for attackers trying to obtain cracking clues from the semantic information of locations. Different methods have different characteristics, and we can select relatively appropriate methods according to our own needs and purposes when using these methods to select dummies to construct dummy location's set.

Table 3. Selection methods of dummy on query probability similarity, physical dispersion and semantic diversity.

Category	Reference	Methods of Selection
Query probability similarity	[7]	avoids dummies with $q_i = 0$
	[42]	dummies have the same probability as the real ones
	[58]	information entropy-based
	[59]	current query probability
Physical dispersion	[8]	virtual circles and virtual grids
	[42]	the product of locational distances
	[60]	the effective distance
	[62]	the road network distance
Semantic diversity	[9]	location semantic tree
	[64]	Euclidian distance
	[65]	the intersection and union of a location's semantic attributes

An overall comparison of random selection and other selection schemes that consider different factors when selecting dummies is shown in Table 4. In Table 4, we observe that different schemes can choose different system structures and take different factors into account to design different schemes according to their own purposes and needs. As a consequence, the types of attacks they can defend against are not the same, and, of course, the corresponding computational overheads are somewhat different. Dummy location selection methods that take into account query probability, physical dispersion, and semantic diversity yield better security than random selection with a relatively small computational overhead. Furthermore, depending on different selection factors, the attacks that can be defended against are varied when selecting a dummy location. When a dummy location is chosen, the more factors are taken into account, the better the privacy protection effects of the scheme are strengthened, while the difference in computing overhead is not readily apparent. As a result, schemes increasingly seek to take more factors into account when selecting dummies. They are no longer always based on a single factor, such as [21,66], which incorporates two factors, and three factors are considered simultaneously in the literature [46]. As the research goes further, new factors are discovered and considered, and new rules are established in [67,68].

Table 4. Summary of dummy selection.

Selection Method	Reference	CO [a]	Architecture		Attack		
			TTP	Non-TTP	AoQ [b]	AoD [c]	AoS [d]
Random Selection	[6]	$O(k \log k)$		√			
Considering Q	[42]	$O(k)$		√	√		
Considering D	[7]	*Null*		√		√	
Considering S	[62]	*Null*	√				√
	[64]	*Null*	√				√
Considering Q+D	[8]	$O(k)$		√	√	√	
	[9]	$O(\log k)$		√	√	√	
	[58]	$O(\alpha \log_2 \alpha)$		√	√	√	
	[60]	$O(k^2 + IJU)$		√	√	√	
Considering D+S	[59]	$O(k)$		√		√	√
Considering Q+S	[65]	$O(It \cdot k)$		√	√		√
All of them	[18]	$O(\log N)$	√		√	√	√

[a] CO: the computation overhead. [b] AoQ: the attack of query probability; Q: Query probability. [c] AoD: the attack of location distribution; D: Location distribution. [d] AoS: the attack of semantic similarity; S: Semantic similarity. Notes: k: the number of dummies; α: $(\omega + m)\log(\omega + m)$, $\omega = (maxtier - 1)(1 - e)m$, m: the number of dummies candidate set, $maxtier$: the max times of iteration; IJ: an area is divided into IJ cells; U: the number of services; It: the times of iteration; N: the total number of users in the region to be clocked.

5. Conclusions

In this article, we provide a review of dummy-based location privacy protection techniques for LBS. First, we distinguished the relationship between the LoP, QoS, and system overhead. At the same time, we made an overall comparison of several representative methods of location privacy protection techniques. We described the merits of dummy-based location privacy protection on LBS. Meanwhile, a summary of the major attacks on dummy-based location privacy protection techniques was also included.

Second, we systematically and comprehensively analyzed and summarizde the ways of selecting dummies on three aspects, namely the query probability, the physical dispersion, and the semantic diversity of locations.

Third, we provided an overview of the methods for achieving query probability, physical dispersion, and semantic diversity while choosing dummies. Furthermore, the different privacy protection advantages of different selection rules when choosing dummies can be seen from a comparative analysis. The results of this comparative analysis can benefit both users and researchers who are studying this field. When the requesting service needs to construct a hiding area that hides their true location, the user can refer to this comparative evaluation to choose a dummy-based location privacy protection method that better meets their needs. Moreover, researchers studying this area can gain a better understanding of dummy-based privacy protection schemes from the results of this comparative analysis. They can also get to know the challenges posed by the expanding background knowledge of attackers and the intersection between LBS and other emerging technologies.

Dummy location selection approaches that took into account new circumstances in the selection of dummies emerged as research progressed. There are still some significant issues to be resolved and perfected in the area of dummy location selection.

First, as new technologies such as social networks, edge computing, and federal learning have been advanced, new privacy concerns have also emerged.

- Because location acquisition technology is becoming more widely available, it is now possible to add geo-information to already-existing social networks, which has facilitated in the emergence and expansion of LBSN. LBSN, a combination of LBS and social networks, involves a range of personal private information, such as shared common locations, personal interests, daily behaviors and activities, etc. [69].

- LBS@E [70] delocalizes LBSs and retrieves local information from nearby edge servers around them instead of the cloud. Consequently, it tackles the location privacy problem innovatively. However, LBS@E brings new challenges to location privacy. Mobile users can still be localized to specific privacy areas jointly covered by edge servers accessed by mobile users. The small privacy area puts the mobile user's location at risk of similarity.
- Ref. [71] uses federated learning to select the best location privacy protection mechanism (LPPM) for each user according to the real location and the user's configuration, which avoids the direct use of the real location information. Nevertheless, it is vulnerable to poisoning attacks and untrusted users who intend to add a backdoor to the model [72] or defend against attacks on model information leakage [73].

Second, existing dummy-based solutions do not account for all aspects of real-world privacy protection [74], and there is a significant gap between theoretical and real-world privacy protection effects. According to Sun et al. [75], attackers can also rule out impossible dummy locations by determining whether users can reach the query location in a reasonable amount of time from their current location.

Third, dummy-based approaches that focus on the spatio-temporal correlation of location are commonly used in trajectory privacy protection, which poses new challenges in trajectory privacy. Zhao [76] assumes that all users(dummies) involved are trustworthy and report their real locations. However, it is often not the case in reality. There are untrusted users who conduct location injection attacks (LIAs) in continuous LBS queries. Zhen [77] found that the trajectory data were published without proper processing. A great amount of work has been devoted to merging one's own trajectories with those of others, without protecting the semantic information about the location. In continuous LBS queries, users can obfuscate their true query location by selecting dummy locations and predicted locations, thus improving their privacy. However, selecting a large number of dummies for each query can increase the query cost of the system and influence the accuracy of the predicted location [78].

Funding: This research was funded by the National Natural Science Foundation of China (No. 61702180), the Research Foundation of Education Bureau of Hunan Province (No. 21B0493), and the Hunan Province Science and Technology Project Funds (2018TP1036).

Institutional Review Board Statement: Not applicable.

Informed Consent Statement: Not applicable.

Conflicts of Interest: The authors declare no conflict of interest.

Abbreviations

The following abbreviations are used in this manuscript:

LTTP	location privacy protection techniques
LoP	the level of privacy protection
RM	representative method
QoS	the quality of service
CO	computation overhead
AoQ	attack of query probability
Q	query probability
AoD	attack of location distribution
D	location distribution
AoS	attack of semantic similarity
S	semantic similarity

References

1. Statista Research Department. Share of U.S. smartphone Owners Using Geosocial and Location-Based Services from 2011 to 2015. 2016. Available online: https://www.statista.com/statistics/224949/mobile-geosocial-and-location-based-service-usage-by-age/ (accessed on 15 May 2022).
2. Auxier, B. Pew Research Center: Internet and Technology-How Americans See Digital Privacy Issues Amid the COVID-19 Outbreak. 2020. Available online: https://www.pewresearch.org/fact-tank/2020/05/04/how-americans-see-digital-privacy-issues-amid-the-covid-19-outbreak/ (accessed on 15 May 2022).
3. Laricchia, F. Bluetooth Location Services Device Shipments Worldwide from 2016 to 2026. 2022. Available online: https://www.statista.com/statistics/1226718/global-bluetooth-location-device-shipment-forecast/#statisticContainer (accessed on 15 May 2022).
4. Security, R.B. Year End Data Breach QuickView Report. 2022. Available online: https://www.riskbasedsecurity.com/2022/02/04/data-breach-report-2021-year-end/ (accessed on 14 May 2022).
5. Association, C.C. Investigation Report on App Personal Information Leakage. 2018. Available online: https://www.cca.org.cn/jmxf/detail/28180.html (accessed on 10 May 2022).
6. Kido, H.; Yanagisawa, Y.; Satoh, T. An anonymous communication technique using dummies for location-based services. In Proceedings of the International Conference on Pervasive Services, 2005, Santorini, Greece, 11–14 July 2005; pp. 88–97.
7. Hara, T.; Suzuki, A.; Iwata, M.; Arase, Y.; Xie, X. Dummy-Based User Location Anonymization Under Real-World Constraints.k *IEEE Access* **2016**, *4*, 673–687. [CrossRef]
8. Niu, B.; Zhang, Z.; Li, X.; Hui, L. Privacy-area aware dummy generation algorithms for Location-Based Services. In Proceedings of the IEEE International Conference on Communications, Sydney, NSW, Australia, 10–14 June 2014; pp. 957–962.
9. Chen, S.H.; Shen, H. Semantic-Aware Dummy Selection for Location Privacy Preservation. In Proceedings of the Trustcom/BigDataSE/ISPA, Tianjin, China, 23–26 August 2016; pp. 752–759.
10. Magkos, E. Cryptographic Approaches for Privacy Preservation in Location-Based Services: A Survey. *Int. J. Inf. Technol. Syst. Approach* **2011**, *4*, 48–69. [CrossRef]
11. Shin, K.G.; Ju, X.; Chen, Z.; Hu, X. Privacy protection for users of location-based services. *IEEE Wirel. Commun.* **2012**, *19*, 30–39. [CrossRef]
12. Chatzikokolakis, K.; ElSalamouny, E.; Palamidessi, C.; Pazii, A. Methods for Location Privacy: A comparative overview. *Found. Trends® Priv. Secur.* **2017**, *1*, 199–257. [CrossRef]
13. Zhang, S.; Wang, G.; Alam, B.; Qin, L. A Dual Privacy Preserving Scheme in Continuous Location-Based Services. *IEEE Internet Things J.* **2018**, *5*, 4191–4200. [CrossRef]
14. Ma, M.; Du, Y.; Li, F.; Liu, J. Review of semantic-based privacy-preserving approaches in LBS. *Chin. J. Netw. Inf. Secur.* **2016**, *2*, 10–13.
15. Wan, S.; Li, F.; Niu, B.; Sun, Z.; Li, H. Research progress on location privacy-preserving techniques. *IEEE Internet Things J.* **2016**, *37*, 18.
16. Zhang, J.; Xu, L.; Tsai, P.W. Community structure-based trilateral stackelberg game model for privacy protection. *Appl. Math. Model.* **2020**, *86*, 20–35. [CrossRef]
17. Qiu, Y.; Liu, Y.; Xxuan, L.; Chen, J. A Novel Location Privacy-Preserving Approach Based on Blockchain. *Sensors* **2020**, *20*, 3519. [CrossRef] [PubMed]
18. Ni, L.; Tian, F.; Ni, Q.; Yan, Y.; Zhang, J. An anonymous entropy-based location privacy protection scheme in mobile social networks. *EURASIP J. Wirel. Commun. Netw.* **2019**, *2019*, 1–19. [CrossRef]
19. Palanisamy, B.; Liu, L. Attack-Resilient Mix-zones over Road Networks: Architecture and Algorithms. *IEEE Trans. Mob. Comput.* **2015**, *14*, 495–508. [CrossRef]
20. Liu, B.; Zhou, W.; Zhu, T.; Gao, L.; Xiang, Y. Location Privacy and Its Applications: A Systematic Study. *IEEE Access* **2018**, *6*, 17606–17624. [CrossRef]
21. Chen, S.; Fu, A.; Shen, J.; Yu, S.; Wang, H.; Sun, H. RNN-DP: A new differential privacy scheme base on Recurrent Neural Network for Dynamic trajectory privacy protection. *J. Netw. Comput. Appl.* **2020**, *168*, 102736. [CrossRef]
22. Farouk, F.; Alkady, Y.; Rizk, R.Y. Efficient Privacy-Preserving Scheme for Location Based Services in VANET System. *IEEE Access* **2020**, *8*, 60101–60116. [CrossRef]
23. Gupta, S.; Arora, G. Use of Homomorphic Encryption with GPS in Location Privacy. In Proceedings of the 2019 4th International Conference on Information Systems and Computer Networks (ISCON), Mathura, India, 21–22 November 2019; pp. 42–45.
24. Sahai, A.; Waters, B. Fuzzy Identity-Based Encryption. *IACR Cryptol. ePrint Arch.* **2005**, *2004*, 86.
25. Cui, Y.; Gao, F.; Li, W.; Shi, Y.; Panaousis, E. Cache-Based Privacy Preserving Solution for Location and Content Protection in Location-Based Services. *Sensors* **2020**, *20*, 4651. [CrossRef]
26. Khodaei, M.J.; Papadimitratos, P. Cooperative Location Privacy in Vehicular Networks: Why Simple Mix Zones are Not Enough. *IEEE Internet Things J.* **2021**, *8*, 7985–8004. [CrossRef]
27. Khoshgozaran, A.; Shahabi, C.; Shirani-Mehr, H. Location privacy: Going beyond K-anonymity, cloaking and anonymizers. *Knowl. Inf. Syst.* **2011**, *26*, 435–465. [CrossRef]
28. Jadallah, H.; Aghbari, Z.A. Spatial cloaking for location-based queries in the cloud. *J. Ambient. Intell. Humaniz. Comput.* **2019**, *10*, 3339–3347. [CrossRef]

29. Xu, C.; Zhu, L.; Liu, Y.; Guan, J.; Yu, S. DP-LTOD: Differential Privacy Latent Trajectory Community Discovering Services over Location-Based Social Networks. *IEEE Trans. Serv. Comput.* **2021**, *14*, 1068–1083. [CrossRef]
30. Li, W.; Li, C.; Geng, Y. APS: Attribute-Aware Privacy-Preserving Scheme in Location-Based Services. *Inf. Sci.* **2020**, *527*, 460–476. [CrossRef]
31. Ni, L.; Yuan, Y.; Wang, X.; Yu, J.; Zhang, J. A Privacy Preserving Algorithm Based on R-constrained Dummy Trajectory in Mobile Social Network. *Procedia Comput. Sci.* **2018**, *129*, 420–425. [CrossRef]
32. Andrés, M.E.; Bordenabe, N.E.; Chatzikokolakis, K.; Palamidessi, C. Geo-indistinguishability: differential privacy for location-based systems. In Proceedings of the 2013 ACM SIGSAC Conference on Computer and Communications Security, Berlin, Germany, 4–8 November 2013; pp. 901–914.
33. Li, H.; Wang, Y.; Guo, F.; Wang, J.; Wang, B.; Wu, C. Differential Privacy Location Protection Method Based on the Markov Model. *Wirel. Commun. Mob. Comput.* **2021**, *2021*, 4696455. [CrossRef]
34. Fung, E.; Kellaris, G.; Papadias, D. Combining Differential Privacy and PIR for Efficient Strong Location Privacy. In Proceedings of the International Symposium on Spatial and Temporal Databases, Hong Kong, China, 26–28 August 2015; pp. 1–18.
35. Russell, P.; Golam, K.M.; Xun, Y.; Elisa, B. Privacy-Preserving and Content-Protecting Location Based Queries. *IEEE Trans. Knowl. Data Eng.* **2014**, *26*, 1200–1210.
36. Domingo-Ferrer, J. Microaggregation for Database and Location Privacy. In Proceedings of the International Workshop on Next Generation Information Technologies and Systems, Kibbutz Shefayim, Israel, 4–6 July 2006; pp. 106–116.
37. Shokri, R.; Theodorakopoulos, G.; Papadimitratos, P.; Kazemi, E.; Hubaux, J.P. Hiding in the Mobile Crowd: Location Privacy through Collaboration. *IEEE Trans. Dependable Secur. Comput.* **2014**, *11*, 266–279.
38. Zhang, S.; Xiong, L.; Tan, Z.; Tao, P.; Wang, G. A caching and spatial K-anonymity driven privacy enhancement scheme in continuous location-based services. *Future Gener. Comput. Syst.* **2019**, *94*, 40–50. [CrossRef]
39. Niu, B.; Li, Q.; Zhu, X.; Cao, G.; Hui, L. Enhancing privacy through caching in location-based services. In Proceedings of the IEEE Conference on Computer Communications (INFOCOM), Hong Kong, 26 April–1 May 2015; pp. 1017–1025.
40. Sweeney, L. k-Anonymity: A Model for Protecting Privacy. *Int. J. Uncertain. Fuzziness -Knowl.-Based Syst.* **2002**, *10*, 557–570. [CrossRef]
41. Beresford, A.R.; Stajano, F. Location Privacy in Pervasive Computing. *IEEE Pervasive Comput.* **2003**, *2*, 46–55. [CrossRef]
42. Niu, B.; Li, Q.; Zhu, X.; Cao, G.; Hui, L. Achieving k-anonymity in privacy-aware location-based services. In Proceedings of the IEEE INFOCOM 2014-IEEE Conference on Computer Communications, Toronto, ON, USA, 27 April–2 May 2014; pp. 754–762.
43. Wernke, M.; Skvortsov, P.; Dürr, F.; Rothermel, K. A classification of location privacy attacks and approaches. *Pers. Ubiquitous Comput.* **2012**, *18*, 163–175. [CrossRef]
44. Chow, C.Y.; Mokbel, M.F.; Liu, X. Spatial cloaking for anonymous location-based services in mobile peer-to-peer environments. *Geoinformatica* **2011**, *15*, 351–380. [CrossRef]
45. Mokbel, M.F. Privacy in Location-Based Services: State-of-the-Art and Research Directions. In Proceedings of the International Conference on Mobile Data Management, Mannheim, Germany, 1 May 2007; p. 228.
46. Shokri, R.; Theodorakopoulos, G.; Boudec, J.Y.L.; Hubaux, J.P. Quantifying location privacy. In Proceedings of the IEEE Symposium on Security and Privacy, Washington, DC, USA, 22–25 May 2011; pp. 247–262.
47. Lee, B.; Oh, J.; Yu, H.; Kim, J. Protecting location privacy using location semantics. In Proceedings of the 17th ACM SIGKDD International Conference on Knowledge Discovery and Data Mining, San Diego, CA, USA, 21–24 August 2011; pp. 21–24.
48. Jiang, H.; Li, J.; Zhao, P.; Zeng, F.; Xiao, Z.; Iyengar, A. Location Privacy-preserving Mechanisms in Location-based Services. *ACM Comput. Surv.* **2021**, *54*, 1–36. [CrossRef]
49. Gedik, B.; Ling, L. Protecting Location Privacy with Personalized k-Anonymity: Architecture and Algorithms. *IEEE Trans. Mob. Comput.* **2007**, *7*, 1–18. [CrossRef]
50. Vu, K.; Rong, Z.; Jie, G. Efficient Algorithms for K-Anonymous Location Privacy in Participatory Sensing. In Proceedings of the IEEE International Conference on Computer Communications, Orlando, FL, USA, 25–30 March 2012; pp. 2399–2407.
51. Yu, L.; Liu, L.; Pu, C. Dynamic Differential Location Privacy with Personalized Error Bounds. In Proceedings of the Network and Distributed System Security Symposium, San Diego, CA, USA, 26 February–1 March 2017; pp. 1–15.
52. Differential Privacy Team, A. Learning with Privacy at Scale. 2017. Available online: https://machinelearning.apple.com/research/learning-with-privacy-at-scale (accessed on 20 April 2022).
53. Xu, M.; Zhao, H.; Ji, X.; Shen, J. Distribution-Perceptive-Based Spatial Cloaking Algorithm for Location Privacy in Mobile Peer-to-Peer Enviroments. *J. Softw.* **2018**, *19*, 1852–1862.
54. Huang, Y.; Huo, Z.; Meng, X.F. CoPrivacy: A Collaborative Location Privacy-Preserving Method without Cloaking Region. *J. Softw.* **2018**, *19*, 1852–1862. [CrossRef]
55. Sun, G.; Cai, S.; Yu, H.; Maharjan, S.; Chang, V.; Du, X.; Guizani, M. Location Privacy Preservation for Mobile Users in Location-Based Services. *IEEE Access* **2019**, *7*, 87425–87438. [CrossRef]
56. Shaham, S.; Ding, M.; Liu, B.; Lin, Z.; Li, J.Y. Privacy Preservation in Location-Based Services: A Novel Metric and Attack Model. *IEEE Trans. Mob. Comput.* **2021**, *20*, 3006–3019. [CrossRef]
57. Wu, D.; Zhang, Y.; Liu, Y. Dummy Location Selection Scheme for K-Anonymity in Location Based Services. In Proceedings of the 2017 IEEE Trustcom/BigDataSE/ICESS, Sydney, NSW, Australia, 1–4 August 2017; pp. 441–448.
58. Li, L.L.; Hua, J.F.; Wan, S.; Zhu, H.; Li, F.H. Achieving efficient location privacy protection based on cache. *J. Commun.* **2017**, *38*, 148–157.

59. Fei, F.; Li, S.; Dai, H.; Hu, C.; Dou, W.; Ni, Q. A K-Anonymity Based Schema for Location Privacy Preservation. *IEEE Trans. Sustain. Comput.* **2019**, *4*, 156–167. [CrossRef]
60. Xu, X.; Chen, H.; Xie, L. A Location Privacy Preservation Method Based on Dummy Locations in Internet of Vehicles. *Appl. Sci.* **2021**, *11*, 4594. [CrossRef]
61. Zhou, C.; Chen, Y.; Tian, H.; Cai, S. Location Privacy and Query Privacy Preserving Method for K-nearest Neighbor Query in Road Networks. *J. Softw.* **2020**, *31*, 471–492.
62. Chen, H.; Qing, X. Location-semantic-based location privacy protection for road network. *J. Commun.* **2016**, *37*, 67–76.
63. Bostanipour, B.; Theodorakopoulos, G. Joint obfuscation of location and its semantic information for privacy protection. *Comput. Secur.* **2021**, *107*, 1–22. [CrossRef]
64. Zeng, H.; Zuo, K.; Wang, Y.; Liu, R. Semantic Diversity Location Privacy Protection Method in Road Network Environment. *Comput. Eng. Appl.* **2020**, *56*, 102–108.
65. Tian, C.; Xu, H.; Lu, T.; Jiang, R.; Kuang, Y. Semantic and Trade-Off Aware Location Privacy Protection in Road Networks Via Improved Multi-Objective Particle Swarm Optimization. *IEEE Access* **2021**, *9*, 54264–54275. [CrossRef]
66. Liao, D.; Huang, X.; Anand, V.; Sun, G.; Yu, H.F. k-DLCA: An efficient approach for location privacy preservation in location-based services. In Proceedings of the IEEE International Conference on Communications (ICC), Kuala Lumpur, Malaysia, 22–27 May 2016; pp. 1–6.
67. Song, D.; Song, M.B.; Shakhov, V.V.; Park, K. Efficient dummy generation for considering obstacles and protecting user location. *Concurr. Comput. Pract. Exp.* **2021**, *33*, e5416. [CrossRef]
68. Kuang, L.; Wang, Y.; Zheng, X.; Huang, L.; et al. Using location semantics to realize personalized road network location privacy protection. *Eurasip J. Wirel. Commun. Netw.* **2020**, *2020*, 1. [CrossRef]
69. Li, J.; Zeng, F.; Xiao, Z.; Jiang, H.; Zheng, Z.; Liu, W.; Ren, J. Drive2friends: Inferring Social Relationships From Individual Vehicle Mobility Data. *IEEE Internet Things J.* **2020**, *7*, 5116–5127. [CrossRef]
70. Cui, G.; He, Q.; Chen, F.; Jin, H.; Xiang, Y.; Yang, Y. Location Privacy Protection via Delocalization in 5G Mobile Edge Computing Environment. *IEEE Trans. Serv. Comput.* **2021**, *9*, 1–12. [CrossRef]
71. Khalfoun, B.; Mokhtar, S.B.; Bouchenak, S.; Nitu, V. EDEN: Enforcing Location Privacy through Re-identification Risk Assessment: A Federated Learning Approach. *Proc. ACM Interact. Mob. Wearable Ubiquitous Technol.* **2021**, *5*, 1–25. [CrossRef]
72. Geyer, R.; Klein, T.; Nabi, M. Differentially Private Federated Learning: A Client Level Perspective. *arXiv* **2017**, arXiv:1712.07557.
73. Bagdasaryan, E; Veit, A; Hua, Y; Estrin, D; Shmatikov, V. How To Backdoor Federated Learning. *arXiv* **2018**, arXiv:1807.00459.
74. Yang, X.; Gao, L.; Wang, H.; Li, Y.; Zheng, J.; Xu, J.; Ma, Y. A User-related Semantic Location Privacy Protection Method In Location-based Service. In Proceedings of the IEEE 27th International Conference on Parallel and Distributed Systems (ICPADS), Beijing, China, 14–16 December 2021; pp. 691–698.
75. Sun, G.; Song, L.; Liao, D.; Yu, H.; Chang, V. Towards privacy preservation for "check-in" services in location-based social networks. *Inf. Sci.* **2019**, *481*, 616–634. [CrossRef]
76. Zhao, P.; Li, J.; Zeng, F.; Xiao, F.; Wang, C.; Jiang, H. ILLIA: Enabling k-Anonymity-Based Privacy Preserving Against Location Injection Attacks in Continuous LBS Queries. *IEEE Internet Things J.* **2018**, *5*, 1033–1042. [CrossRef]
77. Tu, Z.; Zhao, K.; Xu, F.; Li, Y.; Su, L.; Jin, D. Protecting Trajectory from Semantic Attack Considering k-Anonymity, l-diversity and t-closeness. *IEEE Trans. Netw. Serv. Manag.* **2018**, *16*, 264–278. [CrossRef]
78. Zhang, S.; Mao, X.; Choo, K.K.R.; Peng, T.; Wang, G. A trajectory privacy-preserving scheme based on a dual-K mechanism for continuous location-based services. *Inform. Sci.* **2020**, *527*, 406–419.

sensors

MDPI

Article

A Secure and Traceable Vehicles and Parts System Based on Blockchain and Smart Contract

Chin-Ling Chen [1,2], Zhi-Peng Zhu [3], Ming Zhou [3], Woei-Jiunn Tsaur [4,5,*], Chih-Ming Wu [6] and Hongyu Sun [7,8,*]

1 School of Information Engineering, Changchun Sci-Tech University, Changchun 130600, China
2 Department of Computer Science and Information Engineering, Chaoyang University of Technology, Taichung City 413310, Taiwan
3 School of Computer and Information Engineering, Xiamen University of Technology, Xiamen 361024, China
4 Computer Center, National Taipei University, New Taipei City 237303, Taiwan
5 Department of Computer Science and Information Engineering, National Taipei University, New Taipei City 237303, Taiwan
6 School of Civil Engineering and Architecture, Xiamen University of Technology, Xiamen 361024, China
7 Department of Computer Science, Jilin Normal University, Siping 136000, China
8 State Key Laboratory of Numerical Simulation, Siping 136000, China
* Correspondence: wjtsaur@mail.ntpu.edu.tw (W.-J.T.); hongyu@jlnu.edu.cn (H.S.)

Abstract: As society advances, so does the total number of vehicles on the road, creating a massive consumer market for automobiles. According to statistics, a major portion of today's traffic difficulties are caused by accidents caused by subpar cars and auto parts. As a result, each country has, over time, enacted equivalent rules and regulations to prevent such tragedies. However, in the face of profit, some people are desperate enough to employ illegal parts and illegally modified cars, and auto fraud is rampant. As a result, we employ the blockchain of the symmetrical Blockchain's digital ledger and smart contract technology to build a decentralized supply chain system that can identify specific parts. In this study, we design and discuss the proposed system framework by user functions and the flow of parts based on blockchain, and we discuss communication protocols that use the symmetry and asymmetry cryptography, algorithms, properties, and security of the mechanism while providing related analysis and comparing the properties and costs of the system with other studies. Overall, the proposed method has the potential to successfully address the issue of automobile fraud.

Keywords: blockchain; smart contract; automation supply chain; traceability; asymmetry cryptography

1. Introduction

1.1. Background

As of 2020, according to the Bureau of Transportation Statistics (BTS) and National Bureau of Statistics of China statistics, the total number of vehicles in the US is about 276 million, and 280 million in China. In 2020, the world car production grew to 76 million [1–3].

With that many vehicles, a huge vehicle consumer market is produced, and the same with many traffic problems that are due to the vehicles and parts themselves. For example, in the National Motor Vehicle Crash Causation Survey (NMVCCS) [4], an estimated 44,000 crashes are caused by vehicles, which is about 2% of the crashes counted by NMVCSS; additionally, the National Highway Traffic Safety Administration in the literature [5] stated that the critical causes in 10.5% of crashes are steering, suspension, transmission, and engine failures, while about 21% of crashes are caused by various other vehicle failures or defects. Hoque and Hasan [6] stated that: as a percentage of the total number of crashes, vehicle defects caused 16.0% of the crashes and 29.0% of the total casualties by the same factor. It can be seen that unqualified parts would reduce the stability of the car, and then lead to the occurrence of traffic accidents.

Thus, to reduce traffic problems that are caused by a flaw in vehicles or parts, many countries limit illegally modified vehicles and the sale of non-compliant parts or other car equipment and devices by laws or regulations. For example, in the United States, where car control is relatively loose, California law considers it illegal to sell non-compliant car equipment and devices, and other states have similar laws and regulations [7]. Additionally, under the section 75 of the Road Traffic Act 1988 in the UK, it is an offense to alter a vehicle in such a way that the use of the vehicle on a road would be unlawful [8]. This is the same for other countries in the world, such as Japan or China [9,10].

However, some services such as repair and car maintenance require more professional car knowledge. Although the law regulates the sale of car modifications and parts, car repair frauds are still common, because the notion of every car mechanic or car repair company being honest is unrealistic. For example, in some auto repair shops, the owners use counterfeit auto parts instead of high-quality parts to decrease costs [11], and some auto manufacturers privately allow their automakers to modify vehicles privately [12]. In addition, some dealers also sell accident cars or used cars as new cars after modification to make profits [13]. These defective vehicles will increase the probability of traffic accidents, reducing the trust between consumers and car sales-as-a-service providers. This is very detrimental to the safety of life and property of the market and consumers.

Therefore, only having legal constraints is not enough. We need to take practical measures to supervise vehicles and parts to ensure the legality and qualification of vehicles and parts on the road. This in turn minimizes consumer exposure to car fraud and curbs illegal car modifications.

Existing supply chain usage generally involves tagging parts using radio-frequency identification (RFID) and one-dimensional or two-dimensional barcodes and then going to a centralized database for information access. Unfortunately, the data in the system can be easily tampered with or falsified, and it is not easy or even possible to trace the flow of parts. A decentralized blockchain-based system, however, is a superior solution to make the information more reliable and is traceable, immutable, secure and transparent. In addition, the Elliptic Curve Digital Signature Algorithm (ECDSA) [14] is used in our system to ensure data integrity and this system is built in Hyperledger Fabric [15].

All in all, in this study, we proposed a based-blockchain system that will accomplish the following:

(1) Ensure data integrity.
(2) Construct a simple quality identification scheme.
(3) Enable traceable, identifiable parts service with efficiency and mutual trust.

1.2. Related Works

The automotive supply chain (ASC) has been an intricate system due to the various parts used in each vehicle, the need for many part supplies, and the many stakeholders that exist in the ASC. Before this study, lots of scholars on the issue have also combined blockchain with supply chain, as sown in Table 1.

Chen et al. [16] proposed a relatively complete theoretical framework for blockchain-based supply chains by elaborating on their proposed Supply Chain Quality Management (SCQI) and briefly discussing the issues that arise in the context of the case, but there is no mention of arbitration in the study. Sharma et al. [17] proposed a blockchain-based distributed architecture for the smart city automobile industry that examines the entire process from many perspectives and suggests a practical strategy. However, the research does not elaborate on the circulation process of parts and does not address the algorithms necessary to carry out the suggested circulation process. Kim et al. [18] handle the authentication of genuine vehicle parts via both Blockchain Governance Game (BGG) [19] and Fog Computing [20] techniques. However, the studies lack a thorough examination of the roles of the various blockchain tasks and do not suggest a comprehensive service structure. In the study by Miehle et al. [21], the authenticity and tracking and tracing of the source of parts are addressed, access control and licensing systems to secure private license chains are introduced, archiving using external chains and external databases is enabled, and the entry barrier for SMEs to the alliance chain is lowered, thereby effectively improving the supply chain's comprehensiveness and integrity, but the regulation and the stalemate are not addressed. Hao developed a Blockchain-based logistics monitoring system (BLMS) in the study [22], which allows customers, logistics operators, and all other parties in the supply chain to track their parcels and information to ensure fairness and transparency, but not enough for the subsequent regulation of automotive services. Yahiaoui's paper [23] describes a blockchain-based supply chain system and briefly explores the integration of its blockchain supply chain. Li and Ye [24] integrated blockchain technology into the ASC, customizing smart contracts to meet functional requirements, and demonstrating product traceability to consumers and regulators. Wang et al. [25] applied blockchain to auto service to emphasize the importance of component supply chain management, and subsequent service assurance, and offered a blockchain-based Product-Service System (PSS) framework for vehicles and several other application frameworks, but no privacy protection is provided for transactions between supply chain parties, and no specific algorithm or implementation is proposed.

Table 1. Comparison of existing auto parts traceability system.

Authors	Year	Objective	Technologies	Merits	Demerits
Chen et al. [16]	2015	A theoretical framework for combining blockchain and supply chain	Blockchain	Proposed intelligent quality management of supply chain based on the blockchain technology.	There is no discussion on the regulation and analysis of services outside the supply chain.
Sharma et al. [17]	2018	a distributed framework model for the entire life cycle phases of the automotive industry blockchain-based	Blockchain	Analyzing the processes of the automotive industry from multiple perspectives and provided a miner node algorithm.	There is no elaboration on the flow process of the parts and no proposed algorithm to be implemented for the flow process.
Kim et al. [18]	2019	A blockchain-based design for authentication of automotive parts	BGG, Fog Computing	Provide service of authentic certification of auto parts and protection of blockchain.	Lack of analysis of the role of stakeholders in the supply chain.
Miehle et al. [21]	2019	A traceable parts supply chain application built on blockchain and smart contracts	Distributed Ledger, Smart Contract, Blockchain	Introduces access control and licensing systems to secure private license chains, and use external chains and external databases to archive.	There is no solution to the regulation of all parties in the supply chain, and there is no corresponding analysis of the subsequent service of the car.
Helo and Hao [22]	2019	A Blockchain-based logistics monitoring system prototype	JavaScript, Blockchain	All parties on the chain can track and access their package information.	No corresponding solution is proposed for the regulation of subsequent car services.
Yahiaoui et al. [23]	2020	Blockchain and smart contract-based supply chain model	Blockchain	An ASC system based on blockchain and smart contracts is proposed and analyzed.	There is no description of the parties of the ASC, algorithms, and car maintenance services.
Li and Ye [24]	2020	Combines blockchain and ASC for distributed storage of production and sales data	Blockchain, Smart Contract	Ensures the security of ASC data, increases the mutual trust of the parties, and increases that process sensitive data.	No analysis is made for the subsequent service of the car, and no specific algorithm is proposed.
Wang et al. [25]	2020	Blockchain-based Product-Service System service framework for vehicle products	Blockchain, smart-contract	All parties to accurately update and verify vehicle information and easier to verify the condition of vehicles in usage.	no specific algorithm or implementation is proposed.

In this paper, we use a symmetrical copy of the decentralized ledger for all users under the security of asymmetric cryptography. the contents of the other sections are as follows: Section 2 involves some related knowledge of this study. Section 3 describes the communication protocol and algorithm of each phase. We analyzed the characteristics and security issues in Section 4. In Section 5, we make some evaluations for communication costs and computation costs. Lastly, we conclude this paper in Section 6.

2. Preliminary

2.1. Blockchain and Smart Contracts

Blockchain Technology systems came from a paper on the cryptocurrency Bitcoin, "Bitcoin: A Peer-to-Peer Electronic Cash System" [26], proposed by a named Satoshi Nakamoto in 2008. It involves many disciplines, such as mathematics, cryptography, and computer science. In the blockchain, distributed computational storage, public and private keys, real-time broadcasting, and timestamping bring the characteristics of being decentralized, transparently developed, and tamper-proof, and the data structure Merkle tree is used to ensure the traceability of the blockchain. These features make blockchain that can be integrated with various fields.

Smart contacts were proposed by Nick Szabo, a well-known American computer scientist [27]. Smart contacts are codes that run on the blockchain are and automatically executed on the blockchain when conditions are met and cannot be accessed by anyone for execution [28,29]. It is the digital equivalent of traditional contracts, and combined with these blockchains, such as decentralization, tamper-evident, transparent traceability, perpetual operation, and mutual corroboration, smart contracts achieve the effect of decentralization from trusting third-party institutions to trusting the contract itself.

2.2. ECDSA

ECDSA was proposed by Rivest et al. It combines Digital Signature Algorithm (DSA) and Elliptic Curve Cryptography (ECC). Compared with traditional encryption methods, ECDSA has the characteristics of smaller parameters, keys and certificates, stronger key bit strength, and faster operating speed [15,30–32].

Suppose that A wants to send a message M to B. The signature is generated by sender A and verified by receiver B. Firstly, both parties must agree on the elliptic curve (CURVE, G, n), where G is the base point on the curve, n is the order of G, and H is the hash function.

Signature: A chooses a random integer d_A as a private key with values in the range $[0, n-1]$, and generates the public key $Q_A = d_A G$.Computing: $z = h(m)$, $kG = (x_1, y_1)$, $r = x_1 \bmod n$ and $s = k^{-1}(z + rd_A) \bmod n$. Then, the message m and the signature value (r, s) are sent to B.

Verification: B verifies the correctness of the message after receiving the signature value and message m from A. B calculates: $z' = h(m)$, $a_1 = z's^{-1} \bmod n$, $a_2 = rs^{-1} \bmod n$, $(x', y') = a_1 G + a_2 Q_A$. If the equation $r = x' \bmod n$ holds, the verification passes.

2.3. Hyperleader Fabric

Hyperleader Fabric was led by IBM and Linux, a blockchain-based open-source project. It is mainly to establish an enterprise-class distributed ledger system compatible with pluggable consensus mechanisms and supporting identity authentication, which is typical of current federated chains. Additionally, Hyperleader Fabric is modular, scalable, and provides privacy and confidentiality features to enable the platform to give social good, insurance, and finance, as well as supply chain logistics and other industry use cases to provide more effective and novel features.

3. Proposed Scheme

This study uses a symmetrical copy of the blockchain-based ledger technology to build a new automotive parts traceability system by building a Hyperleader Fabric federated chain to implement some functions following text. The system consists of the shareholder's members of the federated chain Parts Manufacturer (PM), Automobile Manufacturer (AM), Car Dealer (CD), Car Owner (CO), and Repair Shop (RS), as well as Competent Authorities

(CA) and Arbitrator (AB) and Blockchain Center (BCC). The system framework is shown in Figure 1.

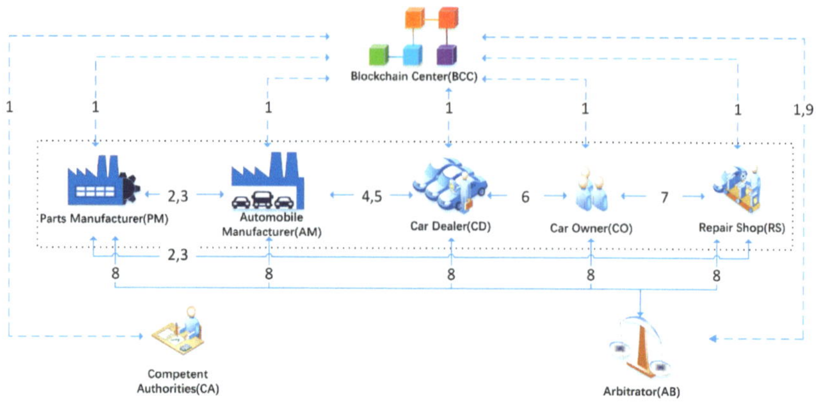

Figure 1. System architecture diagram.

3.1. System Architecture

(1) Parts Manufacturer (PM): PM obtains orders from automobile manufacturers (AM) and Repairers Shop (RS), and then produces the corresponding parts according to the order information and sells them to AM and RS.

(2) Automobile Manufacturer (AM): AM is responsible for the production of research and development of cars, ordering parts from PM for car production. In the meantime, AM also is the seller of car dealers.

(3) Car Dealer (CD): CD is the wholesale vehicle from AM and will sell the vehicle to the consumer (also known as the car owner (CO)).

(4) Car Owner (CO): The end-user of the car, who needs to buy the car from CD, is also the consumer of the Repair Shop (RS) and can go to RS for vehicle repair and parts replacement.

(5) Repair Shop (RS): Order parts from PM to repair the consumer's vehicle.

(6) Competent Authorities (CA): If a member of the alliance chain is unsure of the legitimate source of a part, the auditor has the right to certify any problems with the flow of the part.

(7) Arbitrator (AB): A third-party arbitrator that receives complaints from members of the alliance chain, can find the flow of parts for cars via the Internet, and can find broken parts that are in circulation on the market.

(8) Blockchain Center (BCC): A blockchain that records key information about parts and vehicles as well as information about the distribution process, and the blockchain associates the ID of the recorded part or vehicle with the vehicle or part. The chain code in the BCC can check the status of the part during the transaction. At the same time, each member needs to register with the blockchain center and request a unique ID to be added to the blockchain.

Figure 1 shows the process of a car part passing through the manufacturer of the part to the car manufacturer, then the car manufacturer agrees to assemble it, then it passes through the dealership, the owner, and through the manufacturer of the part to the repair shop and then to the owner. Of course, in reality, there is more than one member in the alliance chain, and the diagram only shows the flow of parts or cars. And the numbers 1–9 of the Figure 1 is correspond to step 1–9. A description of the specific distribution process is as follows.

Step 1. Each role must register an account on BCC; simultaneously, BBC records the specific information of each member and returns a pair of public and private keys.

Step 2. When AM needs to produce a batch of cars or RS needs to receive a batch of parts, it needs to order parts from PM and send the order information to PM.
Step 3. When PM receives the order information, it will produce the parts and engrave the ID number of each part on the part, and send the parts to AM or RS.
Step 4. If the CD is obtaining a batch of cars from the AM, it needs to send the order information to the AM.
Step 5. AM receives the order and delivers the products to CD.
Step 6. CO goes to CD to buy the vehicle and CO needs to provide the identity for the transaction.
Step 7. CO goes to RS to repair the vehicle.
Step 8. If either party disputes the quality or origin of the parts, they may submit a request for arbitration to the AB.
Step 9. Parts and vehicle-related information and circulation process information are recorded on BCC, AB can retrieve and verify the parts and vehicle-related records through BCC.

3.2. Data Definition

Figures 2 and 3 are the basic structure of chain code in our designation. Figure 2 shows the product message structure of parts and vehicles. When the product of a vehicle or a part circulates in every Access Party (AP), its details will disclose this structure. In Figure 3, the left shows the storage structure of AP, and the right shows the definition of roles.

```
type PartInfo struct{                type VehicleInfo struct{
      PUID string                          VUID string
      PName string                         VName string
      PParameter string                    VParameter string
      PAgingStandard string                VAgingStandard string
      PManuName string                     VManuName string
      PProductionDate string               VProductionDate string
      PExfactoryDate string                VExfactoryDate string
      PAging bool                          VPUIDs []string
}                                          VAging bool      }
```

Figure 2. The chaincode structure of the parts and car.

```
type APInfo struct{            type Role struct{
      ID string                      PartManufacturer string
      Name string                    AutomobileManufacturer string
      Detail string                  CarDealer string
      Var roleType Role              CarOwner string
}                                    RepairShop string
                                     CompetentAuthories string
                                     Arbitrator string
                               }
```

Figure 3. Chaincode structure of the accessing party and the enumeration of the role type.

3.3. Registration Phase

All parties who join the system must register an account with BCC. When registration is successful, BCC records its message and returns a pair of public key and private key to the member of the register. The specific registration process is shown in Figure 4.

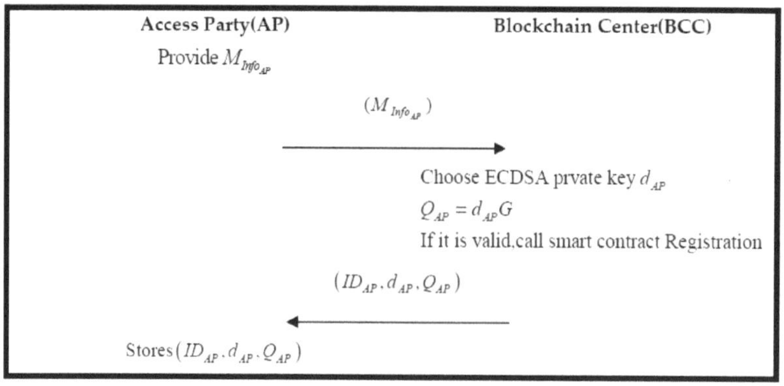

Figure 4. The flowchart of the registration phase.

Step 1. AP sends its message $M_{Info_{AP}}$ (e.g., name, role type, etc.) to the blockchain center for the registration request.

Step 2. BCC uses ECDSA to create a private key d_{AP} using the key to calculate the public key Q_{AP}:

$$Q_{AP} = d_{AP}G \tag{1}$$

If the creation is successful, add the role and trigger smart contact. The algorithm of the smart contract is as follows: Algorithm 1. Then, BCC sends $(ID_{AP}, d_{AP}, Q_{AP})$ to AP.

Step 3. AP receive and storage $(ID_{AP}, d_{AP}, Q_{AP})$.

Algorithm 1: Chaincode Registration of the proposed scheme.

```
func Registration (var Name string, var Detail string, var Role string)(UID string){
    UID = GenerateUID()
    count++
    AP[count].UID = UID
    AP[count].Name = Name
    AP[count].Detail = Detail
    AP[count].Role = Role
    return UID
}
```

3.4. Authentication Phase

Since the actors in the initial stage of the blockchain cannot verify each other's true identity, both parties who need to perform actions need to be authenticated. The "signature" and "verification" are required when using the algorithm ECDSA implemented for authentication. We assume both users A and B need to authenticate. The specific implementation flow is shown in Figure 5. User A generates a random number k_1 and a message M_{A1} and calculates h_{A1}:

$$M_{A1} = (ID_A, ID_B, TS_{A1}, M_{Info_A}) \tag{2}$$

$$h_{A1} = H(M_{A1}) \tag{3}$$

Then, User A calculates the parameter of ECDSA and through "Sign" of Algorithm 2 generates a signature. The specific process of signature shows in Equations (4)–(6):

$$(x_{A1}, y_{A1}) = k_1 G \tag{4}$$

$$r_{A1} = x_{A1} \bmod n \tag{5}$$

$$s_{A1} = {x_{A1}}^{-1}(h_{A1} + r_{A1}d_A) \bmod n \tag{6}$$

Then, A uses B's public key Puk_B to encrypt a message M_{A1}:

$$C_{A1} = E_{Puk_B}(M_{A1}) \tag{7}$$

Finally, A sends the information that is A generating $C_{A1}, (r_{A1}, s_{A1})$ to B.

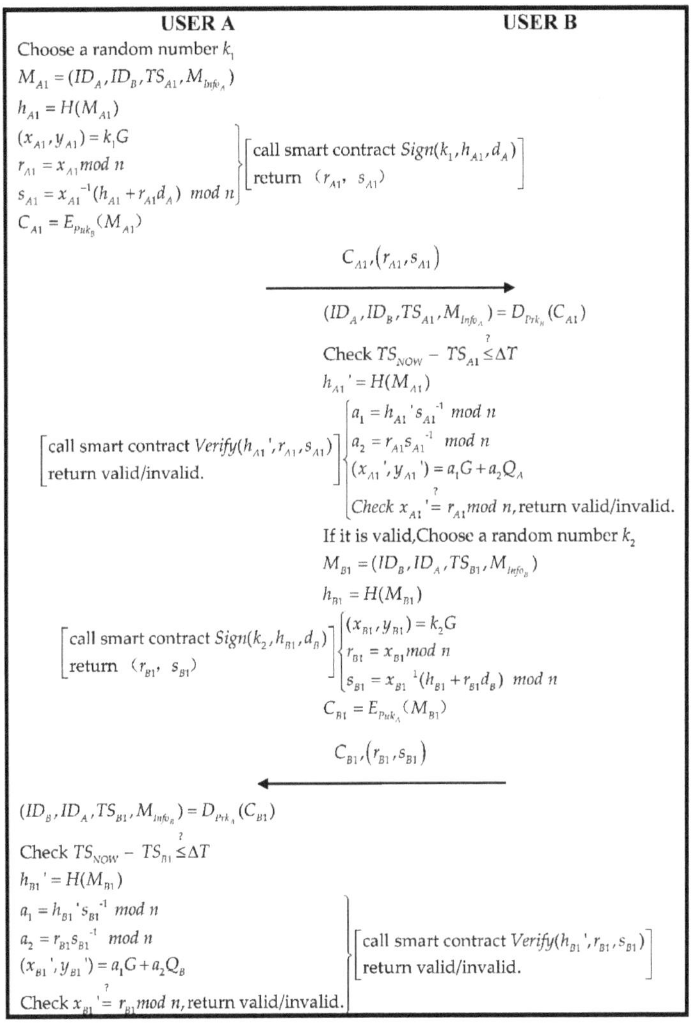

Figure 5. The flowchart of the authentication phase.

Step 1. User B receives a message from A and uses B's private key Prk_B deciphering C_{A1} to acquire the data $(ID_A, ID_B, TS_{A1}, Info_A)$ within the message M_{A1}. In the meantime, determine whether the timestamp is legal or not:

$$(ID_A, ID_B, TS_{A1}, M_{Info_A}) = D_{Prk_B}(C_{A1}) \tag{8}$$

$$TS_{NOW} - TS_{A1} \overset{?}{\leq} \Delta T \tag{9}$$

If Equation (9) is true, the smart contract "*Verify*" of Algorithm 2 will trigger and verify the signature of ECDSA. The specific process of verification is shown in Equations (10)–(14):

$$h_{A1}{}' = H(M_{A1}) \tag{10}$$

$$a_1 = h_{A1}{}' s_{A1}{}^{-1} \bmod n \tag{11}$$

$$a_2 = r_{A1} s_{A1}{}^{-1} \bmod n \tag{12}$$

$$(x_{A1}{}', y_{A1}{}') = a_1 G + a_2 Q_A \tag{13}$$

$$x_{A1}{}' \stackrel{?}{=} r_{A1} \bmod n \tag{14}$$

If Equation (14) is true, the message is from A, which can be confirmed. Then, B generates a random number k_2 and a message M_{B1} and calculates h_{B1}:

$$M_{B1} = (ID_B, ID_A, TS_{B1}, M_{Info_B}) \tag{15}$$

$$h_{B1} = H(M_{B1}) \tag{16}$$

Then, B calculates the parameter of ECDSA and generates a signature through the "Sign" of Algorithm 2. The specific process of signature is shown in (17)–(19).

$$(x_{B1}, y_{B1}) = k_2 G \tag{17}$$

$$r_{B1} = x_{B1} \bmod n \tag{18}$$

$$s_{B1} = x_{B1}{}^{-1}(h_{B1} + r_{B1} d_B) \bmod n \tag{19}$$

Then, B using the public key Puk_A of A encrypts a message M_{B1}:

$$C_{B1} = E_{Puk_A}(M_{B1}) \tag{20}$$

Finally, B sends information $C_{B1}, (r_{B1}, s_{B1})$ to A.

Step 2. When A receives a message from B, it uses its own private key Prk_A to decode C_{B1} and acquire information $(ID_B, ID_A, TS_{B1}, Info_B)$ within M_{B1}. In the meantime, it is verified whether the following timestamp is true or not true:

$$(ID_B, ID_A, TS_{B1}, M_{Info_B}) = D_{Prk_A}(C_{B1}) \tag{21}$$

$$TS_{NOW} - TS_{B1} \stackrel{?}{\le} \Delta T \tag{22}$$

If Equation (22) passes, the smart contract "*Verify*" of Algorithm 2 will trigger and verify the signature of ECDSA. The specific process of verification shows in Equations (23)–(27):

$$h_{B1}{}' = H(M_{B1}) \tag{23}$$

$$a_1 = h_{B1}{}' s_{B1}{}^{-1} \bmod n \tag{24}$$

$$a_2 = r_{B1} s_{B1}{}^{-1} \bmod n \tag{25}$$

$$(x_{B1}{}', y_{B1}{}') = a_1 G + a_2 Q_B \tag{26}$$

$$x_{B1}{}' \stackrel{?}{=} r_{B1} \bmod n \tag{27}$$

If Equation (35) passes, we can confirm the message is A sending to B. The authentication between user A and user B is successful.

Algorithm 2: Chaincode Sign and Verify the proposed scheme

```
func Sign(var k string, var h string, var d string){
    (x, y) = k * G
    r = x % n
    s = (1/k) * (h + r * d) % n
    return (r, s)
}
func Verify(var h string, var r string){
    a1 = (z/s) % n
    a1 = (r/s) % n
    (x, y) = a1 * G + a1 * G
    if x == r
        return "valid"
    else
        return "invalid"
}
```

3.5. Order and Transaction Phase

In the phase, we assume both roles that are User A and User B to simulate order and transaction actions. In this phase, A is the buyer purchasing products, and B is the seller. If the AM needs to perform car production and RS is short of parts for vehicle repair and needs to order parts from PM, then User A is AM and RS and User B is PM. If CD needs to order vehicles for sales activities, then User A is CD, and User B is AM at this time. The flowchart is as follows in Figure 6.

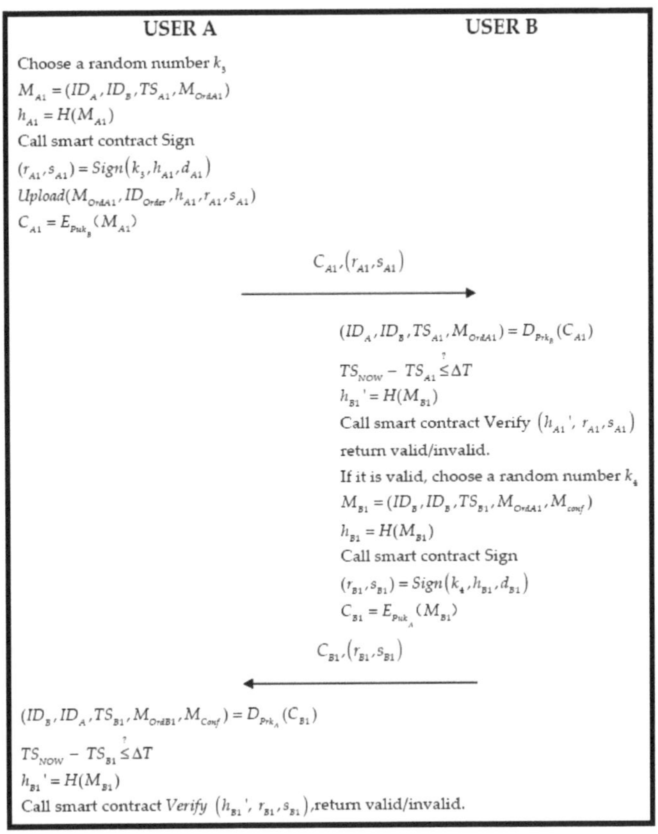

Figure 6. The flowchart of the order phase.

Step 1. User A generates a random number k_3 and message M_{A1} and calculates h_{A1}:

$$M_{A1} = (ID_A, ID_B, TS_{A1}, M_{OrdA1}) \tag{28}$$

$$h_{A1} = H(M_{A1}) \tag{29}$$

Then, User A calculates the parameters of ECDSA, and uses the "Sign" of Algorithm 2 to generate the signature:

$$(r_{A1}, s_{A1}) = Sign(k_3, h_{A1}, d_{A1}) \tag{30}$$

Afterward, User A uploads the order to the blockchain; in the meantime, it uses the public key Puk_B of User B to encrypt a message M_{A1}:

$$Upload(M_{OrdA1}, ID_{Order}, h_{A1}, r_{A1}, s_{A1}) \tag{31}$$

$$C_{A1} = E_{Puk_B}(M_{A1}) \tag{32}$$

Finally, User A delivers $C_{A1}, (r_{A1}, s_{A1})$, which is A generated to User B.

Step 2. User B receives the message from User A and using its private key Prk_B to decrypt C_{A1} to acquire data $(ID_A, ID_B, TS_{A1}, M_{OrdA1})$ of M_{A1}, and verifies that the timestamp holds:

$$(ID_A, ID_B, TS_{A1}, M_{OrdA1}) = D_{Prk_B}(C_{A1}) \tag{33}$$

$$TS_{NOW} - TS_{BR1} \overset{?}{\leq} \Delta T \tag{34}$$

If Equation (34) is established, the smart contract "*Verify*" of Algorithm 2 is triggered to verify that the ECDSA signature is correct:

$$h_{A1}{}' = H(M_{A1}) \tag{35}$$

$$Verify\ (h_{A1}{}', r_{A1}, s_{A1}) \tag{36}$$

If it is correct, we can testify the message is from User A, and then User B generates a random number k_4 and uses order request information M_{conf} and order information M_{OrdA1} to generate a message M_{B1}. The message is sent to A and User B calculates h_{B1}:

$$M_{B1} = (ID_B, ID_B, TS_{B1}, M_{OrdA1}, M_{conf}) \tag{37}$$

$$h_{B1} = H(M_{B1}) \tag{38}$$

Then, User B calculates the parameters of the ECDSA and generates a signature by "Sign" of Algorithm 2:

$$(r_{B1}, s_{B1}) = Sign(k_4, h_{B1}, d_{B1}) \tag{39}$$

Afterward, User A encrypts a message M_{B1} by the public key Puk_A of User B:

$$C_{B1} = E_{Puk_A}(M_{B1}) \tag{40}$$

Finally, B sends $C_{B1}, (r_{B1}, s_{B1})$ to User A.

Step 3. User A receives the message from User B and uses his private key Prk_A to decrypt C_{B1} to acquire data $(ID_B, ID_B, TS_{B1}, M_{OrdA1}, M_{conf})$ within the message M_{B1}, and verifies that the timestamp holds:

$$M_{B1} = (ID_B, ID_B, TS_{B1}, M_{OrdA1}, M_{conf}) \tag{41}$$

$$TS_{NOW} - TS_{PR1} \overset{?}{\leq} \Delta T \tag{42}$$

If Equation (42) is established, the smart contract "*Verify*" of Algorithm 2 is triggered to verify the signature of ECDSA that is correct:

$$h_{B1}{}' = H(M_{B1}) \tag{43}$$

$$Verify\left(h_{B1}{}', r_{B1}, s_{B1}\right) \tag{44}$$

If it is correct, the message is proved to have been sent by User B. Otherwise, the order is voided. At this point, the order is confirmed.

After the order phase mentioned above, both parties to the transaction have completed the task of placing and finalizing the order. In this phase, User B uploads the key information of the generated product to the blockchain. User A receives the product and information from User B and decrypts and verifies the correctness of the information. If it is accurate, the transaction is completed. The specific flowchart is as follows in Figure 7.

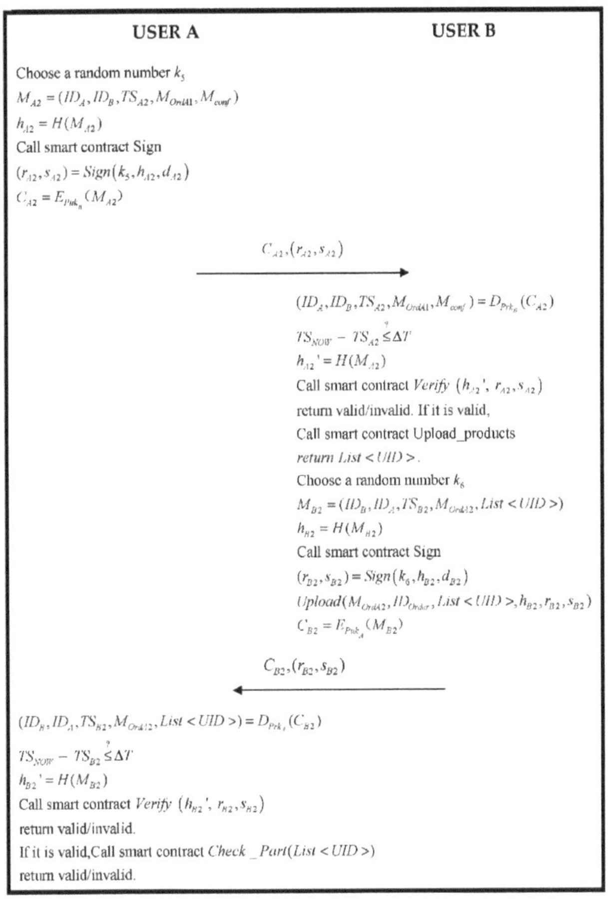

Figure 7. The flowchart of the transaction phase.

Step 1. User A generates a random number k_5, receives the product confirmation M_{conf}, and creates a message M_{A2}. Calculating h_{A2}:

$$M_{A2} = (ID_A, ID_B, TS_{A2}, M_{OrdA1}, M_{conf}) \tag{45}$$

$$h_{A2} = H(M_{A2}) \tag{46}$$

Then, User A calculates the parameter of ECDSA and generates the signature by "Sign" of Algorithm 2:

$$(r_{A2}, s_{A2}) = Sign(k_5, h_{A2}, d_{A2}) \tag{47}$$

After User A uses the public key Puk_B of User B to encrypt M_{A1}:

$$C_{A2} = E_{Puk_B}(M_{A2}) \tag{48}$$

At last, User A sends C_{A2}, (r_{A2}, s_{A2}) to User B.

Step 2. User B receives the message from User A and using his private key Prk_B decrypts C_{A1} to acquire the data $(ID_A, ID_B, TS_{A2}, M_{OrdA1}, M_{conf})$ within M_{A2}, in the meantime verifying if the timestamp is legal:

$$(ID_A, ID_B, TS_{A2}, M_{OrdA1}, M_{conf}) = D_{Prk_B}(C_{A2}) \tag{49}$$

$$TS_{NOW} - TS_{A2} \overset{?}{\leq} \Delta T \tag{50}$$

If (50) is established, the smart contract "*Verify*" of Algorithm 2 is triggered to verify that the signature of ECDSA is correct:

$$h_{A2}' = H(M_{A2}) \tag{51}$$

$$Verify\ (h_{A2}', r_{A2}, s_{A2}) \tag{52}$$

If Equation (52) is correct, it proves that the order information is sent by User A, triggering smart contacts *UploadParts* or *UploadVehicles* within Algorithm 3 or Algorithm 4 to upload the information of products. If it is a transaction among AM, RS, and PM, UploadParts is triggered, and if it is a transaction between CD and AM, $UploadVehicles$ is triggered. In the meantime, the functions $List < UID >$ (UID symbol ID_{Car} or ID_{Part}). Then, User B generates a random number k_6 and uses $List < UID >$, and $Order_{A1}$ generates M_{B1}, which is returned with information of the order. Calculating h_{B1}:

$$M_{B2} = (ID_B, ID_A, TS_{B2}, M_{OrdA2}, List < UID >) \tag{53}$$

$$h_{B2} = H(M_{B2}) \tag{54}$$

Then, User B calculates the parameter of ECDSA and generates a signature by "Sign" of Algorithm 2.

Algorithm 3: Chaincode UploadParts of the proposed scheme

```
var PI []PartInfofunc UploadParts(var pnum int, var PUID string, var PName string, var PParameter string,
var PAgingStandard string, var PManuName string, var PProductionDate string, var PExfactoryDate string,
var PAging bool){
    for (i = 0; i < pnum; i++){
        PI = append(PI, new PartInfo{
        PUID: PUID[i]
        PName: PName
        PParameter: PParameter
        PAgingStandard: PAgingStandard
        PManuName: PManuName
        PProductionDate: PProductionDate[i]
        PExfactoryDate: time.Now
        PAging: false})
        ListPUIDs = append(ListPUIDs, PI[i].ListPUIDs)
        return ListPUIDs
    }
}
```

Step 3. User A acquires the message of User B, uses his private key Prk_A decrypting C_{B2} to obtain data $(ID_B, ID_A, TS_{B2}, M_{OrdA2}, List < UID >)$ within M_{B2}, and verifies if the timestamp is correct:

$$(ID_B, ID_A, TS_{B2}, M_{OrdA2}, List < UID >) = D_{Prk_A}(C_{B2}) \tag{55}$$

$$TS_{NOW} - TS_{B2} \overset{?}{\leq} \Delta T \tag{56}$$

If the verification passes the above, if the above verification holds, "*Verify*" of Algorithm 2 is triggered and checking if the signature of ECDSA is correct:

$$h_{B2}{}' = H(M_{B2}) \tag{57}$$

$$Verify\ (h_{B2}{}', r_{B2}, s_{B2}) \tag{58}$$

If it is true, the system triggers the smart contract Algorithm 5 and proves the information of the product. If it is successful, the transaction finishes.

Algorithm 4: Chaincode UploadVehicles of the proposed scheme

```
var VI []VehicleInfo
func UploadVehicles(var num int, var VUID string, var VName string, var VParameter string, var
VAgingStandard string, var VManuName string, var VProductionDate string, var VExfactoryDate string, var
VAging bool, var VPUIDs []string){
        for (i = 0; i < vnum; i++){
        VI = append(VI, new VehicleInfo{
        VUID: VUID[i]
        VName: VName
        VParameter: VParameter
        VAgingStandard: VAgingStandard
        VManuName: VManuName
        VProductionDate: VProductionDate[i]
        VExfactoryDate: time.Now
        VAging: false
        for(j = 0; i < pnum;j++){
            VPUIDs[j]: VPUIDs[j] }
        })
        ListVUIDs = append(ListVUIDs, VI[i].ListVUIDs)
        return ListVUIDs
    }
}
```

Algorithm 5: Chaincode Check_products of the proposed scheme

```
func CheckParts(var pnum int. ListPUIDs []string){
for(i = 0; i < pnum; i++){
    if(PI[i].PAging == True)
        return "invalid" }
    return "valid"
}
func CheckVehicles(var vnum int. ListVUIDs []string){
    index = searchCar(VI, VUID)
    if(index ! = null)
        return "invalid"
    for(i = 0; i < vnum; i++){
        index2 = searchPID()
        if(PI[i].PAging == True)
    return "invalid" }
        return "valid"
}
```

3.6. Sale Phase

In the phase, CO purchases vehicle in the CD. The specific process is as following Figure 8.

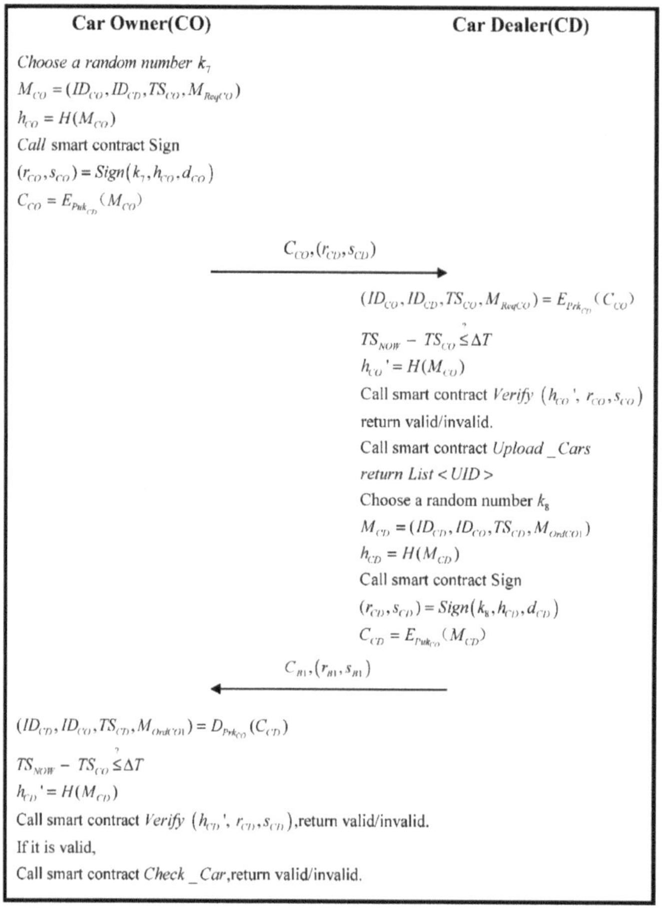

Figure 8. The flowchart of the sale phase.

Step 1. CO choices a product and sends M_{ReqCO} to a CD. First, CO generates a random number k_7 and generates M_{CO}. Calculating h_{CO}:

$$M_{CO} = (ID_{CO}, ID_{CD}, TS_{CO}, M_{ReqCO}) \tag{59}$$

$$h_{CO} = H(M_{CO}) \tag{60}$$

Then, CO calculates the parameter of ECDSA and generates a signature by "Sign" of Algorithm 2, and uses the public key of CD to encrypt:

$$(r_{CO}, s_{CO}) = Sign(k_7, h_{CO}, d_{CO}) \tag{61}$$

$$C_{CO} = E_{Puk_{CD}}(M_{CO}) \tag{62}$$

At last, CO sends $C_{CO}, (r_{CO}, s_{CO})$ to CD.

Step 2. CD receives data $(ID_{CO}, ID_{CO}, TS_{CO}, M_{ReqCO})$ from C_{CO}, and verifies if the timestamp is correct:

$$(ID_{CO}, ID_{CD}, TS_{CO}, M_{ReqCO}) = E_{Prk_{CD}}(C_{CO}) \tag{63}$$

$$TS_{NOW} - TS_{CO} \overset{?}{\leq} \Delta T \tag{64}$$

If (74) is correct, the smart contract "*Verify*" of Algorithm 2 is triggered to verify if the signature of ECDSA is legal or not:

$$h_{CO}{}' = H(M_{CO}) \tag{65}$$

$$Verify\ (h_{CO}{}',\ r_{CO}, s_{CO}) \tag{66}$$

If it is true, it proves the information of the order that sends from CO. Additionally, the system finds the vehicle of the request of the order. CD sends ID_{Car} to CO and a random number k_8 is generated by CD. In the meantime, according to UID_{part} and M_{OrdCO1}, which are created by CO, message M_{CD} is generated. Returning the information of the order to CO calculate h_{CD}:

$$M_{CD} = (ID_{CD}, ID_{CO}, TS_{CD}, M_{OrdCO1}) \tag{67}$$

$$h_{CD} = H(M_{CD}) \tag{68}$$

Additionally, then CD calculates the parameter of ECDSA and generates a signature by "Sign" of Algorithm 2:

$$(r_{CD}, s_{CD}) = Sign(k_8, h_{CD}, d_{CD}) \tag{69}$$

Afterward, the CD using the public key Puk_{CO} of CO encrypts M_{CD}:

$$C_{CD} = E_{Puk_{CO}}(M_{CD}) \tag{70}$$

At last, CD sends $C_{CD}, (r_{CD}, s_{CD})$ to CO.

Step 3. CO receiving the message from CD, using its private key Prk_{CO}, decrypts C_{CD} to acquire data $(ID_{CD}, ID_{CO}, TS_{CD}, M_{OrdCO1})$ within M_{CD}, and it verifies if the timestamp is correct:

$$(ID_{CD}, ID_{CO}, TS_{CD}, M_{OrdCO1}) = D_{Prk_{CO}}(C_{CD}) \tag{71}$$

$$TS_{NOW} - TS_{CO} \overset{?}{\leq} \Delta T_{(2)} \tag{72}$$

If (72) is established, the smart contract "*Verify*" of Algorithm 2 is triggered, in the meantime verifying if the signature of ECDSA is correct or not:

$$h_{CD}{}' = H(M_{CD}) \tag{73}$$

$$Verify\ (h_{CD}{}',\ r_{CD}, s_{CD}) \tag{74}$$

If it is correct, Algorithm 6 is triggered, and the transaction is finished.

3.7. Repair Phase

At this stage, CO goes to RS for vehicle maintenance. The specific process is shown in Figure 9.

Step 1. RS sends ID_{part1} of the old parts and ID_{part2} of new parts that need to be replaced to the CO, and generates random numbers k_9:

$$M_{RS} = (ID_{RS}, ID_{CO}, TS_{RS}, ID_{part1}, ID_{part2}) \tag{75}$$

$$h_{RS} = H(M_{RS}) \tag{76}$$

Then, the user CO calculates the parameters of ECDSA, generates a signature through "Sign" of Algorithm 2, and then encrypts it with the CO's public key:

$$(r_{RS}, s_{RS}) = Sign(k_9, h_{RS}, d_{RS}) \tag{77}$$

$$C_{RS} = E_{Puk_{CO}}(M_{RS}) \tag{78}$$

Finally, RS sends $C_{RS}, (r_{RS}, s_{RS})$, which is generated and sent to CO.

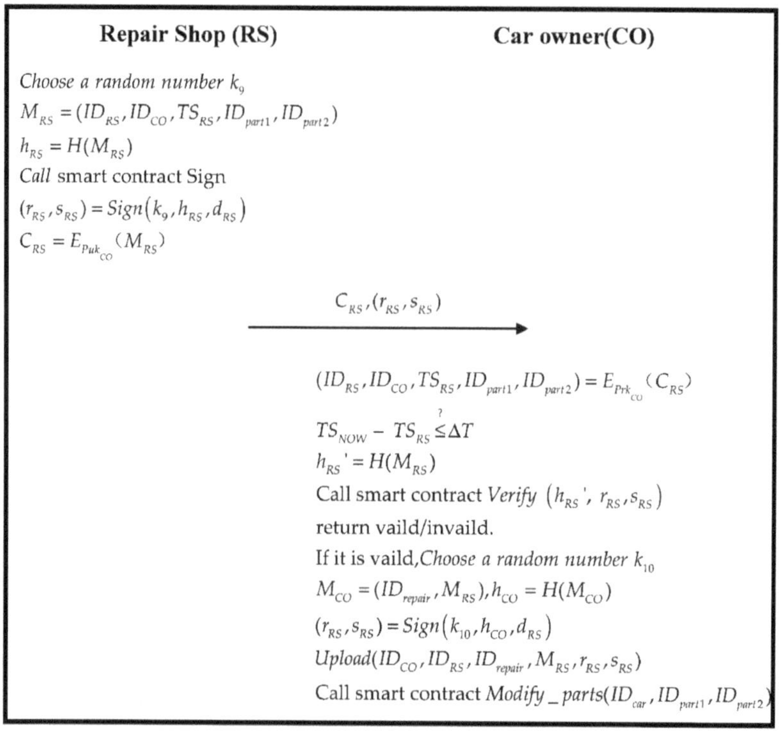

Figure 9. The flowchart of the repair phase.

Step 2. CO receives the data $(ID_{RS}, ID_{CO}, TS_{RS}, ID_{part1}, ID_{part2})$ of the message M_{RS} from RS and verifies whether the timestamp holds:

$$(ID_{RS}, ID_{CO}, TS_{RS}, ID_{part1}, ID_{part2}) = E_{Prk_{CO}}(C_{RS}) \tag{79}$$

$$TS_{NOW} - TS_{RS} \overset{?}{\le} \Delta T \tag{80}$$

If established, it triggers the smart contract "*Verify*" of Algorithm 2 to verify that the ECDSA signature is correct:

$$h_{RS}\prime = H(M_{RS}) \tag{81}$$

$$Verify\ (h_{RS}\prime,\ r_{RS}, s_{RS}) \tag{82}$$

If the verification is passed, a random number k_{10} is generated after confirming the information M_{CO}, a message is generated, and then the maintenance message is signed and uploaded.

$$M_{CO} = (ID_{repair}, M_{RS}) \tag{83}$$

$$h_{CO} = H(M_{CO}) \tag{84}$$

$$(r_{RS}, s_{RS}) = Sign(k_{10}, h_{CO}, d_{RS}) \tag{85}$$

$$Upload(ID_{CO}, ID_{RS}, ID_{repair}, M_{RS}, r_{RS}, s_{RS}) \tag{86}$$

Trigger the smart contract after uploading Algorithm 6.

Algorithm 6: Chaincode Modify_parts of the proposed scheme

```
func ModifyPart(var VUID string, var newPUID string, var oldPUID string){
    index = searchCar(VI, VUID)
    if(index! = null)
        index2 = searchVheiclePUIDs(VI[index].VehiclePUIDs,oldPUID)
        if(index2! = null)
            replace(VI[index].VehiclePUID[index2],newPUID)
            index3 = searchPUID(PI,oldPUID)
            PI[index]. Paging = True
            return "valid"
        else
            return "invalid"
    else
        return "invalid"
}
```

3.8. Arbitration Phase

When either party doubts the validity of a part, they can arbitrate its legitimacy through Arbitration. The process of arbitration is shown in Figure 10, and the numbers 1–4 correspond to step 1–4. The precise details of this process are as follows:

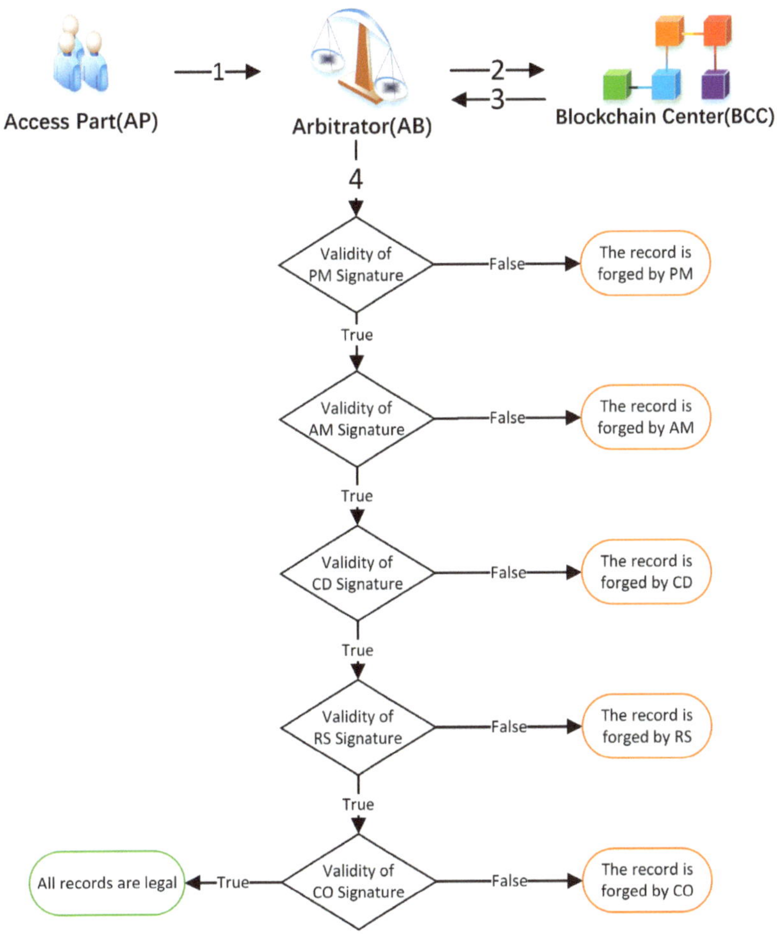

Figure 10. The validation flow in the arbitration phase.

Step 1. AP provides the UID of a specific part to AB.
Step 2. AB sends a TID request message with its signature to BCC.
Step 3. BCC checks the signature of AB, and if the signature is valid, BBC delivers the signature list to AB.
Step 4. AB checks the validity of each signature in the signature list. The order of the checks is as follows.

(a) Verify the signature of PM, if it is not legal, the record is proved to be forged by PM.
(b) Otherwise Verify the signature of AM, if it is not legitimate, the record is forged by AM.
(c) Verify the signature of the CD, if it is not legal, the record is proved to be forged by the CD.
(d) Verify the signature of RS, if it is not legal, the record is proved to be forged by RS.
(e) Verify the signature of CO, if it is not legal, the record is proved to be forged by CO.
(f) If all the above signature is valid, then the process of circulation of the part is proven and verified by AU.

4. Analysis

4.1. Data Integrity

We use ECDSA and hash functions to ensure data integrity. In a blockchain, each participant has a pair of public and private keys. The sender must compute a hash and generate a set of signatures using the receiver's public key before sending the message, and the receiver needs to verify the message and the signatures using his private key to ensure the validity of the message. If the attacker tampers with the data to send to the receiver, then the receiver will verify if the hash value and signature are not passed. All phases' detailed information is listed in Table 2.

Table 2. Verification of the data integrity of the proposed scheme.

Phase	Party		Message	Hash Value	Verification
	Sender	Receiver			
Authentication	USER A	USER B	$M_{A1} = (ID_A, ID_B, TS_{A1}, Info_A)$	$h_{A1} = H(M_{A1})$	$Verify(h_{A1}, r_{A1}, s_{A1})$
	USER B	USER A	$M_{B1} = (ID_B, ID_A, TS_{B1}, Info_B)$	$h_{B1} = H(M_{B1})$	$Verify(h_{B1}, r_{B1}, s_{B1})$
Order and Transaction phase	USER A	USER B	$M_{A1} = (ID_A, ID_B, TS_{A1}, Order_{A1})$	$h_{A1} = H(M_{A1})$	$Verify(h_{A1}', r_{A1}, s_{A1})$
	USER B	USER A	$M_{B1} = (ID_B, ID_B, TS_{B1}, M_{Ord\,A1}, M_{conf})$	$h_{B1} = H(M_{B1})$	$Verify\ (h_{B1}', r_{B1}, s_{B1})$
	USER A	USER B	$M_{A2} = (ID_A, ID_B, TS_{A2}, Order_{A1}, Info_{Confirm})$	$h_{A2} = H(M_{A2})$	$Verify\ (h_{A2}', r_{A2}, s_{A2})$
	USER B	USER A	$M_{B2} = (ID_B, ID_A, TS_{B2}, Order_{A2}, List < UID >)$	$h_{B2} = H(M_{B2})$	$Verify(h_{B2}', r_{B2}, s_{B2})$
Sale phase	Car Owner (CO)	Car Dealer (CO)	$M_{CO} = (ID_{CO}, ID_{CD}, TS_{CO}, Request)$	$h_{CO} = H(M_{CO})$	$Verify(h_{CO}', r_{CO}, s_{CO})$
	Car Dealer (CD)	Car Owner (CD)	$M_{CD} = (ID_{CD}, ID_{CO}, TS_{CD}, Order_{CO})$	$h_{CD} = H(M_{CD})$	$Verify\ (h_{CD}', r_{CD}, s_{CD})$
Repair phase	Repair Shop	Car Owner (CO)	$M_{RS} = (ID_{RS}, ID_{CO}, TS_{RS}, ID_{part_{old}}, ID_{part_{new}})$	$h_{RS} = H(M_{RS})$	$Verify\ (h_{RS}', r_{RS}, s_{RS})$

4.2. Non-Repudiation

In this paper, we use Verify of ECDSA to resolve the repudiation issue. In the blockchain mechanism, all messages transmitted by the sender must sign with their private key, and the receiver using the sender's public key verifies the messages. That ensures messages cannot be denied. Table 3 is the non-repudiation verification of the proposed scheme.

4.3. Traceability and Unforgeability

Based on blockchain characteristics, we learn that all transaction records are stored and chained to the ledger of every peer, and the records are traceable and unforgeable. In the meantime, data can be verified and transparent. For example, AB can trace records to verify whether blockchain data are legal or not. In Figure 10, if the signature cannot pass the verification, the signatures are forged.

Table 3. Non-repudiation verification of the proposed scheme.

Phase	Party		Message	Signature	Verification
	Sender	Receiver			
Authentication	USER A	USER B	$M_{A1} = (ID_A, ID_B, TS_{A1}, Info_A)$	$Sign(k_1, h_{A1}, d_A)$	$Verify(h_{A1}, r_{A1}, s_{A1})$
	USER B	USER A	$M_{B1} = (ID_B, ID_A, TS_{B1}, Info_B)$	$Sign(k_2, h_{B1}, d_B)$	$Verify(h_{B1}, r_{B1}, s_{B1})$
Order and Transaction phase	USER A	USER B	$M_{A1} = (ID_A, ID_B, TS_{A1}, Order_{A1})$	$Sign(k_3, h_{A1}, d_{A1})$	$Verify(h_{A1}', r_{A1}, s_{A1})$
	USER B	USER A	$M_{B1} = (ID_B, ID_B, TS_{B1}, Order_{A1}, Info_{Confirm})$	$Sign(k_4, h_{B1}, d_{B1})$	$Verify\ (h_{B1}', r_{B1}, s_{B1})$
	USER A	USER B	$M_{A2} = (ID_A, ID_B, TS_{A2}, Order_{A1}, Info_{Confirm})$	$Sign(k_5, h_{A2}, d_{A2})$	$Verify\ (h_{A2}', r_{A2}, s_{A2})$
	USER B	USER A	$M_{B2} = (ID_B, ID_A, TS_{B2}, Order_{A2}, List < UID >)$	$Sign(k_6, h_{B2}, d_{B2})$	$Verify(h_{B2}', r_{B2}, s_{B2})$
Sale phase	Car Owner (CO)	Car Dealer (CO)	$M_{CO} = (ID_{CO}, ID_{CD}, TS_{CO}, Request)$	$Sign(k_7, h_{CO}, d_{CO})$	$Verify(h_{CO}', r_{CO}, s_{CO})$
	Car Dealer (CD)	Car Owner (CD)	$M_{CD} = (ID_{CD}, ID_{CO}, TS_{CD}, Order_{CO})$	$Sign(k_8, h_{CD}, d_{CD})$	$Verify\ (h_{CD}', r_{CD}, s_{CD})$
Repair phase	Repair Shop	Car Owner (CO)	$M_{RS} = (ID_{RS}, ID_{CO}, TS_{RS}, ID_{part_{old}}, ID_{part_{new}})$	$Sign(k_9, h_{RS}, d_{RS})$	$Verify\ (h_{RS}', r_{RS}, s_{RS})$

4.4. Man-in-the-Middle Attack

Man-in-the-middle attack (MIMT) generally refers to the attacker intercepting the normal network communication data between the client and the server [33]. In the communication protocol, each communication message on the blockchain uses asymmetric encryption for defense against MIMT, i.e., the receiver's public key encrypts the message when it is sent, and the receiver decrypts the message with his or her private key to ensure that the source of the message is correct.

Scenario: An attacker tampers with the communication messages or eavesdrops between the communicating parties.

Analysis: In the blockchain, the sender uses the public key of the receiver to encrypt messages. Additionally, if the attacker did not use a match private key to decrypt, it did not learn the content of the message. The private key only is known to the receiver.

For example, in the authentication phase, User A encrypts the message M_{A1} with User B's public key Puk_B, then generates a ciphertext C_{A1} and sends it to User B. B then uses his private key Prk_B to decrypt the ciphertext to obtain the original message M_{A1}. The related details are shown as follows:

$$C_{A1} = E_{Puk_B}(M_{A1}) \tag{87}$$

$$M_{A1} = D_{Prk_B}(C_{A1}) \tag{88}$$

Therefore, it is guaranteed that the attacker cannot decrypt the message without the receiver's private key. Each stage of asymmetric encryption and decryption is shown in Table 4.

Table 4. Encryption and decryption to prevent a man-in-the-middle attack.

Phase	Party		Encryption	Decryption
	Sender	Receiver		
Authentication	USER A	USER B	$C_{A1} = E_{Puk_B}(M_{A1})$	$M_{A1} = D_{Prk_B}(C_{A1})$
	USER B	USER A	$C_{B1} = E_{Puk_A}(M_{B1})$	$M_{B1} = D_{Prk_A}(C_{B1})$
Order	USER A	USER B	$C_{A1} = E_{Puk_B}(M_{A1})$	$M_{A1} = D_{Prk_B}(C_{A1})$
	USER B	USER A	$C_{B1} = E_{Puk_A}(M_{B1})$	$M_{B1} = D_{Prk_A}(C_{B1})$
Transaction	USER A	USER B	$C_{A2} = E_{Puk_B}(M_{A2})$	$M_{A2} = D_{Prk_B}(C_{A2})$
	USER B	USER A	$C_{B2} = E_{Puk_A}(M_{B2})$	$M_{B2} = D_{Prk_A}(C_{B2})$
Sale	Car Owner (CO)	Car Dealer (CO)	$C_{CO} = E_{Puk_{CD}}(M_{CO})$	$M_{CO} = D_{Prk_{CD}}(C_{CO})$
	Car Dealer (CD)	Car Owner (CD)	$C_{CD} = E_{Puk_{CO}}(M_{CD})$	$M_{CD} = D_{Prk_{CO}}(C_{CD})$
Repair	Repair Shop	Car Owner (CO)	$C_{RS} = E_{Puk_{CO}}(M_{RS})$	$M_{RS} = D_{Prk_{CO}}(C_{RS})$

4.5. Replay Attack

A replay attack is a type of network attack that uses malicious or fraudulent ways to repeat or delay valid data and the attacker intercepts the message of the communication and retransmits the data to the receiver [34]. In this study, to prevent the replay attack, we add a timestamp to each message, and the receiver needs to calculate the difference of the timestamp when receiving the corresponding message and compare it with the set threshold value, and if the time difference exceeds the threshold value it identifies that the message is being replayed.

Scenario: An attracter listens to messages between sender and receiver and, after that, it re-sends the same message to the receiver.

Analysis: If the receiver receives the ciphertext and decrypts it to acquire the timestamp TS_X of the sender, the receiver verifies that the difference between the current timestamp TS_{NOW} and the timestamp in the message is less than a threshold ΔT. When this does not hold, the communication that suffered a replay attack is confirmed.

For example, in the verification phase, the timestamp TS_{A1} when User A sends the data will be detected $TS_{NOW} - TS_{A1} \stackrel{?}{\leq} \Delta T$ when User B receives the data, and if it passes, it proves that the data are not under replay attack. Table 5 is the timestamp verification for each stage, where the timestamp after the receiver receives the data is collectively called.

Table 5. Timestamp validation to prevent replay attack.

Phase	Party		Send Time	Validation
	Sender	Receiver		
Authentication	USER A	USER B	TS_{A1}	$TS_{NOW} - TS_{A1} \stackrel{?}{\leq} \Delta T$
	USER B	USER A	TS_{B1}	$TS_{NOW} - TS_{B1} \stackrel{?}{\leq} \Delta T$
Order and Transaction	USER A	USER B	TS_{A1}	$TS_{NOW} - TS_{A1} \stackrel{?}{\leq} \Delta T$
	USER B	USER A	TS_{B1}	$TS_{NOW} - TS_{B1} \stackrel{?}{\leq} \Delta T$
	USER A	USER B	TS_{A2}	$TS_{NOW} - TS_{A2} \stackrel{?}{\leq} \Delta T$
	USER B	USER A	TS_{B2}	$TS_{NOW} - TS_{B2} \stackrel{?}{\leq} \Delta T$
Sale	Car Owner (CO)	Car Dealer (CO)	TS_{CO}	$TS_{NOW} - TS_{CO} \stackrel{?}{\leq} \Delta T$
	Car Dealer (CD)	Car Owner (CD)	TS_{CD}	$TS_{NOW} - TS_{CO} \stackrel{?}{\leq} \Delta T$
Repair	Repair Shop	Car Owner (CO)	TS_{RS}	$TS_{NOW} - TS_{RS} \stackrel{?}{\leq} \Delta T$

4.6. Counterfeiting Attack

In this paper, the counterfeiting attack is the behavior of an attacker using falsified and uploaded fake parts' information or disguising as a parts owner to trade on the system. We verify the legitimacy of the data during the transaction process to prevent this attack.

Scenario 1: The attacker fakes and uploads fake parts' information, and uses these parts to trade.

Analysis 1: Uploading parts' information is a unique function of PM. Other users cannot sign and upload parts without a PM private key. Additionally, because the alliance chain is used, each role needs to be authenticated, and the chances of an attacker disguising PM successfully are not possible. At the same time, based on the characteristics of the blockchain, the source of the parts can be traced. Therefore, when that counterfeit part appears on the blockchain, we can quickly locate the attacker.

Scenario 2: Malicious RS or rental car users replace expensive parts with low-cost fake parts.

Analysis 2: In our proposal, the part and the vehicle to which it belongs are bound together and belong to the same owner. As shown in Figure 11, when malicious RS replaced expensive parts reappear on the supply chain and conduct transactions, the buyer of the part will check again whether the source of the part is legitimate. If not, the system will notify the original owner of the part, who can quickly apply for arbitration with an arbitration institution.

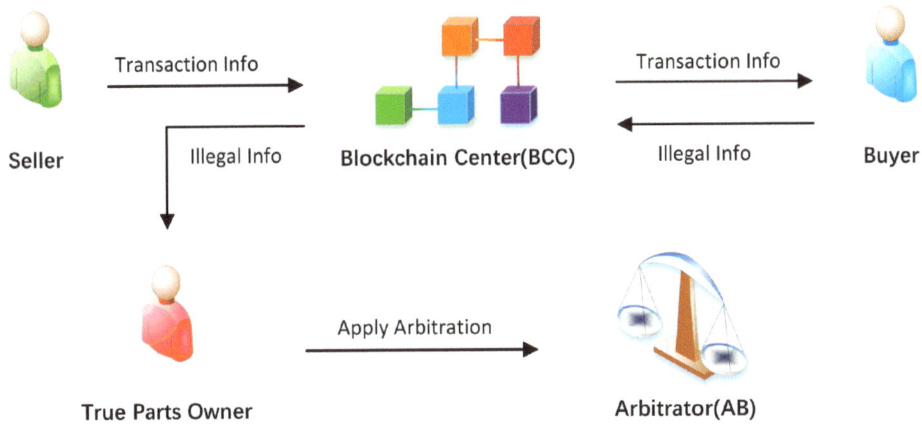

Figure 11. Trading illegal parts handling process.

5. Discussion

5.1. Communication Cost

In this section, we calculate the different communication costs for different network rates as shown in Table 6. Firstly, we assume that the length of the ECDSA key and signature is 160 bits, the length of asymmetrically encrypted data is 1024 bits, and other information (ID, timestamp, etc.) is 80 bits. The total size is 160 bits × 2 + 80 bits × 2 + 1024 bits × 2 = 2588 bits. It takes 0.431 ms in 3G (6 Mpbs), communication environment, 0.026 ms in 4G (100 Mpbs) communication environment, and 0.129 us in 5G (20 Gpbs) communication environment [35].

Table 6. Communication costs of the proposed scheme.

Phase	Message Length	3G (6 M bps)	4G (100 M bps)	5G (20 G bps)
Authentication	2588 bits	0.431 ms	0.026 ms	0.129 us
Order	2588 bits	0.431 ms	0.026 ms	0.129 us
Transaction	2588 bits	0.431 ms	0.026 ms	0.129 us
Sale	2588 bits	0.431 ms	0.026 ms	0.129 us
Repair	1294 bits	0.216 ms	0.013 ms	0.065 us

5.2. Computation Cost

Table 7 shows the computational cost analysis of the roles in each phase. Taking the authentication phase as an example, in this phase both User A and User B need to perform the signature operation, verification operation, encryption, and decryption operation, comparison operation once each, and hash operation twice.

Table 7. Computation costs of the proposed scheme.

Phase	Access Part 1	Access Part 2
Authentication	User A $T_{sig} + T_{ver} + 2T_{hash} + T_{cmp} + 2T_{E/D}$	User B $T_{sig} + T_{ver} + 2T_{hash} + T_{cmp} + 2T_{E/D}$
Order	User A $T_{sig} + T_{ver} + T_{upload} + T_{cmp} + 2T_{E/D} + 2T_{hash}$	User B $T_{sig} + T_{ver} + 2T_{hash} + T_{cmp} + 2T_{E/D}$
Transaction	User A $T_{sig} + T_{ver} + 2T_{hash} + T_{cmp} + 2T_{E/D} + T_{chd}$	User B $T_{sig} + T_{ver} + 2T_{hash} + T_{cmp} + 2T_{E/D} + T_{upload}$
Sale	Car Owner(CO) $T_{sig} + T_{ver} + 2T_{hash} + T_{cmp} + 2T_{E/D} + T_{chd}$	Car Dealer (CO) $T_{sig} + T_{ver} + 2T_{hash} + T_{cmp} + 2T_{E/D}$
Repair	Repair Shop (RS) $T_{sig} + T_{hash} + T_{E/D}$	Car Owner (CO) $T_{sig} + T_{hash} + T_{E/D}$

Note: T_{sig}: Signature operation; T_{ver}: Verify operation; $T_{E/D}$: Encryption/Decryption operation; T_{hash}: Hash function operation; T_{cmp}: Comparison operation; T_{chd}: Check data function; T_{upload}: Upload data operation.

5.3. Function Comparison

Table 8 shows the comparison with the previous researchers. In this paper, we proposed a blockchain-based automotive and parts supply chain service framework, related algorithm, and communication protocol and analyzed related cost and security.

Table 8. Comparison with surveyed related works.

Authors	Year	Objectives	1	2	3	4	5
Chen et al. [16]	2015	A theoretical framework for combining blockchain and supply chain	N	Y	N	Y	N
Sharma et al. [17]	2018	A distributed framework model for the entire life cycle phases of the automotive industry blockchain-based	N	Y	Y	Y	Y
Kim et al. [18]	2019	A blockchain-based design for authentication of parts	N	Y	N	Y	N
Miehle et al. [21]	2019	A traceable parts supply chain application built on blockchain and smart contracts	N	Y	N	Y	N
Helo and Hao [22]	2019	A Blockchain-based logistics monitoring system prototype	N	Y	N	Y	Y
Yahiaoui et al. [23]	2020	Blockchain and smart contract-based supply chain model	N	Y	N	Y	N
Li and Ye [24]	2020	Combines blockchain and ASC for distributed storage of production and sales data	N	Y	N	Y	N
Wang et al. [25]	2020	Blockchain-based Product-Service System service framework for vehicle products	N	Y	N	Y	Y
Our method	2022	Blockchain-based ASC and service framework	Y	Y	Y	Y	Y

Notes: 1: Communication protocol, 2: Blockchain-based architecture, 3: Algorithm, 4: Complete architecture or framework, 5: Analysis, Y: Yes, N: No.

6. Conclusions

The quality of vehicles and parts is closely related to traffic safety. To solve safety hazards caused by flaws in vehicles and parts and information asymmetry between providers and consumers, we proposed an automotive supply chain framework that is based on blockchain and smart contracts, in the meantime also designing communication flows and algorithms in the blockchain. In our analysis and discussion, this study-proposed system has excellent performance and security.

In this blockchain system, all access parties must register with BC to require a pair of public-private keys and a unique ID; in the meantime, both communicating parties should authenticate each other's identities before communicating. In addition, during communication, each role signs and encrypts the information to be sent and uploads it to the chain, and decrypts and verifies the validity of the received message. Furthermore, when a dispute arises with a participant in the system, the participant can apply for arbitration by AB. Additionally, then AB, using the participant, provides a message to acquire blockchain information, confirming the legality.

By the proposed method and framework, we accomplish the features as follows:

(1) Proposed a completely auto supply framework based-blockchain.
(2) Using asymmetrical encryption/decryption to ensure data integrity.
(3) Design some algorithms for simple quality identification of cars and parts.
(4) Analyzing costs of computation and communication.
(5) Parties can verify the legality of an asset by an arbitrator.
(6) Simulate defense against known attacks.

Author Contributions: Conceptualization, C.-L.C. and Z.-P.Z.; methodology, C.-L.C., Z.-P.Z. and M.Z.; validation, W.-J.T., C.-M.W. and H.S.; investigation, C.-L.C., Z.-P.Z. and H.S.; data analysis, W.-J.T., C.-M.W. and H.S.; writing—original draft preparation, C.-L.C. and Z.-P.Z.; writing—review and editing, W.-J.T., C.-M.W. and H.S.; supervision, C.-L.C. and M.Z. All authors have read and agreed to the published version of the manuscript.

Funding: This work was supported in part by the Ministry of Science and Technology in Taiwan (No. MOST 111-2218-E-305-001-MBK and MOST 110-2410-H-324-004-MY2), the Science and Technology Project of Jilin Provincial Department of Education (JJKH20210457KJ), Jilin Province Science and Technology Development Plan Project (20220508038RC), Undergraduate Training Programs for Innovation and Entrepreneurship Project of Jilin Province (J202210203JSJ02) and CERNET Innovation Project (NGII20180315), the National Natural Science Foundation for Young Scientists of China (Grant No. 51808474).

Institutional Review Board Statement: Not applicable.

Informed Consent Statement: Not applicable.

Data Availability Statement: The data used to support the findings of this study are available from the corresponding author upon request.

Conflicts of Interest: The authors declare no conflict of interest.

Notations

q	A k bit prime number
$GF(q)$	Finite group of q
E	Elliptic Curves Defined on Finite Groups q
G	A generating points based on Elliptic Curve E
k_i	The ith random value on the elliptic curve
d_X/Q_X	The ECDSA's private key/public key of the party X
$(x_{Ai}, y_{Ai})/(r_{Ai}, s_{Ai})$	The ith ECDSA/Elliptic curve signature value of User A
TS_{Xi}/TS_{NOW}	The ith timestamp of X/current timestamp
M_{Xi}	The ith message is generated by X
M_{Info_X}/M_{BC_X}	User Info of X/Blockchain Message for X
M_{Conf}/M_{OrdXi}	order Confirmation/The ith order information from X
C_{Xi}	The ith encrypted ciphertext is generated by X
ID_X	Unique ID of User X
$ID_{Car}/ID_{Part}/ID_{Order}/ID_{repair}$	Unique identification code of the vehicle/part/Order/Repair
$H(M)$	One-way hash function
h_{Xi}	The ith hash value of X
Puk_X/Prk_X	X own public/private key that issued by BCC
$E_{Puk_X}(M)/D_{Prk_X}(M)$	Encrypt/Decrypt message M using X public/private key
$A \stackrel{?}{=} B/A \stackrel{?}{\leq} B$	Verify that A is equal to B/Check if A is less than B

References

1. Number of U.S. Aircraft, Vehicles, Vessels, and Other Conveyances. Available online: https://www.bts.gov/content/number-us-aircraft-vehicles-vessels-and-other-conveyances (accessed on 5 July 2022).

2. World Motor Vehicle Production, Selected Countries. Available online: https://www.bts.gov/content/world-motor-vehicle-production-selected-countries (accessed on 5 July 2022).
3. Statistical Bulletin on National Economic and Social Development of the People's Republic of China in 2020. Available online: http://www.stats.gov.cn/xxgk/sjfb/zxfb2020/202102/t20210228_1814159.html (accessed on 5 July 2022).
4. Singh, S. *Critical Reasons for Crashes Investigated in the National Motor Vehicle Crash Causation Survey*; National Center for Statistics and Analysis: Washington, DC, USA, 2015; Available online: https://trid.trb.org/view.aspx?id=1346216&source=post_page (accessed on 6 August 2022).
5. National Highway Traffic Safety Administration. National motor vehicle crash causation survey: Report to congress. *Natl. Highw. Traffic Saf. Adm. Tech. Rep. DOT HS* **2008**, *811*, 59.
6. Hoque, M.S.; Hasan, M.R. Involvement of vehicle factors in road accidents. *J. Civ. Eng.* **2007**, *35*, 17–27.
7. Unlawful Vehicle Modifications: State Laws. Available online: https://www.findlaw.com/traffic/traffic-tickets/unlawful-vehicle-modifications-state-laws.html (accessed on 5 July 2022).
8. Modifying Your Vehicle's Emissions: The Legal, Safety, and Health Implications. Available online: https://www.gov.uk/government/publications/modifying-your-vehicles-emissions/modifying-your-vehicles-emissions-the-legal-safety-and-health-implications (accessed on 5 July 2022).
9. Road Traffic Safety Law of the People's Republic of China. Available online: http://www.gov.cn/banshi/2005-08/23/content_25575.htm (accessed on 5 July 2022).
10. Traffic Safety Measures Basic Law. Available online: https://elaws.e-gov.go.jp/document?lawid=345AC0000000110 (accessed on 5 July 2022).
11. Car Repair Scams. Available online: https://www.fraudguides.com/cars/car-repair-scams/ (accessed on 5 July 2022).
12. Liu, H. The "second-hand" and accident cars are modified and sold. *China Qual. Miles* **2015**, *3*, 27–28.
13. Duboka, Č. Forensic evidence in road accidents caused by vehicle's mechanical failures. In Proceedings of the 26th JUMV International Automotive Conference, Belgrade, Serbia, 19–20 April 2017; pp. 259–268. Available online: https://www.researchgate.net/publication/316740151_FORENSIC_EVIDENCE_IN_ROAD_ACCIDENTS_CAUSED_BY_VEHICLE%27S_MECHANICAL_FAILURES (accessed on 6 August 2022).
14. Johnson, D.; Menezes, A.; Vanstone, S. The elliptic curve digital signature algorithm (ECDSA). *Int. J. Inf. Secur.* **2001**, *1*, 36–63. [CrossRef]
15. Hyperledger. Hyperledger-Fabricdocs. Available online: https://hyperledgerfabric.readthedocs.io/_/downloads/en/release-2.3/pdf/ (accessed on 6 August 2022).
16. Chen, S.; Shi, R.; Ren, Z.; Yan, J.; Shi, Y.; Zhang, J. A blockchain-based supply chain quality management framework. In Proceedings of the 2017 IEEE 14th International Conference on e-Business Engineering (ICEBE), Shanghai, China, 4–6 November 2017; pp. 172–176.
17. Sharma, P.K.; Kumar, N.; Park, J.H. Blockchain-based distributed framework for automotive industry in a smart city. *IEEE Trans. Ind. Inform.* **2018**, *15*, 4197–4205. [CrossRef]
18. Kim, S.-K.; Yeun, C.Y.; Damiani, E.; Al-Hammadi, Y.; Lo, N.-W. New blockchain adoptation for automotive security by using systematic innovation. In Proceedings of the 2019 IEEE Transportation Electrification Conference and Expo, Asia-Pacific (ITEC Asia-Pacific), Seogwipo-si, Korea, 8–11 May 2019; pp. 1–4.
19. Kim, S.-K. The trailer of strategic alliance for blockchain governance game. *arXiv* **2019**, arXiv:1903.11172.
20. Yi, S.; Li, C.; Li, Q. A survey of fog computing: Concepts, applications and issues. In Proceedings of the 2015 Workshop on Mobile Big Data, Hangzhou, China, 21 June 2015; pp. 37–42.
21. Miehle, D.; Henze, D.; Seitz, A.; Luckow, A.; Bruegge, B. PartChain: A decentralized traceability application for multi-tier supply chain networks in the automotive industry. In Proceedings of the 2019 IEEE International Conference on Decentralized Applications and Infrastructures (DAPPCON), Newark, CA, USA, 4–9 April 2019; pp. 140–145.
22. Helo, P.; Hao, Y. Blockchains in operations and supply chains: A model and reference implementation. *Comput. Ind. Eng.* **2019**, *136*, 242–251. [CrossRef]
23. Yahiaoui, S.; Fedouaki, F.; Mouchtachi, A. How blockchain make better the supply chain in the automotive industry. *Int. J. Eng. Adv. Technol.* **2020**, *9*, 2912–2917. [CrossRef]
24. Li, B.; Ye, C. Product traceability system of automobile supply chain based on blockchain. *Comput. Eng. Appl.* **2020**, *56*, 35–42.
25. Wang, X.; Wang, Y.; Liu, A. Trust-driven vehicle product-service system: A blockchain approach. *Procedia CIRP* **2020**, *93*, 593–598. [CrossRef]
26. Nakamoto, S. Bitcoin: A peer-to-peer electronic cash system. *Decentralized Bus. Rev.* **2008**, 21260.
27. Szabo, N. The Idea of Smart Contracts. Nick Szabo's Papers and Concise Tutorials. 1997, Volume 6, p. 199. Available online: http://www.fon.hum.uva.nl/rob/Courses/InformationInSpeech/CDROM/Literature/LOTwinterschool2006/szabo.best.vwh.net/smart_contracts_idea.html (accessed on 5 July 2012).
28. Qiao, L.; Dang, S.; Shihada, B.; Alouini, M.-S.; Nowak, R.; Lv, Z. Can blockchain link the future? *Digit. Commun. Netw.* **2021**, *in press*. [CrossRef]
29. Rivest, R.L.; Hellman, M.E.; Anderson, J.C.; Lyons, J.W. Responses to NIST's proposal. *Commun. ACM* **1992**, *35*, 41–54. [CrossRef]
30. Mehibel, N.; Hamadouche, M.H. A new enhancement of elliptic curve digital signature algorithm. *J. Discret. Math. Sci. Cryptogr.* **2020**, *23*, 743–757. [CrossRef]

31. Pornin, T. Deterministic Usage of the Digital Signature Algorithm (DSA) and Elliptic Curve Digital Signature Algorithm (ECDSA). 2013. Available online: https://www.rfc-editor.org/rfc/rfc6979.html (accessed on 6 August 2022).
32. Kang, B.; Shao, D.; Wang, J. A fair electronic payment system for digital content using elliptic curve cryptography. *J. Algorithms Comput. Technol.* **2018**, *12*, 13–19. [CrossRef]
33. Nayak, G.N.; Samaddar, S.G. Different flavors of man-in-the-middle attack, consequences and feasible solutions. In Proceedings of the 2010 3rd International Conference on Computer Science and Information Technology, Chengdu, China, 9–11 July 2010; pp. 491–495.
34. Malladi, S.; Alves-Foss, J.; Heckendorn, R.B. *On Preventing Replay Attacks on Security Protocols*; Department of Computer Science, Idaho University: Moscow, ID, USA, 2002.
35. Kaur, K.; Kumar, S.; Baliyan, A. 5G: A new era of wireless communication. *Int. J. Inf. Technol.* **2020**, *12*, 619–624. [CrossRef]

sensors

MDPI

Article

A Machine Learning and Blockchain Based Efficient Fraud Detection Mechanism

Tehreem Ashfaq [1], Rabiya Khalid [1], Adamu Sani Yahaya [1,2], Sheraz Aslam [3,4], Ahmad Taher Azar [4,5,6,*] Safa Alsafari [7] and Ibrahim A. Hameed [8,*]

1 Department of Computer Science, COMSATS University Islamabad, Islamabad 44000, Pakistan; tehreemashfaq786@gmail.com (T.A.); rabiyakhalid672@gmail.com (R.K.); asyahaya.it@buk.edu.ng (A.S.Y.)
2 Department of Information Technology, Bayero University Kano, Kano 700006, Nigeria
3 Department of Electrical Engineering, Computer Engineering and Informatics, Cyprus University of Technology, Limassol 3036, Cyprus; sheraz.aslam@cut.ac.cy
4 Automated Systems & Soft Computing Lab (ASSCL), Prince Sultan University, Riyadh 12435, Saudi Arabia
5 College of Computer and Information Sciences, Prince Sultan University, Riyadh 11586, Saudi Arabia
6 Faculty of Computers and Artificial Intelligence, Benha University, Benha 13518, Egypt
7 Department of Computer Science and Artificial Intelligence, College of Computer Science and Engineering, University of Jeddah, Jeddah 23890, Saudi Arabia; sbalsfry@uj.edu.sa
8 Department of ICT and Natural Sciences, Norwegian University of Science and Technology, 7034 Trondheim, Norway
* Correspondence: aazar@psu.edu.sa or ahmad.azar@fci.bu.edu.eg or ahmad_t_azar@ieee.org (A.T.A.); ibib@ntnu.no (I.A.H.)

Abstract: In this paper, we address the problems of fraud and anomalies in the Bitcoin network. These are common problems in e-banking and online transactions. However, as the financial sector evolves, so do the methods for fraud and anomalies. Moreover, blockchain technology is being introduced as the most secure method integrated into finance. However, along with these advanced technologies, many frauds are also increasing every year. Therefore, we propose a secure fraud detection model based on machine learning and blockchain. There are two machine learning algorithms—XGboost and random forest (RF)—used for transaction classification. The machine learning techniques train the dataset based on the fraudulent and integrated transaction patterns and predict the new incoming transactions. The blockchain technology is integrated with machine learning algorithms to detect fraudulent transactions in the Bitcoin network. In the proposed model, XGboost and random forest (RF) algorithms are used to classify transactions and predict transaction patterns. We also calculate the precision and AUC of the models to measure the accuracy. A security analysis of the proposed smart contract is also performed to show the robustness of our system. In addition, an attacker model is also proposed to protect the proposed system from attacks and vulnerabilities.

Keywords: anomaly detection; blockchain; fraud detection; machine learning; random forest; XGboost

1. Introduction

Every industry, including banking, education, health care, and others, has modernized as a result of technological growth. Moreover, with the advent of communication technology, online transactions and means of payment are also being modernized. Through this modernization, traditional currencies are being converted into digital currencies, and all financial transactions are being conducted digitally. However, these transactions are not fully secured and are vulnerable to various digital attacks, such as fraud issues, anomalies, and privacy breaches. Additionally, as the volume of transactions rises, there is an increase in fraud associated with financial transactions. As a result, billions of dollars are lost globally every year [1]. Any suspicious activity on a network that behaves abnormally is called an anomaly. In cybersecurity and digital financial exchange, anomaly detection is used to detect fraud and network invasion. The goal of anomaly detection is to protect the network from illegal and fraudulent activities. In the financial sector, anomaly detection applications have investigated suspicious activity and identified hackers and fraudulent users. However, all anomaly detection methods in traditional financial systems are designed for centralized systems. Therefore, with the development of digital currencies, such as Bitcoin, anomaly detection methods using the blockchain are improving. Despite these advances, there are still many fraud occurrences [2]. Many artificial intelligence (AI) and machine learning techniques have been proposed to detect anomalies and fraud in digital transactions; however, there is no suitable solution for centralized systems. Blockchain is the most advanced and quickly evolving technology in many fields. It first became visible with the appearance of Bitcoin in 2008, which was introduced by Satoshi Nakamoto [3]. It addresses the security issues of centralized systems and provides solutions to external threats. It is a distributed, decentralized, and immutable ledger that time stamps all records and ensures record integrity. However, some participants in the blockchain network behave maliciously [4].

In our work, we apply existing ML techniques (i.e., XGBoost and random forest) to data in the form of blockchain transactions with the goal of detecting fraudulent transactions. To the best of our knowledge, this work is the first to investigate the application of ML to blockchain data with such an objective. The contributions of the study/work are listed below.

- Data-balancing technique and processioning are performed in the proposed system. In pre-processing, the data are divided into a training dataset and a test dataset.
- Machine learning techniques, XGboost and random forest (RF), are used for data classification. They classify the data as fraudulent or non-fraudulent. Both classifiers predict the type of data. These machine learning models are directly connected to the blockchain.
- The machine learning model is linked to the blockchain. A blockchain-based smart contract is written in which the machine learning model is deployed and used to predict the nature of new incoming transactions.
- The blockchain model is used to initiate the transactions, and then machine learning models are used to classify these transactions as malicious or legitimate.
- Two attacker models are also implemented to protect the proposed model from blockchain attacks.

The paper is organized as follows. Related work is discussed in Section 2. The proposed system model along with problem statement is presented in Section 3. Simulation results are discussed in Section 4. A security analysis is given in Section 5. Moreover, the paper's conclusion is described in Section 6. The list of abbreviations is given in Abbreviations.

2. Related Work

Different public and private regions deploy blockchain technologies for various objectives because it is vital to protect and monitor auditing systems. These technologies help to evaluate its repositories and take care of the privacy of auditors. They allow auditors to send their queries in a reliable and accessible manner without exposing their identities to unauthorized users. In [5], consensus algorithms check the legitimacy of the performed transactions. However, it is inefficient to identify the transactions. Therefore, using blockchain as a solution for fraud detection does not completely address the problem. Because of this, new solutions are used to eliminate the vulnerabilities in the existing systems, such as machine learning algorithms. Different supervised machine learning techniques are used to detect fraudulent transactions. Furthermore, a comparative analysis of various machine learning methods is presented [6,7]. In [8], the authors proposed different supervised machine learning solutions to detect fake businesses. Moreover, they also tested over 300,000 accounts using random forest and XGBoost classifiers. The authors in [9] also used XGboost for accurate results. In [10], the authors dealt with the problem of an imbalanced dataset. The dataset belongs to an insurance company and describes the driving patterns of individuals. They use XGboost to predict the performance of drivers along with their telematic information.

According to [11], fraudulent activities are data mining issues because the central server for credit card transactions tells whether a trading transaction is fake or legal. Fraud detection is not a new problem; yet, there are still numerous challenges. The primary reason is that researchers lack real-time data, and banks are unwilling to share their data with researchers because customer data is confidential. At the same time, it is linked to the banks' privacy policies [12]. In [13], the authors used a distributed data mining model to address the problems of slanted delivery of credit cards and non-uniform expenditures. A fraud detection algorithm aws presented in [14], which identifies fraud without relying on any fraudulent historical instances, with a proactive method capable of overcoming the well-known cold-start problem. In [15], The authors suggested and demonstrated the application of the uncertain association law of mining to extract useful data from credit card transactions.

The authors in [16] trainded a Support vector machine model to detect the improper data of credit card transactions. In [17], the authors mixed three different techniques to decrease the wrong beeps in fraud identification. These techniques are Bayesian learning, rule-based learning and Dempster–Shafer theory. In [18], the authors used a transaction aggregation technique to interpret the customer's behavior before any transaction is performed and then used this aggregated data to identify fake transactions. The entire analysis takes place on the behavior of the customers. The primary purpose of the work is to develop a model that can work with unknown datasets and highlight fake datasets in them. Banks give unspecified datasets due to privacy issues. Therefore, the model behaves similarly with all the participant attributes without prioritizing them. The model has also worked on the improper datasets and arranged them in two separate sections: one for legal transactions and the other for fake transactions [18].

In [19], the authors identified the issues of trust, privacy, security and verifiability in centralized-based IoT-driven smart cities. Therefore, the authors proposed a trustworthy privacy-preserving secure framework (TP2SF) for smart cities. The proposed framework comprises three modules: a module for trustworthiness, and two modules that consist of two-layered privacy modules. The trustworthiness module is a blockchain-based reputation system that ensures the system's security. Furthermore, two-layered privacy modules are based on an enhanced proof of work (ePoW) technique and principle component analysis (PCA). These modules transform the data into a reduced shape to prevent the system from poisoning attacks. However, a cloud system is used for data storage, which leads to a centralization problem.

In [20], the authors resolved the issue of privacy preservation through encryption techniques. They also used cryptographic approaches for the computation of data. The proposed system use asymmetric, symmetric and homomorphic encryption techniques to achieve privacy. However, high computational power and time are required to implement these approaches. Cyber attacks and intrusion detection are major problems that cause data privacy issues. Blockchain technology with deep learning algorithms is used to resolve the mentioned in [21]. These models provide security and privacy in virtual machines migrated to the cloud to protect IoT networks. The authors proposed a deep blockchain framework (DBF) model for intrusion detection based on bidirectional long short-term memory (BiLSTM) and blockchain. In [22], the authors identified the issues of centralization and cyber attacks in cloud-based systems. Therefore, they proposed a mixture-of-localization-based outliers (MLO) system with a Gaussian mixture. This collaborative anomaly detection system detects insider and outsider attacks in a cloud-based system. Privacy preservation is highly important for cyber–physical systems (CPSs). In these systems, anomaly detection systems are required to protect the system from inner and outer attacks [23]. Therefore, the authors proposed a new privacy-preserving anomaly detection framework that protects the system from attacks and keeps sensitive information confidential. The proposed method is based on two modules, i.e., the pre-processing module and anomaly detection module that used a Gaussian mixture model (GMM). However, the proposed system is inefficient for tackling modern IoT attacks.

2.1. Adversarial Machine Learning Methods

In adversarial machine learning, some machine learning techniques try to exploit the model's specific vulnerabilities and take advantage of the model's obtained information to generate some malicious attacks [24]. Some adversarial problems are discussed in the following papers. In [25], the authors gave a comprehensive overview of the research conducted in the last decade, considering the pioneering research from the security of non-deep learning algorithms to the advances in this field, i.e., properties of security in deep learning algorithms.

In [26], the authors proposed unsupervised random forest algorithms to reduce the number of fraudulent transactions. Further, this proposed algorithm was used to analyze the detection of credit card fraud. Moreover, the Bayesian network assembles a coordinated non-cyclic chart, further used for the conditional probability distribution for creating a non-cyclic graph. Results show that the random forest-based proposed algorithm performed better than its counterparts. Authors in [27] also proposed a random forest model for detailed feature selection, financial fraud detection, importance measurement of variables, and multidimensional and partial correction analysis. Nevertheless, the authors applied several statistical methodologies, i.e., non-parametric and parametric models, to detect accuracy. They concluded that non-parametric models have less accuracy compared with parametric models. In [28], the authors worked on the problem of intrusion detection in cyber security. They used a dataset which has highly sensitive training data. This type of dataset is vulnerable to cyber attacks. To resolve this issue, they used a random forest algorithm that performs better in detecting cyber attacks. However, there is still room for researchers to improve the detection of cyber attacks. In [29], the authors proposed an effective random forest classifier for anomaly detection in an IoT network. They also compared the performance of an intrusion detection system (IDS) and random forest classifier in terms of accuracy and false alarm rate. However, security is the major issue while implementing an IoT network. In [30], the authors identified the problems of malicious data and manipulation of data by an attacker. Therefore, they implemented the evasion classifier and checked its effectiveness on a test case. The authors analyzed some potential techniques used to increase the robustness of machine learning models against the attacks of data manipulation.

3. Problem Statement and System Model

In this section, we first explain the problem and then present our proposed system model.

3.1. Problem Statement

With the advancement of technology, cyber crime is also increasing day by day, and the financial sector is the most affected sector by cyber crime [5]. The main reason for this problem is security vulnerabilities in financial systems. Anomalies occur in these systems, which are also known as frauds. In traditional financial systems, credit card frauds are the most common frauds, and AI techniques are used to solve these frauds. As a result, the financial industry suffers a loss of billions of dollars each year due to these frauds [1]. In [31], the authors employed unsupervised machine learning techniques to detect the monetary anomalies. However, according to [32], supervised machine learning techniques are more effective for fraud detection. A large amount of learning data and labeled data is good for supervised learning. Therefore, the authors developed a complex model to learn the patterns of anomalies and fraud. However, this model is not able to provide accurate results. Moreover, blockchain innovation solves several fraud problems. It provides security and privacy to the financial sector, as it is decentralized and immutable. However, it does not address such issues as loss of privacy, Sybil attacks, and double-spending attacks.

The purpose of these attacks is to discourage illegal activities and increase financial benefits. Bitcoin is a digital currency based on the concept of proof of work (PoW). In the Bitcoin network, all digital transactions are executed in a distributed manner using digital signatures and hashes via a timestamp service. Bitcoin transactions do not involve a trusted third party to verify the transactions. Therefore, a user can spend the same coin twice, which becomes a fraudulent transaction and is known as a double-spending attack [33]. In [12], the authors discussed the Bitcoin theft known as "all in vain", in which hackers stole nearly 25,000 bitcoins.

To address these issues, we propose a secure and efficient blockchain-based model with the integration of machine learning algorithms. The proposed model detects anomalies and thefts based on the predictive model. In the proposed work, machine learning models are trained on a dataset according to the fraud types and integrated transactions. The proposed model is linked with blockchain to overcome security and threats.

3.2. *Dataset Explanation*

The dataset used in this paper is downloaded from Kaggle [34]. This dataset consists of raw bitcoin transactions. These are the bitcoin transactions from the creation of bitcoin to now. This dataset contains 30 million transactions. However, due to the limited storage and computational power, only 30 thousand transactions are used. The dataset contains the 11 attributes and 30,000 observations. These attributes show the degrees of the bitcoins, mean of out and in degrees and the malicious transactions of these bitcoins. According to the dataset, there are multiple senders and receivers for a single transaction, and a single user can own multiple transaction addresses. In this network, every user is anonymous, as no relevant record is associated with the transaction address [35].

3.3. *Proposed System Model*

The proposed system model consists of two layers: blockchain and machine learning. The blockchain model initiates transactions, and then machine learning models are used to classify these transactions as malicious or legitimate. This is a binary classification. The proposed system model is based on the integration of machine learning and blockchain for fraud and anomaly detection in the financial sector. The anomaly detection system identifies unusual suspicious events that are different from most of the data. A dataset of bitcoin transactions is used for the proposed model. We also use the random forest and XGboost classifiers to classify legitimate and malicious transactions. These classifiers are also used to predict new incoming transactions. The proposed model is trained and tested for legitimate and malicious data patterns using the given dataset. The proposed system model consists of the following steps (discussed in the below subsections).

3.3.1. Data Balancing Using SMOTE

Imbalance of data is a major problem in machine learning, where the distribution of classes is highly imbalanced. The accuracy of machine learning algorithms decreases due to data imbalance. It increases when the number of instances of one class is greater than the other class. Therefore, SMOTE is used to solve this problem, and synthetic samples are randomly generated for the minority class [36]. This technique solves the overfitting problem caused by random oversampling of the data. It is based on random sampling, where a data point is selected from the minority class. Then random weights are assigned to its neighbors, and these neighbors are added to the original samples. The main task of SMOTE is to synthesize the minority class samples. Data balancing improves the effectiveness of machine learning algorithms and helps to achieve better results. In Algorithm 1, SMOTE is used to balance the data, and the class distribution of the data is imbalanced. Lines 1 to 6 show input, output, and initialization of the variables. Lines 7 to 16 show the working mechanism of SMOTE for data balancing. SMOTE works on the pattern of K-nearest neighbor, where the algorithm generates synthetic data. In the first step, SMOTE selects random data from the minority class. In the second step, the

K-nearest neighbors in the dataset are determined. Finally, synthetic data are generated between the randomly selected data by selecting the K-nearest neighbors. Moreover, when we train the model on the imbalance dataset, we check that the data are balanced or not if we are going to balance the data, then we first divide the data into testing and training parts and apply sampling technique only on the training data.

Algorithm 1: Data balancing through SMOTE

Initialization
Inputs: Minority data $M^{(D)} = m_i \in X$, Where i = 1, 2, 3, ..., D
Outputs: Synthetic Data S
Number of minority samples (D)
Percentage of SMOTE (P)
Number of (k) nearest neighbors
for n = 1 to D **do**
Find the K nearest neighbors of D_i
Check $\overline{P} = P/100$
While $\overline{P} \neq 0$ **do**
    Select a random sample m in minority class
    Find neighbor of m
    Pick a random number $\alpha \in [0,1]$
    $\overline{m} = m_i + \alpha(\overline{m} - m_i)$
While Append $\overline{m}$ to S
Check $\overline{P} = P - 1$
end while
end for
End

3.3.2. Detection of Fraudulent Transactions

As more businesses go online, fraud and anomalies in online systems are also on the rise. Fraud detection systems that rely on static rules created by human experts have been used to combat online fraud. For this reason, organizations face a large number of fraudulent activities in online transactions that need to be minimized. In this study, we address fraudulent transactions with Bitcoins. Unusual patterns that do not conform to expected behavior, called outliers, can be detected using anomaly detection. In the proposed model, a dataset of bitcoin transactions is used. This dataset is based on bitcoin transactions in the financial sector. As we know, the transaction pattern of cryptocurrencies of bitcoins and ethers are quite similar. Therefore, we trained our model in the dataset of bitcoins, and it also gives correct prediction on the transactions of ethers. Our proposed model can work efficiently in financial sectors, where blockchain-based cryptocurrencies are used.

3.3.3. XGBoost

XGboost is a boosting algorithm that generates sequential trees. There are multiple trees, and each successive tree aims to reduce the error of the previous tree and update the residual error. Therefore, each new sequential tree has the updated residual error value that is used for boosting. The proposed model uses XGboost to classify legitimate and malicious transactions. Moreover, this algorithm connects to the blockchain smart contract and predicts the new incoming transactions.

Algorithm 2 shows the working of XGboost based on the given dataset. In this algorithm, lines 1 to 3 show the inputs, outputs and the initialization of variables. Lines 4 to 8 show the testing and training of the dataset. The deployment of the model is shown in lines 9 to 11. Blockchain technology is also integrated into this algorithm from lines 12 to 17. These lines show that when a new transaction occurs in the blockchain, it passes to the XGboost to check the transaction's integrity. The notation "if Predictions==0" in line 13 denotes that if the user passes a test sample to the trained XGboost model and it returns '0' in response, then it means the specific test sample belongs to the legitimate class; otherwise, if "if Predictions==1", then it means that it belongs to the malicious class. Furthermore, line number 12 of the algorithm explains the notation predictions. The proposed model predicts the transaction and sends it back to the blockchain with its status. In addition, the performance of the learning algorithms is improved through hyperparameter tuning. A large number of hyperparameters makes XGBoost powerful and scalable; however, it is also difficult to tune because it has a large parameter space.

Algorithm 2: Fraud detection through XGboost

1: **Inputs:** Balanced Dataset S
2: **Outputs:** Transactions in Blockchain B
3: Initialization of Dataset
4: Spliting of S into training and testing
5: $X_{train} \leftarrow$ input variables from dataset
6: $Y_{train} \leftarrow$ target variables to dataset
7: $X_{test} \leftarrow$ input variables from test dataset
8: $Y_{test} \leftarrow$target variables from test dataset
9: Model = XGBClassifier$(n_e stimators = 100)$
10: Model = Model.fit(X_{train}, X_{train})
11: Y_{pred} = Model.predict(X_{test})
12: Predictions = [round(value) for value in Y_{pred}]
13: **if** *Predictions* $==0$ **then**
14: $transaction = legitimate$
15: B.**add** (*transaction*)
16: **else if** *Predictions* $==1$ **then**
17: $transaction = malicious$
18: **end if**
19: **return** B
20: **End**

3.3.4. Random Forest

Random forest is one of the most popular machine learning algorithms that is mainly used for classification. It can be used on both linear and nonlinear data. Random forest is the most productive machine learning algorithm for imbalanced datasets. A single basic classifier cannot solve the problem of an imbalanced dataset. In the proposed system, random forest is used for fraud detection in an unbalanced dataset which has a smaller number of fraud occurrences. In [37], the authors also used random forest on the imbalanced dataset. They used two types of datasets: one with the same number of fraud occurrences and one with a smaller number of fraud occurrences. However, the accuracy of the RF algorithm in the proposed model is better than the previous models. RF integrates several decision trees, where the final outcome is decided based on the majority vote. It also addresses the problem of overfitting. The training sample has a significant imbalance ratio (minority:majority = 0.001:0.999). Under these conditions, conventional classifiers may not be sufficient. In this scenario, RF is used with the benefit of keeping certain essential information about the majority class and using all available information.

3.4. Linkage of Blockchain with Machine Learning in the Proposed Model

Blockchain technology has been used for the past few years to provide security and privacy in various networks. Despite the fascinating features of blockchain, it is still vulnerable to fraudulent activities. The malicious entities may perform invalid and fraudulent transactions using various methods, such as a double-spending attack. In the proposed system, blockchain is combined with machine learning to solve this problem. The database of bitcoin transactions is used in the underlying work, and the proposed machine learning model is trained on the dataset. The pattern of transactions stored in the database is analyzed for further use. In parallel, the transactions are performed on the Ethereum network. The pattern of these transactions is assumed to be similar to the pattern of bitcoin transactions stored in the bitcoin transaction database. Moreover, each new Ethereum transaction is made an input to the machine learning model, and the model is trained on it. The transaction pattern is analyzed and compared with the bitcoin transaction pattern. If the pattern of both transactions matches, the new transaction is classified as legitimate or malicious. To further test the robustness of the proposed system, a double-spending attack is implemented in the underlying work.

In Figure 1, blockchain-based transactions are verified using a machine learning model, and the prediction result shows that the transaction is legitimate or malicious. The prediction of the machine learning model is based on the training and testing of a bitcoin transaction-based dataset.

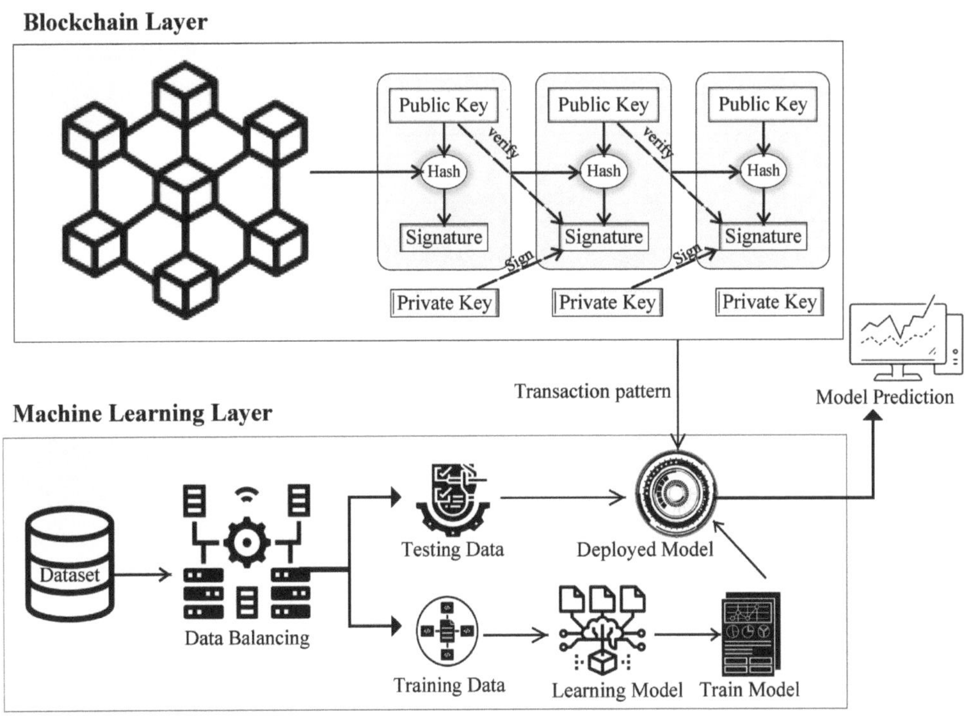

Figure 1. The proposed system mode of blockchain and ML.

4. Results and Discussion

This section first presents the simulation results of our proposed model, then we present the results after inducing modern cyber attacks to the system, i.e., Sybil attack, and double-spending attack.

The selected dataset is highly skewed, as shown in Figures 2 and 3. The classification models are biased toward the majority class due to the imbalance of the data.

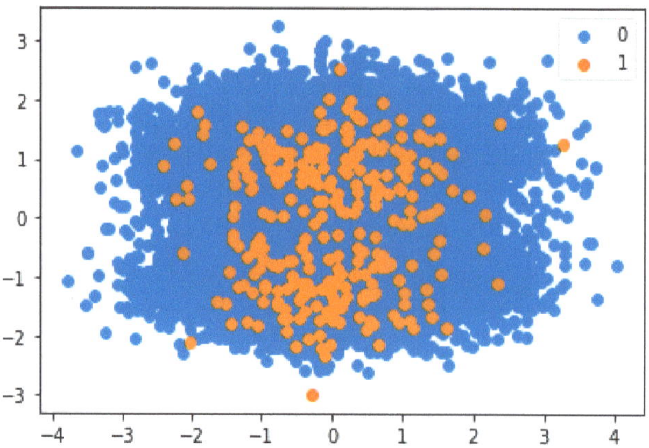

Figure 2. Imbalanced data.

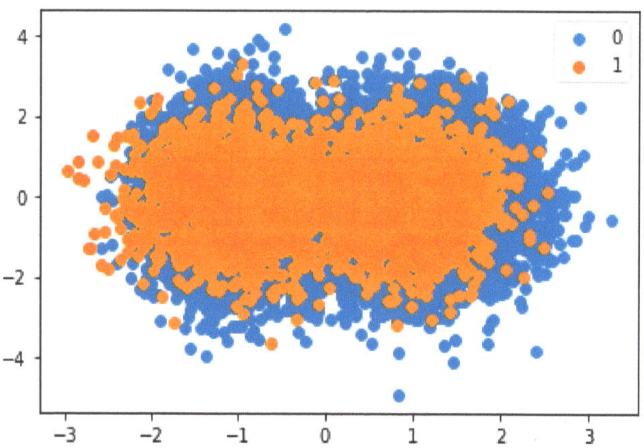

Figure 3. Balanced data.

Figure 2 shows the presence of malicious and honest transactions in the dataset. It can be seen from the figure that the number of honest transactions is higher than the number of malicious transactions. This imbalanced nature of the data leads to a bias in the classification. Synthetic data are used to solve this problem. The malicious entities are oversampled using SMOTE. The synthesized transactions are added to the dataset to limit the bias of the model during classification. The results obtained after using SMOTE are shown in Figure 3.

The observed log loss of XGBoost during training is shown in Figure 4. The log loss is observed for both the training data and the test data. From the figure, it can be seen that at a count of 10 iterations, a drastic drop is observed for both the training and test data. Moreover, the smoothness of the curves indicates that the model efficiently captures the nonlinear patterns of the data. For the test data, the log loss is higher than for the training data. However, the difference is not too large. The smaller difference between the training and test curves indicates that the model is well trained on unseen data. The trained model can be applied to real-world scenarios for anomaly detection in blockchain networks.

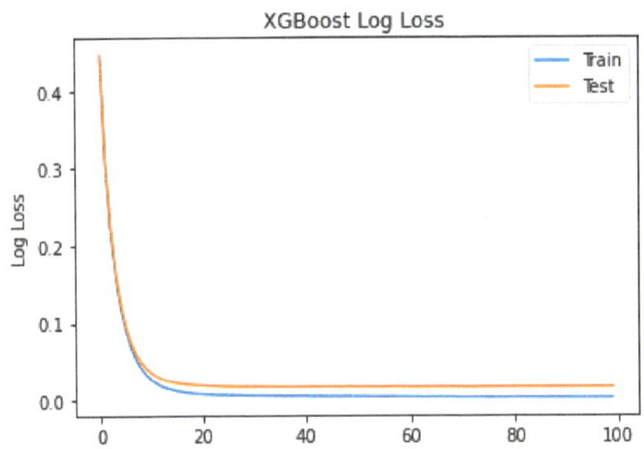

Figure 4. Logloss of XGboost.

Figure 5 shows the correlation between the fraudulent and non-fraudulent class. The correlation value 1 observed for $out_{a}nd_{t}x_{m}alicious$ shows the maximum correlation. Meanwhile, the value almost equal to 0, in the case of $mean_{i}n_{b}tc$, shows the minimum correlation between fraudulent and non-fraudulent.

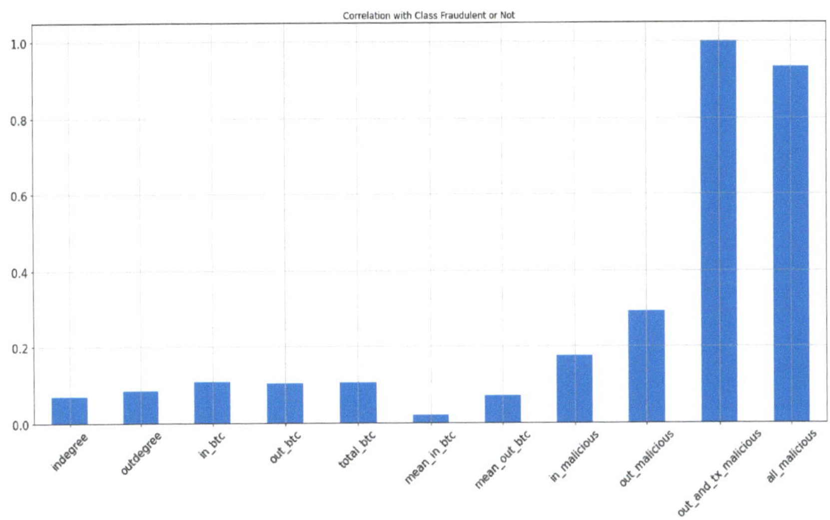

Figure 5. Correlation with class fraudulent or not.

Figure 6 shows the error that occurs when classifying with XGBoost. It shows the error for both training and test data. It can be observed that the classification error decreases as the number of iterations increases. The error is high for training data, and the figure shows a gradual decrease, while it is lower for test data and decreases rapidly.

The precision–recall curve of the XGboost model is visualized in Figure 7. This curve predicts the harmonic mean of both precision and recall. It is seen that a very slight decrease is observed, starting from 1. As soon as the recall value reaches more than 0.9, there is a sudden drop in the precision value. Figure 8 shows the accuracy when XGBoost is used. It shows that the highest peak of 0 to 1 indicates that the model achieves optimal accuracy in classifying blockchain transactions as legitimate or malicious. After reaching the maximum value of 0.9, the accuracy remains constant throughout the training.

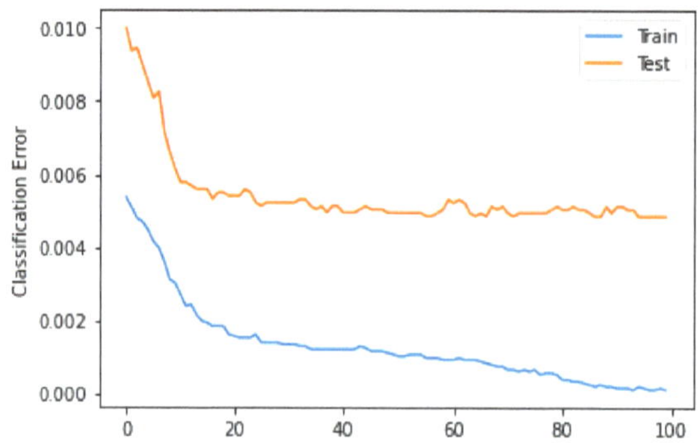

Figure 6. Classification error of XGboost.

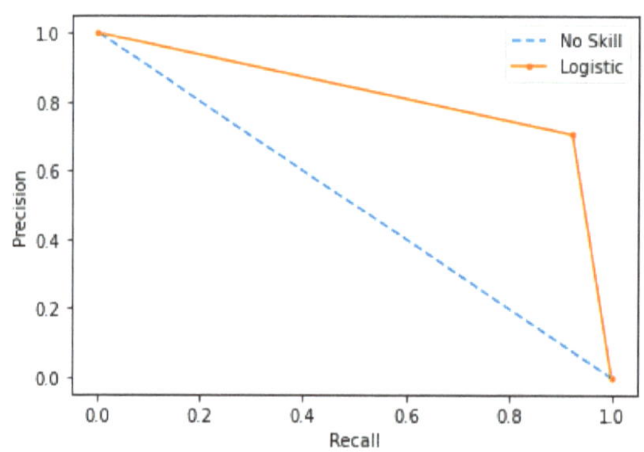

Figure 7. Precision of RF.

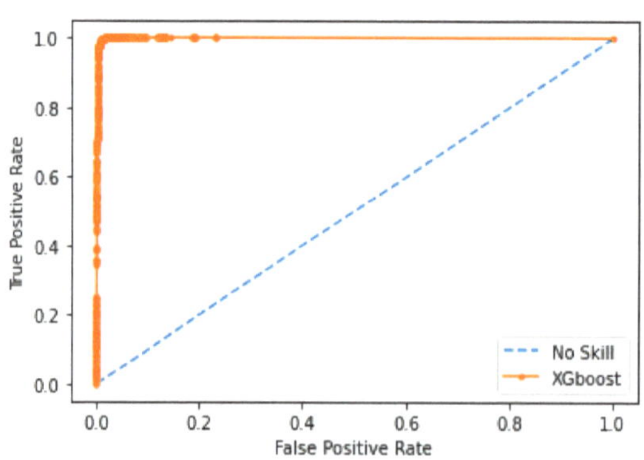

Figure 8. Accuracy of XGboost.

Figure 9 shows the confusion matrix obtained using RF. In this matrix, random forest selects 9014 random samples, correctly identifying 9009 predictions. This means that the proposed model efficiently discriminates between malicious and legitimate transactions. The matrix shows that the highest values are obtained in the case of true negatives, namely 99%. In the other three cases, the number of values is lower. This shows that the proposed model is efficient in detecting true negative transactions. Moreover, the phenomenon of majority voting in the random forest increases the performance of the model during classification. Figure 10 shows the AUC of a random forest. The AUC describes how well the model distinguishes between the positive and negative classes. It can be seen that the value of the AUC increases dramatically at the beginning to almost 0.85. Thereafter, a gradual increase is observed until the maximum value of 0.92 AUC is reached. The random forest model achieves an AUC of 0.92, which means that it performs well in capturing legitimate and malicious transactions.

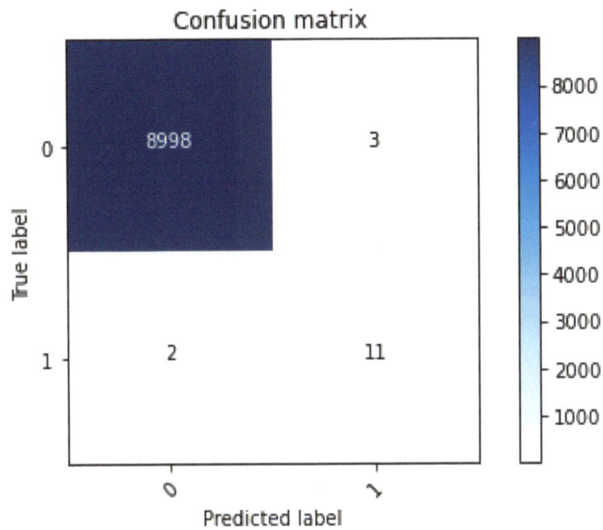

Figure 9. Confusion matrix through random forest.

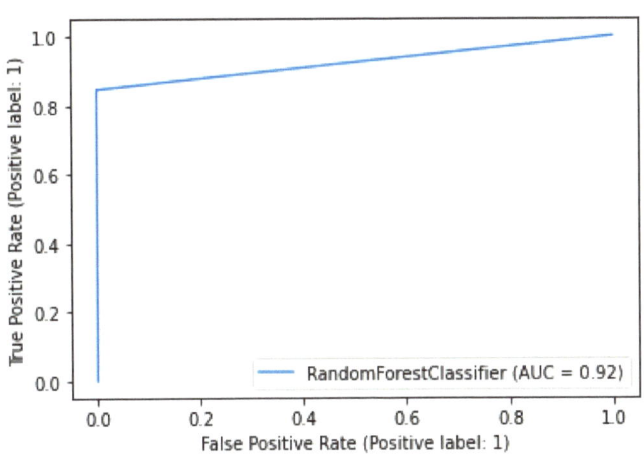

Figure 10. Accuracy of random forest.

Figure 11 shows the transaction and execution costs incurred in executing the functions involved in the blockchain smart contract. The costs are expressed in terms of gas, a basic unit of gas consumption in the blockchain network. From the figure, it can be seen that the transaction costs of all functions remain the same, while the execution costs of the *publish transaction* function are the highest, as mining costs are also included. Overall, the transaction costs are higher than the execution costs for all functions. The reason for this is that the former includes the processing costs of entire transactions, while the latter includes only the execution costs of some operations in a given function.

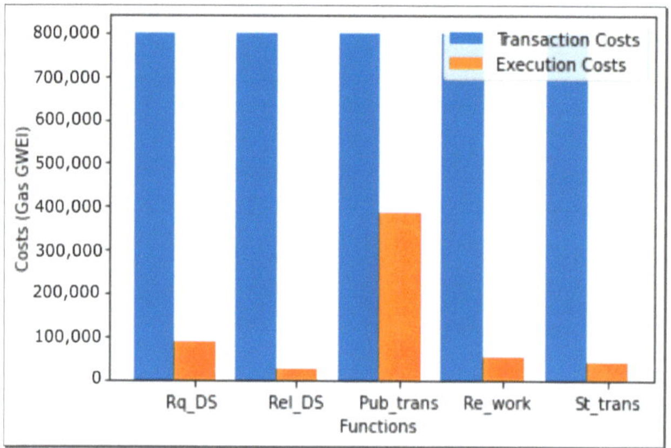

Figure 11. Transactions published and stored on blockchain (where Rq_DS = request dataset, Rel_DS = release dataset, Pub_trans = public transaction, Re_work = reuse work, and St_trans = store transaction).

4.1. *Validation of Proposed Model Based on Modern Cyber Attacks*

Nowadays, blockchain technology is considered the most secure technology for financial transactions due to its advances; however, it is still vulnerable to current cyber attacks. Despite all the advances and security measures, some advanced cyber criminals find strong attacks against the blockchain. The security features of blockchain cannot maintain its security measures against modern cyber attacks, such as selfish mining attacks, Sybil attacks, double-spending attacks, and replay attacks [38]. Therefore, this section explicitly presents results of our proposed model when modern cyber attacks are induced in the system.

4.1.1. Double-Spending Attack

In the blockchain, a transaction is only confirmed after the agreement/verification of all nodes. This verification takes a specific period, which creates a chance for cyber attacks. Double spending is one of these attacks that exploit the transaction verification time. Every transaction on the blockchain takes time for verification, and attackers use this time to their advantage. During the transaction verification delay, the attacker uses the same coin at two places as the verification of both transactions takes place simultaneously. In this way, digital currency is duplicated and falsified easily. In Ref. [33], the authors worked on the two double-spending attacker models. They enhance the two existing attacker models of Satoshi Nakamoto and Rosenfield for double spending. The first proposed model is called the "generalized model", in which authors added a time parameter. This parameter is used to calculate the time advantage of an attacker. The second proposed model is known as the time-based model. This model counts the time when an attacker and honest node mined their last blocks.

The parameters used in both models have the same definitions and use similar notions. The parameters used in the proposed model are given Abbreviations.

The authors discussed the given equations in Ref. [33]. These equations help to evaluate the probability that a double-spending attack can occur in a blockchain network. The probability of a double-spending attack is given in terms of the attacker progressing from 1 block to n blocks and ending up at the difference of $K - n$ blocks. It is given in Equation (1).

$$DS_N(q,K) = \sum_{n=0}^{+\infty} P_N(q,K,n)C_N(q,K-n-1) = 1 - \sum_{n=0}^{K} P_N(q,K,n)(1 - C_N(q,K-n-1)) \quad (1)$$

In Equation (1), C_N is a catch-up function used to define the probability of a double-spending attack. This probability is calculated by the expected branch length of the attacker. Moreover, in the given equation, the catch-up function depends upon a random walk in which the mining reward is given to the honest or attacker node.

$$C(q,z)\begin{cases} (\frac{q}{p})^{z+1} & , if\ q < 0.5 \wedge z > 0 \\ 1 & , otherwise. \end{cases}$$

In the given equation, q defines the computational power of the attacker, and $p = 1 - q$ calculates the probability that an attacker has fewer computational resources. Moreover, z denotes the initial disadvantage of the attacker. K denotes the number of confirmations to declare a block, and n denotes the number of blocks mined by the attacker. The probability that the attacker is successful in mining the block before the honest block is given in Equation (2).

$$\begin{aligned} P(T_q < T_p) &= \int_0^\infty P(T_q = x)P(T_p > x)dx \\ &= \int_0^\infty \frac{q}{\tau} e^{\frac{-q}{\tau}x} e^{\frac{-p}{\tau}x} dx \\ &= q \int_0^\infty \frac{1}{\tau} e^{\frac{-1}{\tau}x} dx \\ &= q \end{aligned} \quad (2)$$

where T_q and T_p are the random variables that are used to calculate the mining time of an honest node and an attacker node, respectively.

The attacker's potential progress function is defined using Equation (3).

$$P(q,m,n,t) = \sum_{z=0}^{n} a(q,t,z)P_N(q,m,n-z) \quad (3)$$

where

$$a(q,t,n) = \begin{cases} 1 & , if\ t = n = 0 \\ 0 & , if\ t <= 0 \\ \frac{(qt)^n}{n!}e^{-qt}, & , otherwise \end{cases}$$

In Equation (3), the $P(q,m,n,t)$ is used to calculate the probability of in how much time an attacker can mine the nth block before the honest node mines the mth block. Furthermore, P_N shows the potential progress function, and $a(q,t,n)$ is used to calculate the probability of mining the nth block in $t\tau$ seconds.

In the proposed work, the impact of a double-spending attack is assessed using the time advantage, computational power, and the number of pre-mined blocks. The number of pre-mined blocks is utilized as an input in Figure 12. The double-spending attack occurs after only a few blocks are created for values of q greater than 40%. It means that as the value of q rises, the probability rises with it, and once an attacker has control over the network, the chances of a double-spending attack become high. The probabilistic values close to 0 indicate that the double-spending attack will fail, while values close to 1 indicate a more significant success percentage for the double-spending attack.

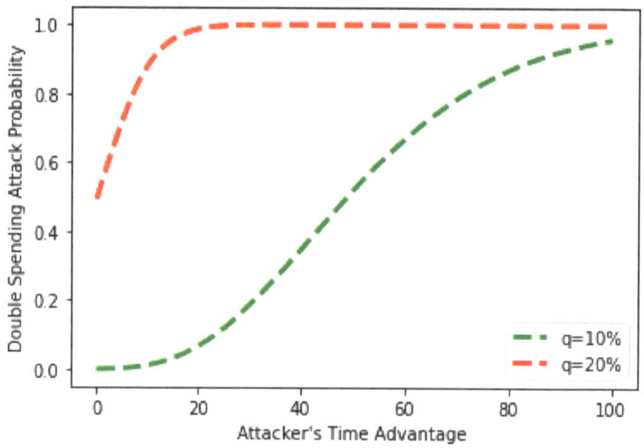

Figure 12. Double spending against time advantage of the attacker.

4.1.2. Sybil Attack

Blockchain has become the most secure platform for digital currency transactions. However, it is vulnerable to blockchain-based attacks, such as the Sybil attack. In a Sybil attack, a user creates multiple identities (IDs) to receive more rewards from the network or to rate himself highly. In the network, some malicious users are present and act maliciously at some point. Fake IDs are used by malicious users to obtain high ratings and deceive the network's legitimate users. It also manipulates the network and its data. All users in the proposed system are registered and have an account. When a registered user engages in bad behavior, several false IDs that are not registered on the network are created. In [39], the authors proposed an equation related to the probability of a Sybil attack, which is given below:

$$P(w) = \frac{\binom{ns}{w}\binom{N-1}{N-w}}{\binom{N+ns-1}{N}} \tag{4}$$

In the given equation, N represents the number of honest nodes' identities, and ns represents the successful Sybil node's identities. Suppose at the initial stage, w is the total identities in the network, which is calculated by using this $w = N + ns - 1$. The probability of an attack is increased when the number of successful Sybil identities is increased in the network. On the other hand, the attacker fails to implement the Sybil attack if the Sybil identities are less than the honest identities. The mentioned equations are hypergeometric equations.

In Figure 13, the evaluating parameters of Sybil attack are given, such as different Sybil identities ns = 12 and 24, number of nodes, and the computational power of the attacker node. The given figure shows the probability and impact of different Sybil identities in the network. It is observed from the figure that when the number of Sybil identities is 12, and computational resources are 0, then the probability of a Sybil attack is zero. However, the probability of a Sybil attack is increased when the computational resources are increased from 100 with 12 Sybil identities. It shows that if the attacker increases the computational resources, the probability of a Sybil attack becomes high. Moreover, when the Sybil identities are increased up to 24 with the computation resources equal to 125, the probability of a Sybil attack is zero. However, when the computational resources of Sybil identities are increased beyond 125, the probability of an attack is also increased. The graph depicts that the probability of a Sybil attack becomes high when the number of Sybil identities and computational resources is high. The findings reveal that the number of Sybil identities established by hostile people determines the likelihood of a Sybil assault.

The mathematical definition of the probability of a Sybil attack's success is shown in Equation (4).

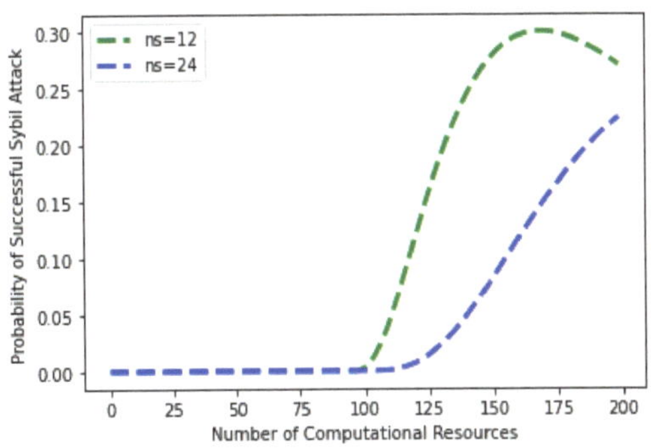

Figure 13. Probability of Sybil attack versus number of Sybil identities.

The idea of a Sybil attack was proposed in [39] to prevent the networks from this attack. The chance of a Sybil assault is calculated in this attacker model, utilizing several characteristics, such as computational power, the number of honest nodes, and the number of fake IDs. When both the number of fake IDs and computational power increase, the likelihood of the Sybil attack increases. In a Sybil attack, the following parameters are employed.

$$P(w) = \frac{\binom{g}{h}\binom{Q-1}{N*-h}}{\binom{g+h-1}{h*}} \tag{5}$$

- Q: number of population
- g: number of items in the population that are classified as success
- h: number of items in the sample that are classified as successes
- c: number of computational power of sample
- N^*: number of items in the sample

The relationship between the attack's probability and processing power is depicted in Figure 14. The graphical representation shows that the probability of an attack increases as the computational power employed by malevolent users and fake IDs increases. When malicious users use less processing power, the likelihood of an attack decreases, and vice versa. Equation (5) gives the mathematical description of the chance of a Sybil assault succeeding against computational power.

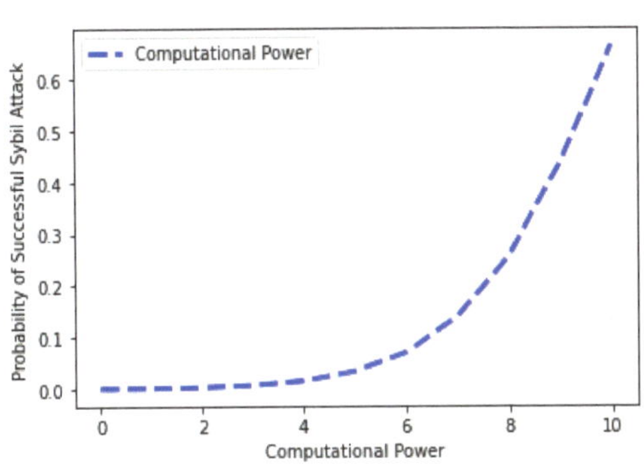

Figure 14. Sybil attack against computing power.

5. Security Analysis

In this section, we analyze the vulnerabilities of the proposed smart contracts. The security analysis of the proposed system is discussed in detail. For the security analysis, we used Oyente software, an open-source tool developed by the authors of [40]. It analyzes the smart contract using symbolic execution techniques based upon the execution of step-wise functions [41]. Oyente software provides a flexible environment, which directly works with the Ethereum Virtual Machine (EVM) and does not require access to high-level representations, such as Solidity and Serpent [42]. Moreover, it is also used to analyze smart contracts against the following significant vulnerabilities:

- Re-entrancy vulnerability;
- Timestamp dependency;
- Callstack depth vulnerability;
- Transaction ordering dependency;
- Parity multisig bug;
- Integer overflow;
- Integer underflow.

Figure 15 shows the security analysis of the smart contract involved in the proposed model. From the figure, it is observed that the outputs of all results in the analysis report are "False", which means that the smart contract used in the proposed system model is robust against well-known vulnerabilities. All of the results being false means the proposed model is secure and robust against these attacks.

```
INFO:symExec:    ============ Results ===========
INFO:symExec:      EVM Code Coverage:                          99.5%
INFO:symExec:      Integer Underflow:                          False
INFO:symExec:      Integer Overflow:                           False
INFO:symExec:      Parity Multisig Bug 2:                      False
INFO:symExec:      Callstack Depth Attack Vulnerability:       False
INFO:symExec:      Transaction-Ordering Dependence (TOD):      False
INFO:symExec:      Timestamp Dependency:                       False
INFO:symExec:      Re-Entrancy Vulnerability:                  False
INFO:symExec:    ====== Analysis Completed ======
```

Figure 15. Security analysis of the proposed smart contract.

Security Features

In this section, we discussed the solutions of our security model, and how it deals with the security threats and ensures the security of the system. The proposed solution consists of blockchain features. These features are decentralization, integrity, non-repudiation, availability and trust. This system is protected against replay attacks and man-in-the-middle (MITM) attacks.

Integrity: is an important feature which is used to ensure that there is no occurrence of data modification. The immutability of blockchain ensures data integrity and exchange messages between all participants and generates logs and events.
Availability: it makes sure that the deployed smart contract in the blockchain is always available for all participants. Availability also ensures that all services are always available. It also protects the system against denial of service (DoS) attacks because all transactions are stored in a distributed ledger of Ethereum. Therefore, there is no fear of hacking, failure and compromise. The ledger of Ethereum is highly robust against the DoS attack because thousands of trusted mining nodes protect this ledger.
Confidentiality: the requirement of confidentiality is achieved using a permissioned or private blockchain, e.g., Hyperledger or private Ethereum networks. The proposed system is based on a permissioned blockchain network in the proposed scenario.

6. Conclusions

Nowadays, blockchain is the latest and most secure technology that covers various research areas related to security. Blockchain development is based on digital currencies and is used to secure digital financial transactions. It protects financial systems from fraudulent attacks. Therefore, a blockchain-based machine learning algorithm is proposed to secure digital transactions. The proposed model predicts whether the incoming transaction in the blockchain is fraudulent or not. The proposed machine learning algorithms are trained and tested on a bitcoin-based dataset based on bitcoin transactions and predict the behavior of the incoming transactions. The given dataset is based on 30,047 entities, with smaller numbers of fraudulent entities. Due to the small amount of fraudulent data in the dataset, good results cannot be obtained because of the data imbalance problem. Therefore, we generate synthetic malicious data points through SMOTE to achieve better results. We use XGboost and random forest to classify the model and calculate the confusion matrix. This classification allows the model to distinguish between fraudulent and real data. The simulation results show that the proposed algorithm works adequately to find transaction fraud. Moreover, two attacker models are implemented to check the efficacy of the system against bugs and attacks. The proposed system is robust against double-spending and Sybil attacks.

A major limitation of our proposal is that it can be affected by the adversarial attack described in Section 2.1; we leave it to future work to address such a threat.

Author Contributions: Conceptualization, S.A. (Safa Alsafari); Data curation, R.K. and S.A. (Safa Alsafari); Formal analysis, R.K., A.S.Y., S.A. (Sheraz Aslam) and S.A. (Safa Alsafari); Investigation, T.A., R.K., A.T.A. and I.A.H.; Methodology, T.A. and S.A. (Safa Alsafari); Project administration, S.A. (Sheraz Aslam) and A.T.A.; Resources, I.A.H.; Supervision, S.A. (Sheraz Aslam), I.A.H. and A.T.A.; Validation, A.S.Y.; Visualization, A.T.A. and I.A.H.; Writing—original draft, T.A., A.S.Y., R.K. and S.A. (Safa Alsafari); Writing—review & editing, T.A., R.K., A.S.Y., S.A. (Sheraz Aslam), S.A. (Safa Alsafari), A.T.A. and I.A.H. All authors have read and agreed to the published version of the manuscript.

Funding: This research was funded by Norwegian University of Science and Technology.

Institutional Review Board Statement: Not applicable.

Informed Consent Statement: Not applicable.

Data Availability Statement: Not applicable.

Acknowledgments: The authors would like to acknowledge the support of Norwegian University of Science and Technology for paying the Article Processing Charges (APC) of this publication. Special acknowledgment to Automated Systems & Soft Computing Lab (ASSCL), Prince Sultan University, Riyadh, Saudi Arabia. In addition, the authors wish to acknowledge the editor and anonymous reviewers for their insightful comments, which have improved the quality of this publication.

Conflicts of Interest: The authors declare no conflict of interest.

Abbreviations

List of abbreviations:

Abbreviation	**Full Form**
AI	Artificial Intelligent
ANN	Artificial Neural Network
CPS	Cyber–Physical System
DBF	Deep Blockchain Framework
DTR	Decision Tree Regression
ePoW	enhanced Proof of Work
GMM	Gaussian Mixture Model
IoT	Internet of Things
LSTM	Long Short-Term Memory
MLO	Mixture Localization-based Outliers

MLP	Multi Layer Perceptron
PCA	Principle Component Analysis
RFE	Recursive Feature Elimination
SDA	Stacked De-noising Autoencoders
SMOTE	Synthetic Minority Oversampling Technique
SVD	Singular Value Decomposition
SVM	Support Vector Machine
XGboost	eXtreme Gradient Boosting

List of acronyms:

Abbreviation	**Full Form**
C_N	Catch-up function
K	Number of confirmation to declare a block
m	Honest nodes mine the block
n	Attackers mine the block
P_N	Potential progress function
q	Probability of attack
T	Time needed for mining
t	Time advantage for the attackers
τ	Average time for the mining of block
x	Available computational power in network
z	Initial disadvantage of attacker

References

1. Staudemeyer, R.C.; Voyiatzis, A.G.; Moldovan, G.; Suppan, S.R.; Lioumpas, A.; Calvo, D. Smart cities under attack. In *Human-Computer Interaction and Cybersecurity Handbook*; CRC Press: Boca Raton, FL, USA, 2018.
2. Podgorelec, B.; Turkanović, M.; Karakatič, S. A machine learning-based method for automated blockchain transaction signing including personalized anomaly detection. *Sensors* **2020**, *20*,147. [CrossRef] [PubMed]
3. Nakamoto, S. Bitcoin: A Peer-to-Peer Electronic Cash System. 2008. Available online: https://bitcoin.org/bitcoin.pdf (accessed on 21 March 2020).
4. Farrugia, S.; Ellul, J.; Azzopardi, G. Detection of illicit accounts over the Ethereum blockchain. *Expert Syst. Appl.* **2020**, *150*, 113318. [CrossRef]
5. Ostapowicz, M.; Żbikowski, K. Detecting fraudulent accounts on blockchain: A supervised approach. In Proceedings of the International Conference on Web Information Systems Engineering, Hong Kong, China, 19–22 January 2020; Springer: Cham, Switzerland, 2020; pp. 18–31.
6. Aziz, A.S.A.; Hassanien, A.E.; Azar, A.T.; Hanafy, S.E. Genetic Algorithm with Different Feature Selection Techniques for Anomaly Detectors Generation. In Proceedings of the 2013 Federated Conference on Computer Science and Information Systems (FedCSIS), Kraków, Poland, 8–11 September 2013.
7. Hassanien, A.E.; Tolba, M.; Azar, A.T. Advanced Machine Learning Technologies and Applications: Second International Conference, AMLTA 2014, Cairo, Egypt, 28–30 November 2014. In *Communications in Computer and Information Science*; Springer: Berlin/Heidelberg, Germany, 2014; Volume 488, ISBN 978-3-319-13460-4.
8. Khan, H.; Asghar, M.U.; Asghar, M.Z.; Srivastava, G.; Maddikunta, P.K.R.; Gadekallu, T.R. Fake review classification using supervised machine learning. In Proceedings of the International Conference on Pattern Recognition, Virtual Event, 10–15 January 2021; Springer: Cham, Switzerland, 2021; pp. 269–288.
9. Shahbazi, Z.; Hazra, D.P.; Park, S.; Byun, Y.C. Toward Improving the Prediction Accuracy of Product Recommendation System Using Extreme Gradient Boosting and Encoding Approaches. *Symmetry* **2020**, *12*, 1566. [CrossRef]
10. Pesantez-Narvaez, J.; Guillen, M.; Alcañiz, M. Predicting motor insurance claims using telematics data—XGBoost versus logistic regression. *Risks* **2019**, *7*, 70. [CrossRef]
11. Li, J.; Gu, C.; Wei, F.; Chen, X. A Survey on Blockchain Anomaly Detection Using Data Mining Techniques. In Proceedings of the International Conference on Blockchain and Trustworthy Systems, Guangzhou, China, 7–8 December 2019; Springer: Singapore, 2019.
12. Reid, F.; Harrigan, M. An analysis of anonymity in the bitcoin system. In *Security and Privacy in Social Networks*; Springer: New York, NY, USA, 2013; pp. 197–223.
13. Ngai, E.W.T.; Hu, Y.; Wong, Y.H.; Chen, Y.; Sun, X. The application of data mining techniques in financial fraud detection: A classification framework and an academic review of literature. *Decis. Support Syst.* **2011**, *50*, 559–569. [CrossRef]
14. Saia, R.; Carta, S. Evaluating Credit Card Transactions in the Frequency Domain for a Proactive Fraud Detection Approach. In Proceedings of the 14th International Conference on Security and Cryptography (SECRYPT 2017), Madrid, Spain, 26–28 July 2017; pp. 335–342.

15. Sánchez, D.; Vila, M.A.; Cerda, L.; Serrano, J.M. Association rules applied to credit card fraud detection. *Expert Syst. Appl.* **2009**, *36*, 3630–3640. [CrossRef]
16. Gyamfi, N.K.; Abdulai, J.D. Bank fraud detection using support vector machine. In Proceedings of the 2018 IEEE 9th Annual Information Technology, Electronics and Mobile Communication Conference (IEMCON), Vancouver, BC, Canada, 1–3 November 2018; pp. 37–41.
17. Panigrahi, S.; Kundu, A.; Sural, S.; Majumdar, A.K. Credit card fraud detection: A fusion approach using Dempster–Shafer theory and Bayesian learning. *Inf. Fusion* **2009**, *10*, 354–363. [CrossRef]
18. Shi, F.B.; Sun, X.Q.; Gao, J.H.; Xu, L.; Shen, H.W.; Cheng, X.Q. Anomaly detection in Bitcoin market via price return analysis. *PLoS ONE* **2019**, *14*, e0218341. [CrossRef]
19. Kumar, P.; Gupta, G.P.; Tripathi, R. TP2SF: A Trustworthy Privacy-Preserving Secured Framework for sustainable smart cities by leveraging blockchain and machine learning. *J. Syst. Archit.* **2021**, *115*, 101954. [CrossRef]
20. Zhao, Y.; Tarus, S.K.; Yang, L.T.; Sun, J.; Ge, Y.; Wang, J. Privacy-preserving clustering for big data in cyber-physical-social systems: Survey and perspectives. *Inf. Sci.* **2020**, *515*, 132–155. [CrossRef]
21. Alkadi, O.; Moustafa, N.; Turnbull, B.; Choo, K.K.R. A deep blockchain framework-enabled collaborative intrusion detection for protecting IoT and cloud networks. *IEEE Internet Things J.* **2020**, *8*, 9463–9472. [CrossRef]
22. AlKadi, O.; Moustafa, N.; Turnbull, B.; Choo, K.K.R. Mixture localization-based outliers models for securing data migration in cloud centers. *IEEE Access* **2019**, *7*, 114607–114618. [CrossRef]
23. Keshk, M.; Sitnikova, E.; Moustafa, N.; Hu, J.; Khalil, I. An integrated framework for privacy-preserving based anomaly detection for cyber-physical systems. *IEEE Trans. Sustain. Comput.* **2019**, *6*, 66–79. [CrossRef]
24. Kurakin, A.; Goodfellow, I.; Bengio, S. Adversarial machine learning at scale. *arXiv* **2016**, arXiv:1611.01236. [CrossRef]
25. Biggio, B.; Roli, F. Wild patterns: Ten years after the rise of adversarial machine learning. *Pattern Recognit.* **2018**, *84*, 317–331.
26. Xuan, S.; Liu, G.; Li, Z.; Zheng, L.; Wang, S.; Jiang, C. Random forest for credit card fraud detection. In Proceedings of the 2018 IEEE 15th International Conference on Networking, Sensing and Control (ICNSC), Zhuhai, China, 27–29 March 2018; pp. 1–6. [CrossRef]
27. Liu, C.; Chan, Y.; Alam Kazmi, S.H.; Fu, H. Financial fraud detection model: Based on random forest. *Int. J. Econ. Financ.* **2015**, *7*, 178–188.
28. Apruzzese, G.; Andreolini, M.; Colajanni, M.; Marchetti, M. Hardening random forest cyber detectors against adversarial attacks. *IEEE Trans. Emerg. Top. Comput. Intell.* **2020**, *4*, 427–439. [CrossRef]
29. Primartha, R.; Tama, B.A. Anomaly detection using random forest: A performance revisited. In Proceedings of the 2017 International Conference on Data and Software Engineering (ICoDSE), Palembang, Indonesia, 1–2 November 2017; pp. 1–6. [CrossRef]
30. Laskov, P. Practical evasion of a learning-based classifier: A case study. In Proceedings of the 2014 IEEE Symposium on Security and Privacy, San Jose, CA, USA, 18–21 May 2014; pp. 197–211.
31. Pham, T.; Lee, S. Anomaly detection in bitcoin network using unsupervised learning methods. *arXiv* **2016**, arXiv:1611.03941.
32. Martin, K.; Rahouti, M.; Ayyash, M.; Alsmadi, I. Anomaly detection in blockchain using network representation and machine learning. *Secur. Priv.* **2022**, *5*, e192.
33. Pinzón, C.; Rocha, C. Double-spend attack models with time advantange for bitcoin. *Electron. Notes Theor. Comput. Sci.* **2016**, *329*, 79–103. [CrossRef]
34. Bitcoin Network Transactional Metadata. Available online: https://www.kaggle.com/datasets/omershafiq/bitcoin-network-transactional-metadata (accessed on 14 September 2022). [CrossRef]
35. Shafiq, O. Anomaly Detection in Blockchain. Master's Thesis, Tampere University, Tampere, Finland, 2019.
36. Chawla, N.V.; Bowyer, K.W.; Hall, L.O.; Kegelmeyer, W.P. SMOTE: Synthetic minority over-sampling technique. *J. Artif. Intell. Res.* **2002**, *16*, 321–357.
37. Sadaf, K.; Sultana, J. Intrusion detection based on autoencoder and isolation Forest in fog computing. *IEEE Access* **2020**, *8*, 167059–167068. [CrossRef]
38. Eyal, I.; Sirer, E.G. Majority is not enough: Bitcoin Mining is vulnerable. In Proceedings of the International Conference on Financial Cryptography and Data Security, Christ Church, Barbados, 3–7 March 2014; pp. 436–454; Springer: Berlin/Heidelberg, Germany, 2014. [CrossRef]
39. Landa, R.; Griffin, D.; Clegg, R.G.; Mykoniati, E.; Rio, M. A Sybilproof indirect reciprocity mechanism for peer-to-peer networks. In Proceedings of the IEEE INFOCOM 2009, Rio De Janeiro, Brazil, 24 April 2009; pp. 343–351.
40. Luu, L.; Chu, D.-H.; Olickel, H.; Saxena, P.; Hobor, A. Making smart contracts smarter. In Proceedings of the 2016 ACM SIGSAC Conference on Computer and Communications Security, Vienna, Austria, 24–28 October 2016.
41. Nizamuddin, N.; Hasan, H.; Salah, K.; Iqbal, R. Blockchain-based framework for protecting author royalty of digital assets. *Arab. J. Sci. Eng.* **2019**, *44*, 3849–3866.
42. Halo Block, Medium. How To Use Oyente, a Smart Contract Security Analyzer—Solidity Tutorial. 2020. Available online: https://medium.com/haloblock/how-to-use-oyente-a-smart-contract-security-analyzer-solidity-tutorial-86671be93c4b (accessed on 13 April 2020). [CrossRef]

sensors

MDPI

Review

Adaptation of IoT with Blockchain in Food Supply Chain Management: An Analysis-Based Review in Development, Benefits and Potential Applications

Amanpreet Kaur [1], Gurpreet Singh [2], Vinay Kukreja [1], Sparsh Sharma [3], Saurabh Singh [4] and Byungun Yoon [4,*]

1 Chitkara University Institute of Engineering & Technology, Chitkara University, Rajpura 173212, Punjab, India
2 Department of Computer Science & Engineering, Punjab Institute of Technology, Rajpura 140401, Punjab, India
3 Department of Computer Science and Engineering, National Institute of Technology, Srinagar 190001, Jammu and Kashmir, India
4 Department of Industrial and Systems Engineering, Dongguk University, Seoul 04620, Korea
* Correspondence: postman3@dongguk.edu

Abstract: In today's scenario, blockchain technology is an emerging area and promising technology in the field of the food supply chain industry (FSCI). A literature survey comprising an analytical review of blockchain technology with the Internet of things (IoT) for food supply chain management (FSCM) is presented to better understand the associated research benefits, issues, and challenges. At present, with the concept of farm-to-fork gaining increasing popularity, food safety and quality certification are of critical concern. Blockchain technology provides the traceability of food supply from the source, i.e., the seeding factories, to the customer's table. The main idea of this paper is to identify blockchain technology with the Internet of things (IoT) devices to investigate the food conditions and various issues faced by transporters while supplying fresh food. Blockchain provides applications such as smart contracts to monitor, observe, and manage all transactions and communications among stakeholders. IoT technology provides approaches for verifying all transactions; these transactions are recorded and then stored in a centralized database system. Thus, IoT enables a safe and cost-effective FSCM system for stakeholders. In this paper, we contribute to the awareness of blockchain applications that are relevant to the food supply chain (FSC), and we present an analysis of the literature on relevant blockchain applications which has been conducted concerning various parameters. The observations in the present survey are also relevant to the application of blockchain technology with IoT in other areas.

Keywords: blockchain; food supply chain; IoT; cloud computing

1. Introduction

The growth of agricultural crops and the management of logistics in food must be stringently monitored. The supply chain of agricultural products and crops is a critical aspect related to product safety; the risk of food spoilage and potential poisoning has led to the increased focus on traceability enhancement. Agricultural food and products are particularly vulnerable, and consumers are quite concerned about the quality, nutritive value, and safety of the food they consume. In fact, the food crisis was ranked as the seventh-highest risk in the year 2018 by the World Economic Forum [1]. Furthermore, with increasing globalization, the international trade market has flourished, and food grains and related products are being traded across multiple countries, requiring intensive tracking. In supply chain management (SCM), the traceability of agricultural products and crops requires robust information management, communications, and the collection of data related to products (e.g., origins and crop exchange information) over their entire life cycle, which can be challenging. The Public Health Organization has observed that product traceability is an essential policy tool for food quality management and monitoring. Dabbene and Gay [2] emphasized enabling product traceability through various tools such as RFID and Bar Codes [3]. In SCM, RFID [4] technology has been employed to facilitate tracking, reduce food wastage, increase operational efficiency, collect data, control aspects such as temperature and humidity, and prevent risks related to shipping and picking.

Cloud computing is now being used for storing information such as the details of food products, customers, and retailers, which can be accessed by users from various websites or barcode scans using mobile phones or other gadgets [5]. Cloud computing also provides instant or small messaging services relating to agricultural products, such as government subsidies, alerts of disease outbreaks, weather conditions, and pesticide information [6]. Cloud computing can be applied at the granular level in the process of keeping food safe by providing the opportunity to analyze and observe the product status from source to destination, i.e., end-to-end delivery. The use of cloud computing and big data is revolutionizing the entire food industry; the cloud structure provides the backbone to analyze and collect data throughout the food supply chain, right from the fields where the crops grow, the warehouses where food is stored, the containers that ship it, to the consumer. They provide a possible alternative to expensive investments relating to hardware and software, allowing the industry to react faster to shifting environments in the marketplace and to gain a competitive advantage. Currently, traceability in the agriculture supply chain suffers from data fragmentation and centralized controls, which cause challenges in data modification and management. Identifying the source and swiftly isolating the product from the supply chain requires close coordination among multiple stakeholders. Individual stages in food supply chains often have good traceability, but the exchange of data and information between stages proves to be difficult to capture and time-consuming.

Many researchers have presented integrations of the food supply chain and blockchain with emerging technologies such as the Internet of things, cloud computing [6], big data, analysis through case studies [7], and survey techniques [8]. These approaches resulted in certain enhancements such as improving traceability efficiency and enhancing transparency in supply chains among users, as well as certain challenges including scalability, immature technologies, lack of legislation, and so on [9]. Blockchain technology is considered a digital ledger that is distributed and organized by a network of various computing devices and machines. A blockchain preserves essential data or information in small chunks called blocks that are secure and cryptographically immutable. Blockchain was first reported in 2008 and is the brainchild of Satoshi Nakamoto. Specifically, the concept of decentralization of the peer-to-peer ledger in 2008 was introduced by Nakamoto. Blockchain technology allows users and suppliers to check transaction details in a real-time environment, as it is used to collect data relating to all transactions occurring within a specific period. Thus, blockchain creates a digital footprint for the verification and validation of data and information [10]. Each blockchain contains several blocks, and each block contains information about the succeeding block in chronological order, as well as the hash of the previous block. Blockchain technology is associated with artificial intelligence (AI) technologies that resolve issues relating to trust, traceability, security, and collaboration in SCM. It is employed in various applications, such as VeChain for certification, Walton chain for apparel supply chains, Ambrosus for food and medicine supply chains, and Modem, exclusively for pharma supply chains. Blockchain technology has been used in the financial domain as the foundation of fully distributed cryptocurrencies such as Bitcoin and in peer-to-peer (or point-to-point) electronic cash systems. It has also aroused interest in various other areas including the food supply chain, medicine, education, e-commerce, real estate, voting systems, and so on.

1.1. Blockchain-Related Components

Given below are the main Blockchain Architecture components also shown pictorically in Figure 1:

- **User or node**—end user or node within the blockchain.
- **Transactions (deals)**—smallest chunks or building blocks for the blockchain transaction system.
- **Blocks or chunks**—a data structure used for preserving a set of transactions, which is distributed to all nodes in the present network.
- **Purchase Chain**—a sequence of blocks related to a purchase order.
- **Miners**—a specific type of user who performs the block verification and validation processes.
- **Consensus**—blockchain operations carried out after verification according to the set of rules and arrangements.

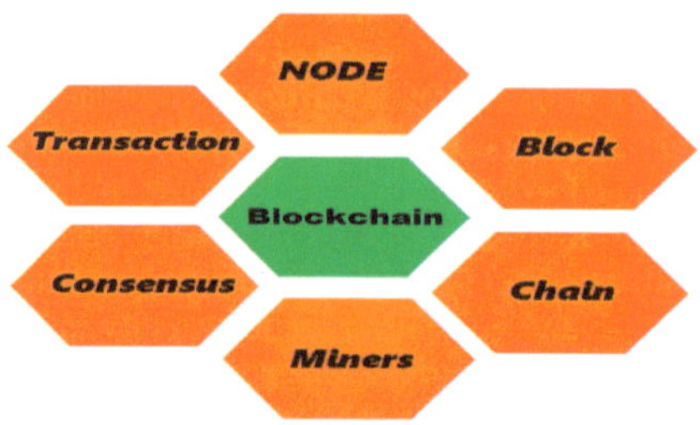

Figure 1. Components of Blockchain.

1.2. Features of Blockchain

- **Immutability:** Blockchain technology provides the essential benefit that once a user enters information or data into the blockchain, it cannot be updated or modified during the entire transaction process. This characteristic is called immutability, and it has made blockchain technology very popular. Consequently, blockchain is being used in all sectors where data integrity, data security, and data protection are of utmost importance.
- **Autonomy:** Blockchain provides the ability to take decisions individually without intervention by others. It allows the manufacturing and delivery of devices smartly, with IoT-based devices for quick and autonomous decisions in transactions.
- **Decentralization:** All transactions of authorized users can be completed over the internet and accessed without any previous intervention. Every registered user has the same ability to monitor and observe the transaction and prepare copies of all transactions [11]. This information will never be changed without other users being intimated [12]. In contemporary internet-based systems, the entire information of a transaction is not saved only on one single server; copies of the transaction data are saved in distributed computers that are considered "nodes" in the blockchain without any supervisory central authority. Then, all the computers are connected to the blockchain network, which is called a distributed ledger because of the distributed data.
- **Smart Contracts:** A smart contract works as a digitalized contract, and after certain agreements, it operates automatically [13]. In actual fact, a smart contract is a computerized transaction protocol that enhances trust and speeds up transactions [14,15]. For example, once a product is developed and received at the warehouse, payment is made automatically. Using smart contracts, developers can reduce processing time, manpower, paperwork, and other resources. In a new observation, Maersk observed that more than 30 people and organizations were involved in the shipping of containers containing roses, avocados, and other perishable goods from Kenya to the Netherlands in the year 2014 [16]. The entire task was managed using smart contracts only; no human intervention was required, and the entire process took ten days. Smart contracts cannot be changed by humans; they are based on the agreements between partners.

Blockchain reduces the risk of transactions at all levels and increases the supply chain visibility, reliability, and transparency while protecting stakeholders' benefits.

- **Transparency:** Blockchain technology provides a clear and transparent environment. No third party is required as a mediator to provide trust between different parties related to data transactions. Furthermore, even the identities of those involved are hidden with the help of a complex cryptography technique.

This review covers aspects such as how blockchain has been used in the food supply chain and how it can help to address food security and humidity-effect issues. The following questions have been considered:

Question 1: What studies have been conducted on the blockchain with IoT adoption in food supply chain management (FSCM)?
Question 2: What are the benefits of using blockchain in the food supply chain?
Question 3: What are the various challenges of blockchain adoption in FSCM?
Question 4: How does blockchain provide control over FSCM?

The above questions have been answered by collating related papers and summarizing an analysis of the literature. These papers apply a content-analysis-based literature review methodology. Here, we introduce a literature review to provide some background information about the key concepts followed by our research methodology. We then present and discuss the findings, and finally, conclude and discuss future research directions.

2. Literature Review

Many researchers have discussed the application of blockchain technology to the food supply chain (FSC). Wang et al. state that FSCs provide a way to design, manage, develop, transition, and systematically organize the food system. Bosona and Gebresenbet [17] explored the notion of food traceability with regard to collecting, storing, transmitting, and preserving food product information throughout the various levels of the FSC using blockchain technology. This further provides control over food quantity, quality, and safety in an FSC. Shih et al [18] described how trading partners for an FSC could preserve a record relating to food transactions. All transactions would be controlled by the trading partners. Lin et al. [19] introduced a traceability system relating to food safety and security, which is based on blockchain technology and a GS1 global standard enabling interoperability (EPCIS). It is specifically used for the process of acquisition, management, and the exchange of product information over the internet. With the help of this system, customers can trace food information through the consumer traceability client application. He et al. [20] developed a nonreversible and decentralized data storage approach related to food. Alonso et al. [21] introduced a unique platform that joins artificial intelligence with IoT, using blockchain technology and edge computing for the management of farming. After using blockchain technology, maintaining food traceability from one end to the other and fragmentation are no longer a challenge. It creates and generates a common platform for data collection. In blockchain technology, consumers can rapidly trace the food forward as well as backward using all the blocks related to products. In another research, "a set of interdependent companies that work closely together to manage the flow of goods and services along the value-added chain of agricultural and food products, to realize superior customer value at the lowest possible costs" was studied. We start the comparison with other companies, then find food products such as producing, by Folkerts and Ko horse [22].

Today, FSCs are centralized and rely on central powers for handling the data and information flow related to food. Several studies have commented on the lack of continuous observation of the FSC and its inability to predict the freshness of the food [23]. Similarly, the conventional food supervision system suffers from various factors, such as inconsistencies in data, interoperability with insufficient resources, fragmentation, and lack of transparency. To mitigate the above concerns, FSC practitioners have adopted various applications related to blockchain technology in the food industry. Near-infrared spectroscopy can be used as a speedy scheme for the evaluation of physicochemical changes in stored foods such as soybeans [24]. Near-infrared spectroscopy (NIRS) is an effective approach for the chemical characterization and screening of agricultural crops. David et al. [25] depict the importance of blockchain technology feasibility in FSCM. With the help of this, organizations can achieve integrity among connected nodes, such as proof of work maintenance, needs, traceability, innovation to reduce intermediaries, and so on. The papers mention a literature review related to sustainable supply chain management (SSCM) that includes the critical factors and performance for developing a comprehensive model of sustainable supply chain management (SSCM) in the food supply chain management industry [26,27]. This paper elaborates on the state-of-the-art relating to blockchain and its applications in supply chain financing and trading. It emphasizes the areas of the blockchain where it may identify the value of trading and supply chain finance [28]. Blockchain technology is a highly growing field in supply chain management, cyber security, banking, and healthcare [29].

Based upon the literature survey, we have identified the few parameters which have been discussed by various papers and it is shown in Table 1. Blockchain is impacting upcoming supply chain practices as well as policies by providing visibility and traceability. Blockchain has the potential to improve as well as enhance the traditional supply chain processes after imposing its own rules, and governance mechanism. In another paper [30], the author discussed how blockchain and IoT-based systems promote value transfer in smart and small-scale agricultural farms. Authors in [31] introduced a method to increase transparency and automate the use of blockchains in agriculture. In another paper [32], the author introduced the various challenges in the implementation of blockchains in the dairy industry. In another paper [33], the author discussed traceability and, with the help of Hyperledger, improved the traceability of a blockchain.

Table 1. Literature review for the supply chain.

Research Paper	Customer Intervention	Implementation System	Froud Detection	Traceability	Price Transparency	Original Database	Based on IoT
[33]	√			√	√	√	
[34]		√	√	√	√	√	√
[35]	√			√		√	√
[36]		√	√	√			
[37]					√		√
[38]	√		√	√		√	
[39]			√	√		√	√
[40]	√		√	√		√	√
[41]	√	√		√			
[42]					√		√
[43]		√	√			√	
[44]				√			
[45]	√	√	√	√			
[46]	√				√		
[47]				√		√	√
[48]	√	√		√		√	
[49]	√			√			√
[50]	√	√					√
[1]	√		√	√		√	√
[51]	√			√		√	
[52]		√		√		√	
[25]	√			√		√	

The estimated acceleration in the population of the world and the associated requirement of food from fields to market has also now been sensed with the Internet of underground things (IOUT) [53], by using sensor devices and IoT techniques. The challenge of merging such technologies also requires smart communication between these devices [54]. Due to the fast evolution of IoT, low-power wide-area technologies are becoming popular due to the power concern of these devices. Material conscious information networks (MCIN) are the newly developing techniques that define smart agriculture architecture and also deal with concerns such as management and commerce which affect the supply chain system [55].

3. Traditional Supply Chain Management

Traditional supply chain management is very simple and smooth because of clear requirements in terms of design, plans, manufacture, and delivery. Traditional supply chain management (TSCM) is the best practice for the synchronization of the flow of importing and transferring raw materials from supplier to consumer. There is a route from supplier to consumer, e.g., supplier to manufacturer to wholesaler to retailer to consumer. This complete route involves the following stages: generation of order, order collection, gathering information, and timely distribution of goods and services to the consumer. There are 5 V's (volatility, volume, velocity, visibility, and veracity) relating to SCM to improve the outcomes with several objectives relating to aspects such as services, support, total expenditure, and so on. These objectives can be achieved by the supply manager after applying new digital techniques in the enhancement of supply chain technologies. Furthermore, supply chain managers create additional sources of revenue by providing new access to markets for the creation of smart products. The main function of supply chain centers are movements related to the transactions of raw materials, capital, and finished goods from one place to another. However, the traditional supply chain has the following limitations:

- Traditional supply chains have a limited view of work.
- Delays and unsynchronized responses because of variations in planning.
- Delayed information while passing through each organization.
- The entire chain has limited visibility.
- As information flows, the end customer demands distortion.

Supply chain management involves maintaining information relating to transactions such as the exchange of money, time, and physical materials. An FSC involves various processes which take food from the farm to the dinner table. This includes multiple stages such as the manufacturing, administration, utilization, supplying, and discarding of food products. Food products travel from manufacturers to consumers via workers who work in various stages of the supply chain. At every level of supply chain operations, man-made resources are required to pass the food item toward its destination. Thus, it is essential to streamline the entire supply chain process to prevent high costs, discrepancies, or inefficiencies. The six stages of the food supply chain include:

- **Seed purchasing:** Various food seeds are purchased from seed companies for sale to farmers.
- **Farming**: Ingredients, fruits, meat, vegetables, and beverages originate and are purchased.
- **Processing**: Plants and animals are converted into edible forms.
- **Distributing:** Retailers and suppliers purchase the food in its final form and further transport it. Distributors sell items, manage inventories, reduce costs, and maintain ledgers to give value to food products.
- **Retailing**: The food product is delivered to the final consumers.
- **Food product purchasing**: The final stage of FSC where the consumer purchases the finished product from the retailer.

In the Figure 2, a simple supply chain is shown. Seed companies will provide raw materials like seeds, plants, and so on. Farmers [56] purchase raw materials and use them for cultivation. After harvesting, the foodstuffs will further move toward processing. During processing, factories process foodstuffs and forward the useful portions for distribution in the next phase. Distributors sell this food to retailers, and customers then purchase items from retailers.

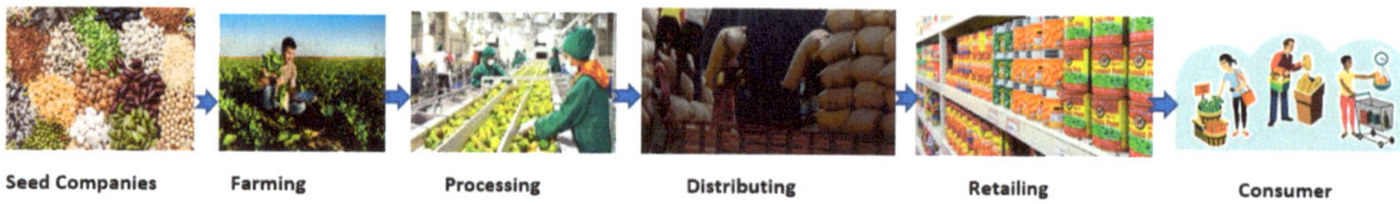

Figure 2. Supply chain management.

4. Blockchain with IoT Devices

The Internet of things (IoT) is an intelligent, reliable, and high-speed information network that connects objects for data collection and transmission. IoT includes radio-frequency identification (RFID), a global positioning system (GPS), a geographic information system (GIS), a wireless sensor network (WSN), and so on. IoT provides the facility of automatic recording through IoT sensors which collect information such as temperature, voice, and humidity. In SCM, IoT provides real-time data collection for fresh food products, which is used to identify the quality of the product concerning the external environment [57]. IoT devices help to eradicate human error and increase the efficiency of monitoring and capturing information. IoT in conjunction with blockchain enables smooth transactions and the monitoring of data transfer through blocks. There are several emerging areas of IoT with blockchain.

4.1. IoT for Healthcare

In recent years, healthcare services have undergone drastic changes in response to increased demand. There are several wearable devices [46] that can identify the medical state of patients. Wearable devices can determine a patient's blood pressure, blood glucose, breathing problems, heartbeat, and so on [58]. However, while wearable devices can be used to transfer and collect data only in hospitals, IoT devices must be used to monitor patients remotely, for which remote monitoring systems are needed. Some of the devices used for healthcare monitoring are:

- Stationary medical devices;
- Medical embedded devices;
- Medical wearable devices;
- Wearable health monitoring devices;
- Glucose monitoring devices;
- Hand hygiene monitoring devices;
- Depression and mood monitoring.

Stationary medical devices are physically installed in specific locations. Embedded devices are placed inside the patient's body. Medical wearable devices, prescribed by doctors, and wearable health monitoring devices are used to monitor the health of the patient and then relay that information to the concerned persons. Blockchain technology works with IoT-enabled systems to maintain the medical records of patients. A blockchain [39] maintains the ledger of data relating to a patient, and their doctor can use this ledger to extract the data.

4.2. IoT for Smart Homes

Smart Homes provide an automated and intelligent system to protect homes from outsiders. Through the Internet of things (IoT), smart electronic devices such as Smart TVs, LED lights, microwaves, AC, and refrigerators can be connected through the internet, and humans can control various household activities. IoT-based smart homes feature a one-to-one communication system between various devices without human intervention. In IoT-based smart homes, the gateway for communication can be configured using blockchain technology which can be used to store and exchange data in the form of chunks or blocks. Blockchain technology [41] also provides decentralization support to overcome issues arising in traditional centralized architecture.

The number of smart homes is increasing every year, and the price for the development of IoT-based networks is also increasing. Surveys claim that worldwide consumer spending in the smart home sector added up to nearly $90 billion in 2014 and $213 billion in 2019, and it is expected to grow at 10% CAGR to $525 billion over the forecasted period from 2019 to 2025 as can be seen in Table 2. Figure 3 shows the year-wise increase in the count of the smart IoT and blockchain based smart devices.

Table 2. Annual count increases and annual spending.

Years	Annual Increase	Annual Spending Per Count
2014	90	$48
2015	120	$60
2016	145	$72
2017	164	$80
2018	189	$96
2019	213	$108
2020	235	$121
2021	255	$132
2022	274	$143
2023	293	$155
2024	340	$167
2025	525	$201

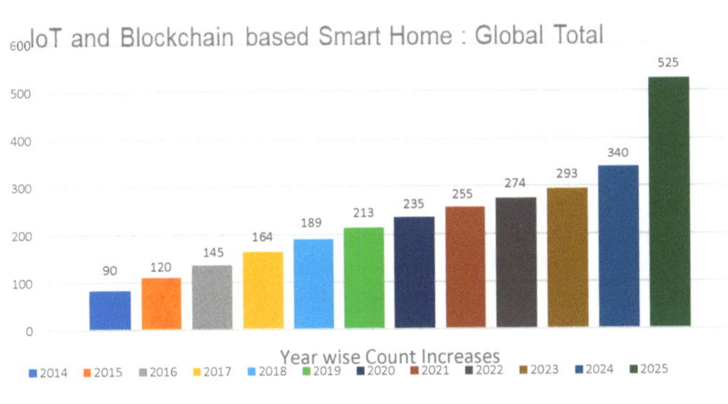

Figure 3. IoT and Blockchain-based Smart Homes.

As an example, in smart homes, we might have a smart refrigerator that may indicate the status of the left-over food in the refrigerator, such as a small egg tray, and may send the status signal of left-over eggs to the owner or to the shopkeeper from where the materials are normally purchased. Similarly, a milk bottle may indicate the amount of milk left in the bottle to the shopkeeper for the supply of another bottle(s). As another example, the smart storage food tray may indicate the freshness status of vegetables stored in it and this way items may be supplied directly to the home.

4.3. IoT for Supply Chain Management

At present, SCM [31,34] is an emerging area of blockchain technology. SCM can be effective if it provides reliable visibility into materials and goods, from the production level to the delivery of the product. IoT has enabled manufacturing companies to achieve such visibility by using smart SCM which is IoT-enabled to optimize production as well as good transmission. Smart SCM is improving the customer delivery service as well. In such systems, multiple IoT-enabled sensors work in various stages of the supply chain to provide functionalities such as real-time monitoring throughout the supply chain, which can be used to monitor raw goods and materials quality. These sensors identify and respond according to real-time changes happening in an environment. Sensors receive inputs from a variety of sources such as light, temperature, motion, and pressure. Sensor technology has made it easy for businesses to transition to IoT-enabled supply chains:

- **Customer satisfaction as per requirements**: At the time of manufacturing, companies identify food that requires special arrangements for shipping and transport. Smart SCM maintains regulatory compliance and high quality throughout the shipment transport.
- **Safeguarding:** Smart IoT food supply chain monitoring systems offer quality identification, consumer assurance, and food validation after meeting customer expectations.
- **Cost-effectiveness**: Smart SCM provides a cost-effective IoT-based system to identify the food quality and shipment from manufacturing to delivery.
- **Maintaining Integrity:** While manufacturing and distributing, certain foods, products, and pharmaceuticals require special preservation to prevent harm. An IoT-based SCM works to maintain the integrity of perishable food. Smart supply chains preserve human health and eliminate waste through stringent monitoring and alerts.
- **Trackability:** IoT devices trace the food freshness level at every transaction point. These collect the food condition information and transmit it to the cloud for further storage. The blockchain collects this information and maintains a ledger for the particular food. Consequently, the transporter and seller can check the condition of the food at every stage.

In Figure 4 shown that modern supply chains have numerous entities, such as devices, automation, transportation, onboard units, services, and so on, with proper components and capacity. Each level of the SCM has been identified and coordinated, and transparency between them is provided. Traditional supply chains work independently at each stage: manufacturing, marketing, distribution, planning, and finally, organizational purchasing. However, individual components of supply chains can be directly connected using IoT devices. IoT devices receive data or information from sensors installed at various stages in the supply chain. This data is then transmitted to a cloud [29] for permanent storage, as shown in the above figure. Researchers observed that combining blockchain technology with IoT effectively provides scalability, security, efficiency, auditing, quality, and interoperability. A self-organized blockchain and IoT-based agriculture system that worked without human intervention was introduced by Lin et al. [47]. In IoT and blockchain environments, tracking initializes from business partners to the end consumer. This whole task is achieved through cloud-based technology where various aspects such as events, status, location, and conditions are monitored. Then, status updates are stored in the status table for further perusal. The location and condition of an event can also be tracked by the cloud throughout the business onboarding, to consumer purchasing and receiving. IoT devices include sensors that record conditions such as temperature and humidity. The collected information is stored in a cloud that is remote and secure. Cloud-based technology associated with blockchain and IoT performs various operations such as maintaining food pedigree and safety, multi-tier logistics network visibility, supply chain integrity, real-time inventory management, real-time planning predictions, and so on. The blockchain creates ledgers for the data and maintains a record of food. However, blockchain technology has certain challenges, e.g., there is no way to change or update data in a blockchain if mistakes occur. It is difficult to manipulate and amend data or information in blockchains. Various security transaction schemes do exist but are difficult to verify. Installed IoT devices can be hacked by hackers.

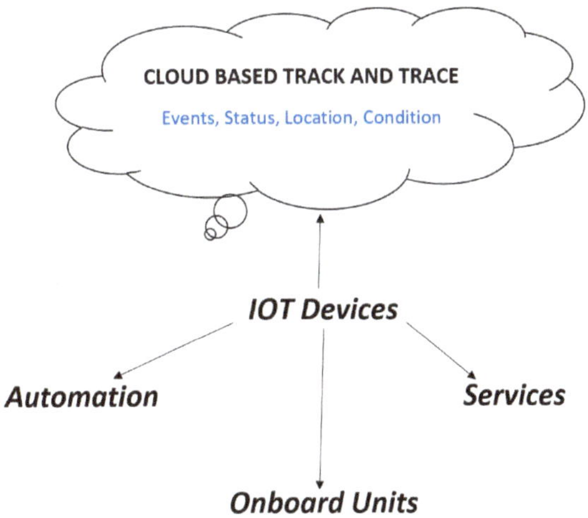

Figure 4. SCM based on IoT and blockchain.

5. Conclusions

A brief literature review is essential to provide a quick analysis of the published research. This review can help researchers determine future directions in developing applications involving IoT with blockchain technology in FSCM. Researchers can observe the challenges and potentials of blockchain in the FSC. In this paper, 60 research papers (journal articles, conference papers, and book chapters) relating to blockchain and IoT have been reviewed and some observations have been presented. Applications of IoT and blockchain in other fields, such as healthcare and smart homes, have also been presented. Blockchain and smart contracts are used to track and trace the performance of business transactions after eliminating intermediaries. At each stage of transactions, IoT devices observe and record the food condition. This information is further stored on the cloud and the blockchain creates a ledger. Blockchains have improved food traceability and enhanced FSC trading activities through collaborative relationships and by maximizing operational efficiencies. However, the study findings are that the blockchain falls under four basic categories: Member name, technical issue, organizational effect, and regulatory barriers. These all include some issues, such as scalability, security, and privacy in the blockchain. This paper shows how the blockchain with IoT influences the food supply chain and the fundamental concepts, and the understanding of blockchain with IoT technology. It is a guide toward the new and relevant research areas of blockchain concerning the food supply chain. This paper includes the technological adoption of blockchain as well as the adoption of challenges for the food chain with IoT devices.

Author Contributions: Conceptualization, A.K. and G.S.; methodology, A.K. and S.S. (Sparsh Sharma); writing—original draft preparation, A.K., V.K. and S.S. (Saurabh Singh); writing—review and editing, A.K., V.K., S.S. (Sparsh Sharma) and S.S. (Saurabh Singh); supervision, B.Y.; funding acquisition, S.S. (Saurabh Singh) and B.Y. All authors have read and agreed to the published version of the manuscript.

Funding: This research was supported by a grant (21163MFDS502) from the Ministry of Food and Drug Safety in 2022.

Institutional Review Board Statement: Not applicable.

Informed Consent Statement: Not applicable.

Data Availability Statement: Not applicable.

Conflicts of Interest: The authors declare no conflict of interest.

References

1. Queiroz, M.M.; Telles, R.; Bonilla, S.H. Blockchain and supply chain management integration a systematic review of the literature. *Supply Chain Manag. Int. J.* **2019**, *25*, 241–254. [CrossRef]
2. Dabbene, F.; Gay, P. Food traceability systems: Performance evaluation and optimization. *Comput. Electron. Agric.* **2011**, *75*, 139–146. [CrossRef]
3. Bazarsukh, R. Global Risks 2018: Fractures, Fears and Failures. Available online: https://reports.weforum.org/global-risks-2018/global-risks-2018-fractures-fears-and-failures (accessed on 10 May 2022).
4. Marsden, T.; Murdoch, J.; Morgan, K. Sustainable agriculture, food supply chains and regional development: Editorial introduction. *Int. Plan. Stud.* **1999**, *4*, 295–301. [CrossRef]
5. Nychas, G.-J.E.; Panagou, E.Z.; Mohareb, F. Novel approaches for food safety management and communication. *Curr. Opin. Food Sci.* **2016**, *12*, 13–20. [CrossRef]
6. Srivastava, H.S.; Wood, L.C. Cloud Computing to Improve Agri-Supply Chains in Developing Countries. In *Encyclopedia of Information Science and Technology*, 3rd ed.; IGI Global: Hershey, PA, USA, 2015; pp. 1059–1069; ISBN 978-1-4666-5888-2.
7. Verhoeven, P.; Sinn, F.; Herden, T.T. Examples from Blockchain Implementations in Logistics and supply chain management exploring the mindful use of a new technology. *Logistics* **2018**, *2*, 20. [CrossRef]
8. Hackius, N.; Petersen, M. Blockchain in logistics and supply chain: Trick or Treat? In Digitalization in Supply Chain Management and Logistics. 2017. Available online: https://tore.tuhh.de/bitstream/11420/1447/1/petersen_hackius_blockchain_in_scm_and_logistics_hicl_2017 (accessed on 24 March 2019).
9. Wang, Y.; Han, J.H.; Davies, P. Understanding blockchain technology for future supply chains: A systematic literature review and research agenda. *Supply Chain Manag. Int. J.* **2018**, *24*, 62–84. [CrossRef]
10. Gay, P.; Piccarolo, P.; Aimonino, D.R.; Tortia, C. Livestock identification and farm management by RFID systems. In Proceedings of the International Conference on Agricultural Engineering, Hersonissos, Greece, 23–25 June 2008; p. P-025.
11. Kouhizadeh, M.; Sarkis, J. Blockchain Practices, Potentials, and Perspectives in Greening supply chains. *Sustainability* **2018**, *10*, 3652. [CrossRef]
12. Nizamuddin, N.; Hasan, H.; Salah, K. IPFS- Blockchain-based Authenticity of Online Publications. In *International Conference on Blockchain (ICBC 2018)*; Springer LNCS: Seattle, WA, USA, 2018.
13. Zhao, J.; Fan, S.; Yan, J. Overview of business innovations and research opportunities in blockchain and introduction to the special issue. *Financ. Innov.* **2016**, *2*, 28. [CrossRef]
14. Jeppsson, A.; Olsson, O. Blockchains as a Solution for Traceability and Transparency. Master Thesis, Lund University, Lund, Sweden, 2017.
15. Park, K. Cryptocurrencies, Blockchain Is about to Revolutionize the Shipping Industry, Bloomberg. 18 April 2018. Available online: https://www.bloomberg.com/news/articles/2018-04-18/drowning-in-a-sea-of-paper-world (accessed on 22 September 2022).
16. Lee, Y.; Rathore, S.; Park, J.H.; Park, J.H. A blockchain-based smart home gateway architecture for preventing data forgery. *Hum.-Cent. Comput. Inf. Sci.* **2020**, *10*, 9. [CrossRef]
17. Bosona, T.; Gebresenbet, G. Food traceability as an integral part of logistics management in food and agricultural supply chain. *Food Control* **2013**, *33*, 32–48. [CrossRef]
18. Shih, D.-H.; Lu, K.-C.; Shih, Y.-T.; Shih, P.-Y. A simulated organic vegetable production and marketing environment by using Ethereum. *Electronics* **2019**, *8*, 1341. [CrossRef]
19. Lin, Q.; Wang, H.; Pei, X.; Wang, J. Food Safety Traceability System Based on Blockchain and EPCIS. *IEEE Access* **2019**, *7*, 20698–20707. [CrossRef]
20. He, X.; Chen, X.; Li, K. A decentralized and non-reversible traceability system for storing commodity data. *KSII Trans. Internet Inf. Syst. (TIIS)* **2018**, *13*, 619–634.
21. Alonso, R.S.; Sittón-Candanedo, I.; García, Ó.; Prieto, J.; Rodríguez-González, S. An intelligent Edge-IoT platform for monitoring livestock and crops in a dairy farming scenario. *Ad Hoc Netw.* **2020**, *98*, 102047. [CrossRef]
22. Folkerts, H.; Koehorst, H. Challenges in international food supply chains: Vertical co-ordination in the European agribusiness and food industries. *Supply Chain. Manag. Int. J.* **1997**, *2*, 11–14. [CrossRef]
23. Tian, F. An Agri-food supply chain traceability system for China based on FRID & Blockchain technology. In Proceedings of the 13th International Conference on Service Systems and Service Management (ICSSSM 2016), Kunming, China, 24–26 June 2016.
24. Bazoni, C.H.; Ida, E.I.; Barbin, D.F.; Kurozawa, L.E. Near-infrared spectroscopy as a rapid method for evaluation physicochemical changes of stored soybeans. *J. Stored Prod. Res.* **2017**, *73*, 1–6. [CrossRef]
25. David, A.; Kumar, C.G.; Paul, P.V. Blockchain Technology in the Food Supply Chain: Empirical Analysis. *Int. J. Inf. Syst. Supply Chain. Manag. (IJISSCM)* **2022**, *5*, 1–12. [CrossRef]
26. Mastos, T.; Gotzamani, K. Sustainable Supply Chain Management in the Food Industry: A Conceptual Model from a Literature Review and a Case Study. *Foods* **2022**, *11*, 2295. [CrossRef]
27. Mastos, T.D.; Nizamis, A.; Terzi, S.; Gkortzis, D.; Papadopoulos, A.; Tsagkalidis, N.; Ioannidis, D.; Votis, K.; Tzovaras, D. Introducing an application of an industry 4.0 solution for circular supply chain management. *J. Clean. Prod.* **2021**, *300*, 126886. [CrossRef]
28. Ioannou, I.; Demirel, G. Blockchain and supply chain finance: A critical literature review at the intersection of operations, finance and law. *J. Bank. Financ. Technol.* **2022**, *6*, 83–107. [CrossRef]

29. Chang, A.; El-Rayes, N.; Shi, J. Blockchain Technology for Supply Chain Management: A Comprehensive Review. *FinTech* **2022**, *1*, 191–205. [CrossRef]
30. Awan, S.H.; Ahmed, S.; Nawaz, A.; Sulaiman, S.; Zaman, K.; Ali, M.Y.; Najam, Z.; Imran, S. Blockchain with IoT, an emergent routing scheme for smart agriculture. *Int. J. Adv. Comput. Sci. Appl.* **2020**, *11*, 420–429. [CrossRef]
31. Zhao, G.; Liu, S.; Lopez, C.; Lu, H.; Elgueta, S.; Chen, H.; Boshkoska, B.M. Blockchain technology in agri-food value chain management: A synthesis of applications, challenges and future research directions. *Comput. Ind.* **2019**, *109*, 83–99. [CrossRef]
32. Ronaghi, M.H. A blockchain maturity model in agricultural supply chain. *Inf. Process. Agric.* **2021**, *8*, 398–408. [CrossRef]
33. Kamilaris, A.; Fonts, A.; Prenafeta-Boldv, F.X. A rise of blockchain technology in agriculture and food supply chains. *Trends Food Sci. Technol.* **2019**, *91*, 640–652. [CrossRef]
34. Umamaheswari, S.; Sreeram, S.; Kritika, N.; Prasanth, D.R.J. BIoT: Blockchain based IoT for agriculture. In Proceedings of the 2019 11th International Conference on Advanced Computing (ICoAC), Chennai, India, 18 December 2019; pp. 324–327.
35. Panarello, A.; Tapas, N.; Merlino, G.; Longo, F.; Puliafito, A. Blockchain and IoT Integration: A systematic survey. *Sensors* **2018**, *18*, 2575. [CrossRef]
36. Salah, K.; Nizamuddin, N.; Jayaraman, R.; Omar, M. Blockchain-based soybean traceability in agricultural supply chain. *IEEE Access* **2019**, *7*, 73295–73305. [CrossRef]
37. Vangala, A.; Das, A.K.; Kumar, N.; Alazab, M. Smart secure sensing for IoT-based agriculture: Blockchain perspective. *IEEE Sens. J.* **2021**, *21*, 17591–17607. [CrossRef]
38. Yadav, V.S.; Singh, A.R. A systematic literature review of blockchain technology in agriculture. In Proceedings of the International Conference on Industrial Engineering and Operations Management, Pilsen, Czech Republic, 23–26 July 2019; pp. 973–981.
39. Rejeb, A.; Keogh, H.G.; Treiblmaier, H. Leveraging the internet of Things and blockchain technology in supply chain management. *Future Internet* **2019**, *11*, 161. [CrossRef]
40. Tian, F. A supply chain traceability system for food safety based on HACCP, blockchain & Internet of things. In Proceedings of the 2017 International Conference on Service Systems and Service Management, Dalian, China, 16–18 June 2017.
41. Caro, M.P.; Ali, M.S.; Vecchio, M.; Giaffreda, R. Blockchain-based traceability in Agri-Food supply chain management: A practical implementation. In Proceedings of the 2018 IoT Vertical and Topical Summit on Agriculture-Tuscany (IOT Tuscany), Tuscany, Italy, 8 May 2018; pp. 1–4.
42. Kim, M.; Hilton, B.; Burks, Z.; Reyes, J. Integrating blockchain, smart contract-tokens, and IoT to design a food traceability solution. In Proceedings of the 2018 IEEE 9thAnnual Information Technology, Electronics and Mobile Communication Conference (IEMCON), Vancouver, BC, Canada, 1 November 2018; Volume 1, pp. 335–340.
43. Devi, M.S.; Suguna, R.; Joshi, A.S.; Bagate, R.A. *Design of IoT Blockchain Based Smart Agriculture for Enlightening Safety and Security*; Springer: Singapore, 2019; Volume 985.
44. Kamble, S.S.; Gunasekaran, A.; Sharma, R. Modeling the blockchain enabled traceability in agriculture supply chain. *Int. J. Inf. Manag.* **2020**, *52*, 101967–102016. [CrossRef]
45. Bumblauskas, D.; Mann, A.; Dugan, B.; Rittmer, J. A blockchain use case in food distribution: Do you know where your food has been? *Int. J. Inf. Manag.* **2020**, *52*, 102008–102010.
46. Mao, D.; Wang, F.; Hao, Z.; Li, H. Credit evaluation system based on blockchain for multiple stakeholders in the food supply chain. *Int. Environ. Res. Public Health* **2018**, *15*, 1627. [CrossRef] [PubMed]
47. Pournader, M.; Shi, Y.; Seuring, S.; Koh, S.C.L. Blockchain applications in supply chains, transport and logistics: A systematic review, of the literature. *Int. J. Prod. Res.* **2019**, *58*, 2063–2081. [CrossRef]
48. Reyna, A.; Martín, C.; Chen, J.; Soler, E.; Díaz, M. On blockchain and its integration with IoT. Challenges and opportunities. *Future Gen. Comput. Syst.* **2018**, *88*, 173–190. [CrossRef]
49. Available online: https://www.greyb.com/smart-home-market/1-feb-2019 (accessed on 22 September 2022).
50. Lin, J.; Shen, Z.; Zhang, A.; Chai, Y. Blockchain and IoT based food traceability for smart agriculture. In Proceedings of the 3rd International Conference on Crowd Science and Engineering, Singapore, 28 July 2018.
51. Folinas, D.; Manikas, I.; Manos, B. Trace- ability data management for food chains. *Br. Food J.* **2006**, *108*, 622–633. [CrossRef]
52. Hasan, H.; Salah, K. Blockchain-based Proof of Delivery of Physical Assets with Single and Multiple Transporters. *IEEE Access* **2018**, *6*, 46781–46793. [CrossRef]
53. Vuran, M.C.; Salam, A.; Wong, R.; Irmak, S. Internet of underground things: Sensing and communications on the field for precision agriculture. In Proceedings of the IEEE 4th World Forum on Internet of Things (WF-IoT'18), Singapore, 5–8 February 2018; pp. 586–591.
54. Cambra, C.; Sendra, S.; Lloret, J.; Garcia, L. An IoT service-oriented system for agriculture monitoring. In Proceedings of the IEEE International Conference on Communications (ICC'17), Paris, France, 21–25 May 2017; pp. 1–6.
55. Gu, X.; Chai, Y.; Liu, Y.; Shen, J.; Huang, Y.; Nan, Y. A MCIN-based architecture of smart agriculture. *Int. J. Crowd Sci.* **2017**, *3*, 237–248. [CrossRef]
56. Kim, H.; Laskowski, M. Sustainable Solutions for Food, Farmers, and Financing. Blockchain Research Institute 2018, 3028164, Canada. Available online: https://papers.ssrn.com/sol3/papers.cfm?abstract_id=3028164 (accessed on 22 September 2022).

57. Astill, J.; Dara, R.A.; Campbell, M.; Farber, J.M.; Fraser, E.D.G.; Sharif, S.; Yada, R.Y. Transparency in food supply chains: A review of enabling technology solutions. *Trends Food Sci. Technol.* **2019**, *91*, 240–247. [CrossRef]
58. Ratta, P.; Kaur, A.; Sharma, S.; Shabaz, M.; Dhiman, G. Application of Artificial Intelligence and Internet of Things (IoT) in Healthcare and Medical Sector: Applications, Challenges and Future Perspective. *J. Food Qual.* **2021**, *2021*, 7608296. [CrossRef]

sensors

MDPI

Article

Securing the Cyber Resilience of a Blockchain-Based Railroad Non-Stop Customs Clearance System

Sungbeen Kim and Dohoon Kim *

Department of Computer Science, Kyonggi University, Suwon-si 16227, Republic of Korea
* Correspondence: karmy01@kyonggi.ac.kr

Abstract: Current railroad customs clearance systems are problematic in that the movement of trains is occasionally restricted for extended periods during inspections to verify cargo integrity at customs clearance. Consequently, significant human and material resources are consumed to obtain customs clearance to the destination, considering different processes exist for cross-border trade. Therefore, we developed a cross-border blockchain-based non-stop customs clearance (NSCC) system to address these delays and reduce resource consumption for cross-border trains. The integrity, stability, and traceability of blockchain technology are used to construct a stable and reliable customs clearance system to address these problems. The proposed method connects diverse trade and customs clearance agreements in a single blockchain network, which ensures integrity and minimal resource consumption, and includes railroads, freight vehicles, and transit stations in addition to the current customs clearance system. The integrity and confidentiality of customs clearance data are protected using sequence diagrams and the blockchain to strengthen the resilience of the NSCC process against attacks; the blockchain-based NSCC system structurally verifies the attack resilience based on matching sequences. The results confirm that the blockchain-based NSCC system is time- and cost-efficient compared with the current customs clearance system and offers improved attack resilience.

Keywords: blockchain; railroad; customs clearance; non-stop; attack resilience; sequence diagram; integrity

1. Introduction

Cross-border freight transport by rail needs to undergo customs clearance at the border. The clearance process is time-consuming and requires considerable human and material resources. Furthermore, the rail freight agreements between different countries differ from each other [1], with several different consultative bodies responsible for handling the consultation, data integrity, and border delay problems associated with customs clearance processes. In this study, we developed a blockchain-based [2–4] border non-stop customs clearance (NSCC) system as a solution to these issues. Trains and freight cars need equipment to operate in a network because NSCC is required to ensure the integrity of goods. In this study, we refer to a setting in which the Internet of Things (IoT) is implemented on trains and freight vehicles [5,6]. Specifically, the cross-border NSCC system overlays a blockchain network on top of the current customs clearance system and uses the combined system as a blockchain platform [2–4] within the IoT context. Blockchain technology can address the issue of potential data breaches in cross-border customs clearance because it offers integrity, reliability, and traceability, while also providing the ability to track freight and trains for customs clearance purposes and ensure the reliability of data [7–9]. A suitable consensus algorithm for the border NSCC system developed in this study was selected based on previous research [10,11]. A basic attack scenario based on MITRE ATT&CK [12–14] was also built to create the NSCC system attack and defense functionality using blockchain technology. Consequently, measures for safeguarding the integrity and confidentiality of the NSCC process were developed, with cyberattacks to which the NSCC is vulnerable actively pre-empted. The structural benefits of NSCC are demonstrated by developing blockchain-based attack resilience [15] and a sequence diagram [16].

The existing customs clearance process transmits customs clearance data using the server–client method, and human resources directly validate cargo integrity. However, NSCC transmits data in a P2P system, with cargo integrity easily checked using the hash function. In addition, compared to the existing customs clearance system, which requires trains to stop at stations, NSCC improves efficiency because it inspects cargo integrity while the train is underway, making this operation the first example of NSCC procedure using blockchain for international rail freight transportation. In addition to resolving the problems associated with the current customs clearance system, the issue of data integrity while the data are being transferred from customs clearance can also be structurally resolved. Furthermore, a blockchain can be effectively applied to other domains because the developed blockchain-based NSCC system offers resilience to common attack scenarios and structural advantages. By utilizing blockchain-based NSCC, a plan was proposed to ensure cyber resilience against attacks targeting station and train nodes that participated in existing customs clearance procedures. In this study, we experimented with the cyber resilience environment that blockchain structurally provides and applied it to existing customs procedures, which enabled us to confirm the structural advantages of blockchain-based NSCC by constructing attack–response scenarios.

The remainder of this article is organized as follows: The structural resilience and consensus process of a blockchain is briefly described in Section 2. The consensus algorithm selected for utilization in this study is discussed, and MITRE ATT&CK is described to structurally prove the attack resilience of the NSCC system. Section 3 presents the designed blockchain-based border NSCC system, and Section 4 discusses the use of a sequence diagram to demonstrate the attack resilience of NSCC, which is the main topic of this study. Section 5 provides the experimental results of NSCC using Simulation of Urban Mobility (SUMO) [17], Docker, and Ethereum [18]. A summary table that explains the current customs clearance system and blockchain-based NSCC system is also presented. Finally, a discussion and the conclusion are presented in Sections 6 and 7, respectively.

2. Related Work

In this study, we develop a blockchain-based NSCC system and demonstrate its structural attack resilience using a blockchain, its consensus algorithm, and an attack sequence diagram created with MITRE ATT&CK. In this section, we discuss relevant prior studies and present a method that applies to this study.

2.1. Structural Attack Resilience of Blockchain

In this section, we describe the structural resilience of a blockchain, including the structural benefits of using a blockchain with the NSCC system, as well as the background of this study. A blockchain is a distributed ledger technology in which every node connected to the blockchain network owns the same ledger and is structurally resilient to data forgery, denial of service (DoS) [19], and availability attacks. Consequently, if data forging occurs at one node, accessing another node allows the original ledger to be restored, and data remain preserved until no further nodes remain. Specifically, data robustness can be maintained even if only one node participates in the blockchain network [20].

One drawback of blockchain technology is that it cannot be utilized in various applications currently in operation in the industry. At present, efforts are ongoing in numerous domains to introduce blockchain technology into conventional industries; however, introducing a blockchain structurally is problematic. Incorporating a blockchain in conventional industries is time-consuming because of the high initial introduction cost. Moreover, an attack can target a blockchain network while it is transmitting, receiving, and distributing data when blockchain technology and older systems are combined. However, in this study, we assume that trains, freight cars, and customs clearance stations comprise one IoT node environment. Moreover, we utilize the structural advantages of a blockchain to demonstrate resilience to various attacks targeting industrial networks. In addition, we identify methods to utilize blockchain technology in various industrial domains.

2.2. Comparison of Consensus Algorithms and Their Application Domains

In this study, we develop a blockchain-based NSCC system. An integrated higher-level decision-making process is necessary because the NSCC system is a new consensus body created by the collaboration of numerous consensus bodies. To choose the best blockchain consensus algorithm for NSCC, we compare four primary consensus algorithms: proof of work (PoW), proof of activity (PoA), proof of stake (PoS), and delegated PoS (DPoS). Table 1 lists the existing consensus algorithms identified by referring to articles that compared and analyzed several consensus techniques. Table 1 presents PoS, chosen as the consensus technique to be implemented in the proposed blockchain-based NSCC. One drawback of PoW is that it is challenging to introduce in different nations owing to the significant reliance of NSCC on hardware. Additionally, PoA and DPoS are consensus algorithms investigated to compensate for the drawbacks of PoW and PoS, although they are less scalable than PoS. PoS with high scalability is more appropriate for NSCC because it requires the participation of numerous nations and councils. PoS must be required by some validation nodes on the blockchain network. Hence, each node that participates in NSCC must be a validation node, e.g., nations and councils.

Table 1. Quantitative indicator analysis of consensus algorithms [11,21].

Category	PoW	PoA	PoS	DPoS
Latency (response time)	10 min	5 min	1 min	3 s
transaction per second (TPS)	≥7 TPS	≥14 TPS	≥300 TPS	≥500 TPS
Computing overhead	High	Low	Medium	Medium
Scalability	Low	Medium	High	Medium
Decentralized level	High	Low	Medium	Medium
Hardware dependency	Yes	No	No	No
Security (in application)	Low	Medium	Medium	Medium
Consensus method	Hash rates	Activity-based	Stake	Stake votes
Reference	[22–24]	[25–27]	[28–31]	[32,33]
Adequacy	X	△	O	△

To construct a blockchain-based NSCC system, this study chooses the PoS consensus algorithm and obtains consensus from each nation and consultative body. Railway cooperation organizations, such as the Organization for Cooperation of Railways (OSJD) [34] and the Organization for International Carriage by Rail (OTIF) [35], also have freight transport agreements, known as the Agreement on the International Goods Transport by Rail and the Uniform Rules Concerning the Contract of International Carriage of Goods by Rail [36]. The consensus algorithm of a blockchain is similar to a cargo transportation agreement because it is a mechanism that moves forward with customs clearance by evaluating the interests of each country. All consensus algorithms were introduced in various application domains except the railroad industry. Furthermore, because a blockchain uses a consensus algorithm to make decisions, an attack directed at the current network can occur while disseminating the consensus results, instead of the consensus algorithm. Therefore, in this study, we develop a strategy to introduce PoS into the railroad industry, structurally construct a blockchain-based NSCC system, and combine many railroad cooperation groups, such as the OSJD and OTIF, into one consensus system. The NSCC attack–response sequence for a network and cyber threats outlined in Section 2.3 are defined and detailed in Section 4.

2.3. Using MITRE ATT&CK

The MITRE organization created MITRE ATT&CK in response to the expansion of the influence of and harm inflicted by cross-border cyberattacks [12–14]. The adversarial tactics, techniques, and common knowledge (ATT&CK) framework is a phase of the cyber kill chain model [37] internally designed and arranged based on actual attack cases in

MITRE. MITRE ATT&CK is a database consisting of standard data produced by analyzing the adversarial behaviors of attackers from the standpoint of attack tactics and techniques, achieved by observing actual cyberattacks and then classifying and cataloging the attack techniques of various attack groups. MITRE ATT&CK is the result of patterning threatening tactics and techniques to improve the detection of intelligent attacks, taking a slightly different perspective from the traditional cyber kill chain concept. At MITRE, development of the ATT&CK framework initially began by recording tactics, techniques, and procedures (TTPs) relating to hacking attacks employed in the Windows corporate network environment before evolving into a framework that can recognize the behavior of an attacker by mapping TTP information based on studying consistent attack behavior patterns generated by attackers.

In this study, we build a blockchain-based NSCC system attack and defense scenario and demonstrate its attack resilience by applying attack scenarios and sequences created using matrices of MITRE ATT&CK. In conventional security studies, the attack life cycle is depicted as a sequence diagram. We demonstrate the attack resilience of the NSCC developed in this study using this attack sequence diagram. Attack sequence diagrams are frequently used in academic research to support and demonstrate the reliability of networks and systems. By employing a DoS attack against the voice-over-internet protocol [38] environment, for instance, the robustness of the environment, in terms of availability to provide services, can be demonstrated [39].

In this study, an attack sequence diagram is established and used for process analysis to demonstrate the attack resilience of the developed blockchain-based NSCC system. A sequence is constructed in accordance with the basic NSCC process, while the attack life cycle is developed by selecting a random attack point in the sequence.

3. Non-Stop Customs Clearance Using Blockchain

The border NSCC system operates in areas in the vicinity of stations located on the borders between different countries. The system is implemented by configuring the blockchain network and enabling data transmission between trains and transit stations that belong to different networks. As shown in Figure 1, a train travels from country A to country B, with network interworking between base stations a and b assumed to be automatic in this process. Go-Ethereum (Geth) blockchain network interworking, required for the border NSCC system to progress, uses a Docker container [40] in the machine of every transit station and train node. Geth software is needed to function as the Ethereum node in the Ethereum network [41]. Considering the Simulation of Urban Mobility (SUMO) framework can simulate the actual traffic environment, it was used to simulate the role of the train in this study. SUMO runs inside the Docker container of the train, with the customs clearance process conducted using communication linking the IP address and port number between Docker containers.

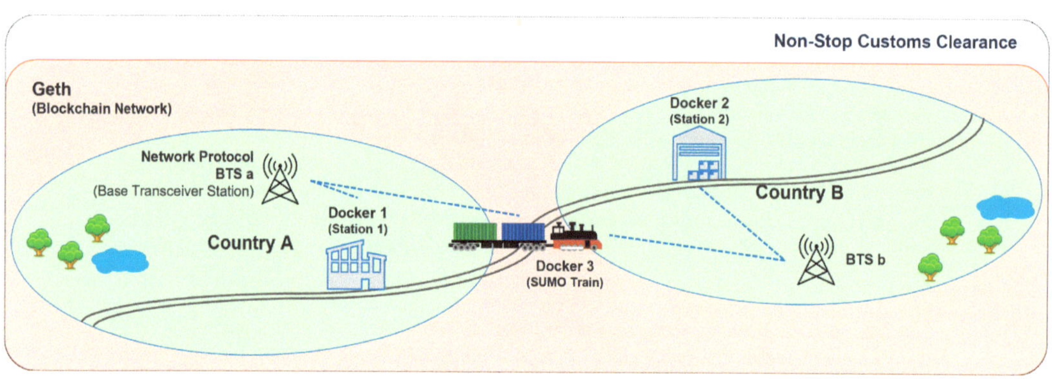

Figure 1. Railway border non-stop custom clearance in real-world simulation.

The process developed in this study enables trains to proceed through customs clearance without stopping between transit stations; in addition, the attack resilience is structurally demonstrated. To proceed through Station 2 in country B, a train that already passed through Station 1 communicates the information required for NSCC. After receiving it, Station 2 checks the integrity of the data by comparing it to the hash value [42,43] of the data stored in the current blockchain network. The hash value for the cargo specs is broadcast to the blockchain network and other transit stations during the departure process once the cargo is loaded on the train at the original departure point. Cargo integrity is examined by contrasting the hash value transmitted with the hash value of the data propagated throughout the process of passing through each transit station. The hash value of the data is compared with the hash value recorded on the blockchain, and if no discrepancies are found, the NSCC process continues. If a problem arises, the train and its cargo are inspected using the existing customs clearance process.

The operation of the NSCC system is described in Section 3.1, with the network setup needed to use the blockchain defined and explained in Section 3.2.

3.1. *Procedure of Blockchain-Based Non-Stop Customs Clearance System*

The NSCC process is divided into five individual steps, as shown in Figure 2, where each step is described. NSCC uses distributed storage as its data storage process because it utilizes a blockchain network [44]. Data are logged using the distributed storage system known as the interplanetary file system [45–47], and data comparison and verification are conducted. Data are recorded and stored using distributed storage, a network of distributed nodes. Consequently, every train node and transit station node involved in the blockchain network participates in the distributed storage system. For the comparison–verification process, the customs clearance data are uploaded to the distributed storage and encrypted using the hash function. The respective steps of the process are shown in Figure 2.

- **(Step 1)** Enter and transact: A train node approaches a station node by this process to conduct the NSCC process. The customs clearance data (raw data) are processed by the train node and sent to the station node for customs clearance. This process employs a security network (e.g., a virtual private network) that utilizes the base station of each country [48,49].
- **(Step 2)** Receive and hash: Data from the train node are relayed to the station node, which hashes the data using a hash function. The calculated hash value is compared and validated in Step 3. The hash function to be used at this point is chosen from SHA-256 [50] or Keccak-256 [51] and applied throughout the customs clearance process.
- **(Step 3)** Compare: The station node compares the hash value of the hashed data with that of the initial customs clearance data generated when the cargo was initially loaded. The hash value uploaded to the distributed storage is currently compared with that produced by the station node based on the transaction recorded in the blockchain. The results of the comparison are broadcast in Step 4.
- **(Step 4)** Broadcast: The success of the NSCC process is determined by comparing the hash value produced by the station node to that in the distributed storage. Subsequently, the train node decides on whether to proceed. If the hash value of the distributed storage that already exists differs from that generated in the relevant station node, the train proceeds in accordance with the existing customs clearance procedures. If the two hash values correspond, indicating that no irregularities exist with the data or cargo, the train node passes through without stopping. The passing information is broadcast to other stations and train nodes.
- **(Step 5)** Dashboard: A dashboard displays the NSCC-related data. The visualized data can be examined and subsequently analyzed. The corresponding dashboard of each node allows users to view information about the blockchain network and hardware resources.

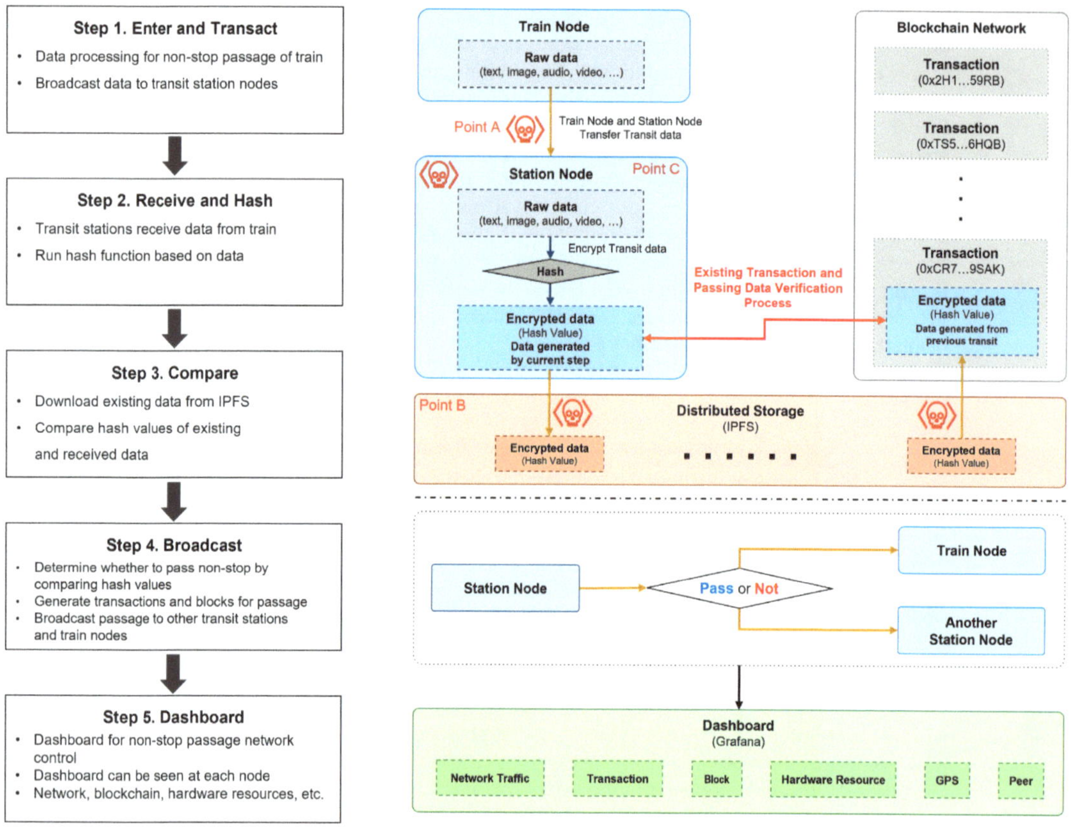

Figure 2. Process overview of non-stop customs clearance system and attack points.

Potential attack points for each node, component, interface, and layer that constitute NSCC are shown in Figures 2 and 3. Points A, B, and C represent potential weak points vulnerable to attacks, which can be attacked by hostile attacker nodes intending to damage the NSCC network and systems. Attacks on NSCC-related data and communications are possible through these points. In this study, we use the properties of the blockchain to structurally demonstrate the attack resilience of points A, B, and C. In Figure 3, target, network, and storage are the three potential attack layers, with an attack scenario created by setting an attack sequence diagram of the respective points. The data shown in Figure 3 can be breached and stolen by the blockchain-based NSCC based on the configured scenario.

For instance, if the station in Figure 2 is attacked, data relevant to customs clearance can be compromised, making the customs data verification process vulnerable to attacks. In the event of an attack, significant issues, such as time delays and misjudgment can occur during the customs clearance process. Moreover, sensitive data can be compromised because the customs clearance process follows an international consensus procedure. However, the attack resilience of the customs clearance node is structurally proven by the developed NSCC sequence diagram, with the method for securing resilience explained using security elements as an example.

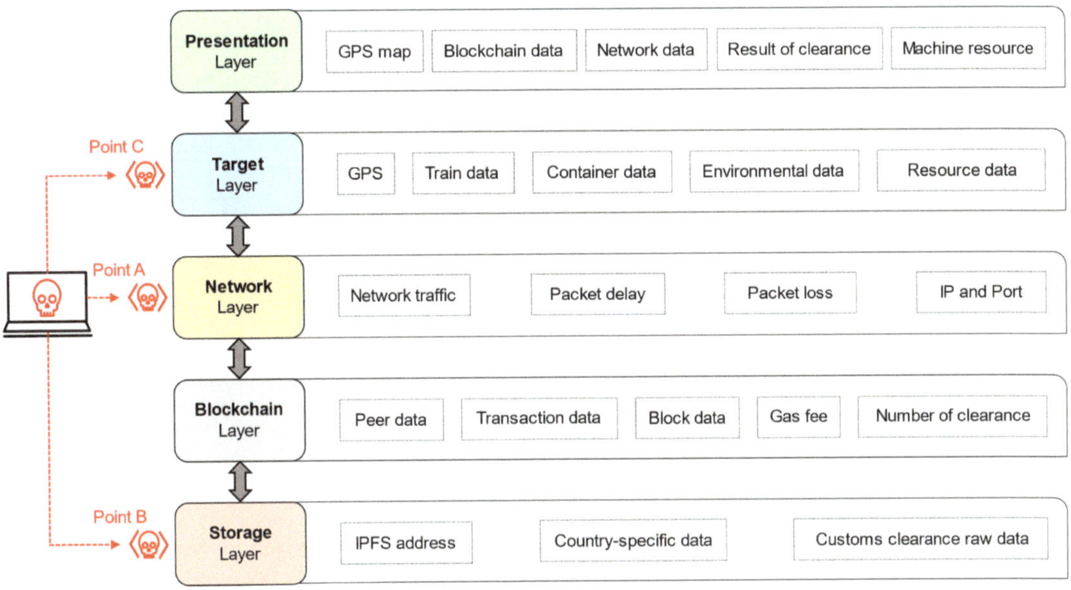

Figure 3. Data derived for each component layer.

3.2. Network Configuration of Blockchain-Based Non-Stop Customs Clearance System

This section describes the organizational structure of the blockchain network, and the manner data are sent to and received from the network via an existing railway network. The structure of the blockchain-based NSCC network is depicted in Figure 4, with the NSCC sequence technique from the perspective of each node summarized in Table 2. The detailed explanation is as follows:

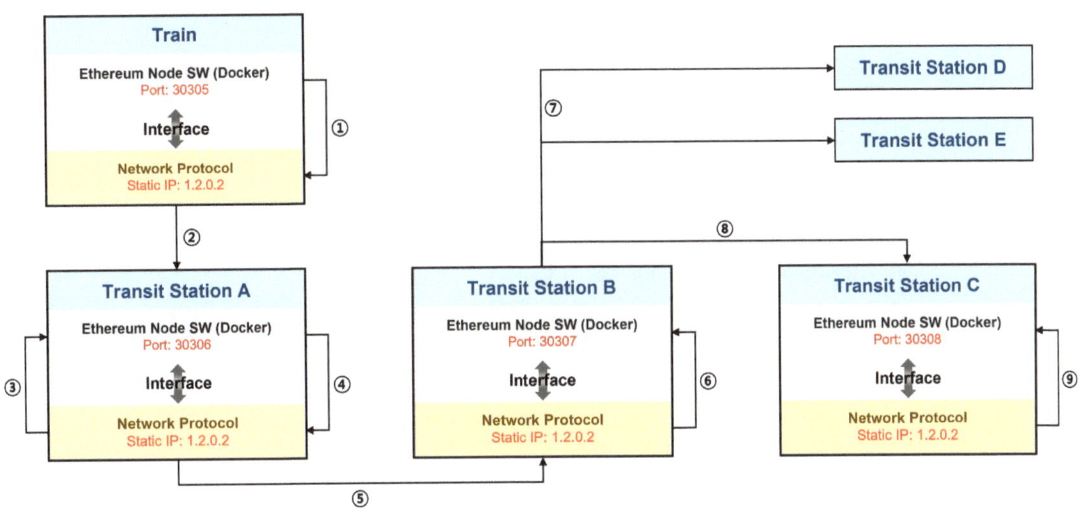

Figure 4. Non-stop customs clearance network implemented with Ethereum.

Table 2. Process of non-stop customs clearance on blockchain network.

Process No.	Description of Each Process
①	Communication data cleaning and communication protocols are accessed to transmit data from trains to transit stations.
②	Data transmitted to transit station using communication protocol of machine.
③	Transit station that received data through communication protocol accesses Ethereum node to verify data.
④	After verification process, communication protocol to deliver data to another transit station is accessed.
⑤	Verified results broadcast to other transit stations (blockchain network) using communication protocol.
⑥	Verification data received from other transit stations through blockchain network are checked.
⑦	Data forwarded to other transit station nodes that do not directly participate in this customs clearance process, and data are verified.
⑧	Verification-related data are transmitted to transit stations on future train routes.
⑨	Transit stations other than those that received data check whether transaction information of blockchain network matches the verification result.

The network for NSCC based on a blockchain, shown schematically in Figure 4, indicates that each IP address uses the same subnet mask because when setting up the experimental environment, the network is configured utilizing many Docker containers inside a single machine. The IP information of each node is expressed differently during the actual NSCC application process. For example, Stations A and B have static IP addresses of 242.42.25.65 and 103.132.54.12, respectively. Furthermore, the port number increases sequentially, as shown in Figure 4; however, when NSCC is applied, appropriate port numbers, such as 9090 and 7897, can be assigned to each node.

The number of connected nodes is also changed if NSCC is applied to customs clearance. The network includes the customs clearance nodes from 29 OSJD and 51 OTIF member countries, assuming that the present customs clearance offices in border areas are participating (as of December 2022) [52]. As more member nodes join the blockchain network, it becomes more stable. Therefore, the NSCC network has high robustness, and the maturity of each node increases with the number of consultative bodies and countries participating in NSCC.

A summary of each process depicted in Figure 4 is provided in Table 2, as identified by the process number. This process is more difficult than the current railroad customs clearance system because data broadcast from a train to a transit station uses network connection protocols and interfaces. An attacker can target the network, trains, and transit stations in this process. Sequence diagrams are used in this study to describe the basic flow of this process. In addition, each attack–defense phase is defined and the structural resilience of the blockchain-based NSCC system is demonstrated.

4. Attack Resilience in Blockchain-Based Railway

A sequence diagram for basic customs clearance is defined and systematically discussed in Section 4.1. In addition, the potential attack time and method, which are the focal elements of this study, are presented in Section 4.2, where the attack–response sequence diagram is defined. The procedures for attack, response, and analysis are described. The attack–response sequence diagram constructed based on potential attack points A, B, and C is depicted in Figures 2 and 3.

4.1. Basic Sequence of Blockchain-Based Non-Stop Customs Clearance

The basic flow of the developed blockchain-based NSCC is shown in Figure 5. When the initial cargo information is transmitted to the blockchain network at a shipping point, the transaction status and block in the blockchain are returned. Once the cargo is recorded in the blockchain network, the train departs for the transit station. When the train passes through this transit station, data pertaining to customs clearance are broadcast to the station and verified. The verification process corroborates the integrity of the cargo based on the data already recorded in the blockchain network. A hash function is used to readily and rapidly substantiate the customs data, as shown in Figure 2.

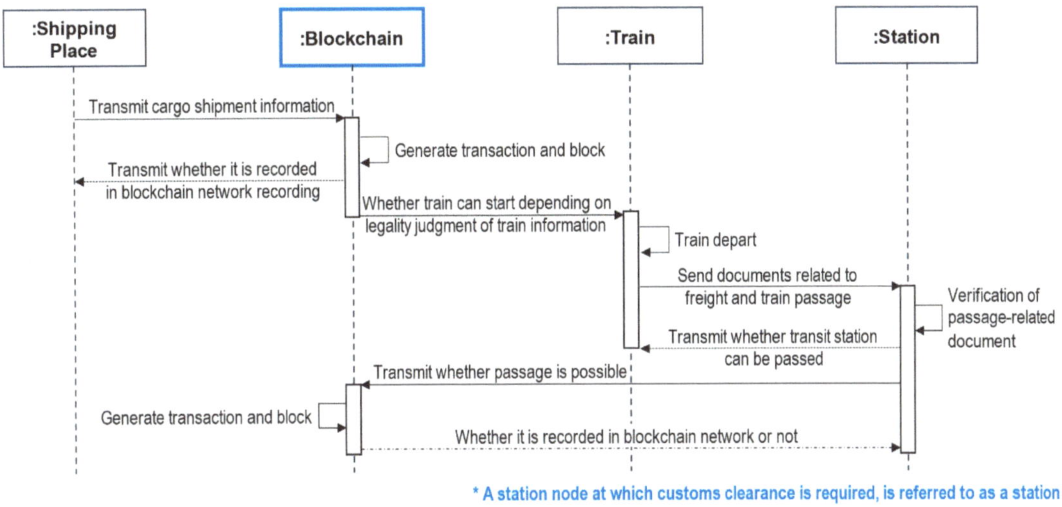

Figure 5. Basic sequence diagram of NSCC.

The sequence diagram shown in Figure 5 only depicts the basic customs clearance process of NSCC, with additional details of each step omitted for clarity. Furthermore, all 'Station' mentioned below are categorized as customs clearance station nodes according to Figure 5. As the diagram is intended to indicate the steps involved in the automatic generation of data recorded throughout the transaction and block creation process of each blockchain and the verification process at each transit station, the processes related to distributed storage are not shown here.

Given that a blockchain is a platform that offers integrity and reliability, the NSCC process can be conducted by relying on these features of a blockchain for customs clearance in trains and transit stations. As presented in Section 4.2, an attack–response sequence is added to the basic NSCC sequence to reinforce the structural robustness of the NSCC and ensure the cyber resilience of the blockchain-based NSCC.

4.2. Attack Sequence of Blockchain-Based Non-Stop Customs Clearance with Attack Resilience

The attack–response sequence diagram for the potential attack points of the blockchain-based NSCC system, which is the main concept of this research, is presented in this section. The entire sequence demonstrates that the blockchain structurally has attack resilience, with an attacker performing an attack sequence against a victim and the victim responding in correspondence with the sequence. As defined in Figure 5, a customs clearance station node is referred to as a Station.

4.2.1. Attack Sequence A: Attacking Clearance Station Node Using DoS

Blockchain-based NSCC technology is robust and has attack resilience in terms of the availability of customs clearance. In this sequence, an attacker targets Station A with a DoS [19] attack. We utilize the structural benefits of the blockchain to defend the system against this attack.

- **Attack sequence:** One of the potential attack points indicated in Figures 2 and 3 is transit Station A, through which the train is expected to pass. An attacker prepares a DoS attack against this target. In addition to overloading the network communication of transit Station A by packet fragmentation, the attacker sends a request to establish socket communication to transit Station A [53]. Accordingly, the train waits at the station without transmitting a request to pass through after receiving data relating to customs clearance and completing the verification process. Thus, the attacker keeps transit Station A overloaded to perform DoS attacks and delay customs clearance.

- **Corresponding sequence:** As the train node cannot receive permission to pass through transit Station A, it sends a request to other nearby transit station nodes for customs clearance. When transit Station B receives a request for customs clearance, the train is granted permission to pass through Station B, and the train is processed for passage through transit Station A in accordance with the existing customs clearance sequence.
- **Analysis and discussion:** A flowchart based on the attack–response scenarios for DoS attacks is shown in Figure 6. As all trains and transit station nodes are connected to the blockchain network, no problems occur when clearance is requested from and processed by transit Station B. If the integrity of the data transmitted from the train can be verified, the train can pass through the customs clearance station without stopping. Thus, the system is designed to enable other trusted customs clearance nodes to handle the data verification process. The attacker targets the availability of the NSCC; however, it offers resilience against these attacks because only one blockchain network is used. As the blockchain network is structurally designed to ensure the reliability and integrity of recorded transactions, even if additional nodes participate in the verification process, the reliability of the verification is ensured.

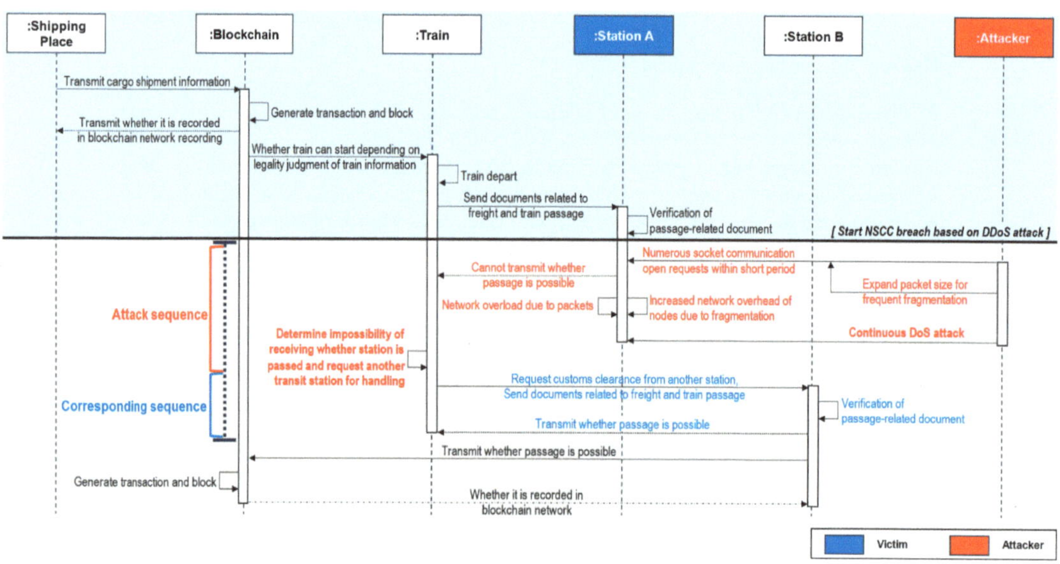

Figure 6. Attack sequence diagram using DoS.

4.2.2. Attack Sequence B: Attacking Distributed Storage Using Spoofing Attack

In terms of the data integrity and reliability of customs clearance, the developed blockchain-based NSCC is attack-resilient and robust. In this sequence, an attacker conducts a spoofing attack [54] targeting the distributed storage. The structural benefits of the blockchain are utilized to defend against this attack.

- **Attack Sequence:** Figure 7 shows the approach followed to target and attack the distributed storage system of the blockchain-based NSCC. An attacker confounds the sender by spoofing the domain address and routing details to connect to the distributed storage at B, a potential attack point, as depicted in Figures 2 and 3. Customs documents are sent to transit Station A by train. Transit Station A utilizes the distributed storage and transaction data on the blockchain network to verify them. During this process, the attacker transfers arbitrary data while changing the routing table of transit Station A to enable the attacker to appear as the distributed storage. Data inconsistency occurs because transit Station A undertakes the verification process based on the data sent by the attacker; consequently, the NSCC process cannot be implemented.

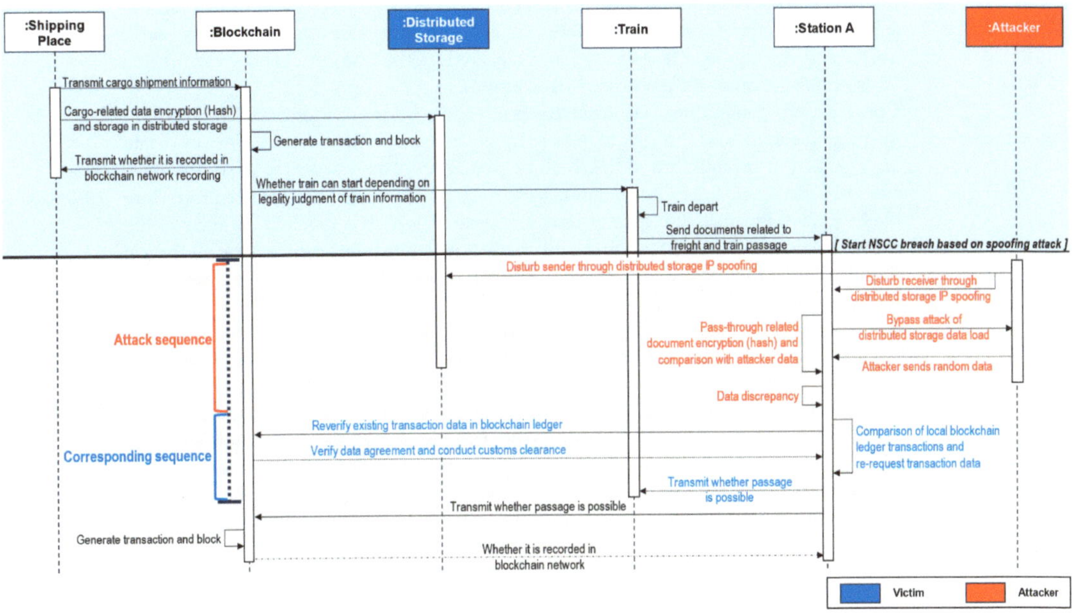

Figure 7. Attack sequence diagram using spoofing.

- **Corresponding sequence:** Transit Station A analyzes the distributed storage data for inconsistencies and compares them to transactions [55] on its own local blockchain ledger. After the first verification, a secondary verification is conducted with the transaction data of the actual blockchain network because the hash value of the customs clearance data is present in the transaction data. Transit Station A updates the routing table and broadcasts permission for the train to pass through after verifying that the data from the blockchain network correspond with the customs clearance data. The sequence is completed after the customs clearance is recorded on the blockchain network.
- **Analysis and discussion:** An attack that targets the routing database occurs when a transit station is proceeding with verification. The distributed storage and blockchain network transactions contain the data needed for verification, and any data inconsistencies can be determined in the event of an attack directed against the distributed storage. In this case, the blockchain network is accessed to verify data because every participating node has the same ledger. The blockchain platform structurally ensures integrity, reliability, and traceability because all participating nodes share the same ledger. These features of the blockchain can be used to safely conduct the data verification process.

4.2.3. Attack Sequence C: Attacking Clearance Station Nodes Using Advanced Persistent Threat and Backdoor Attacks

In terms of data integrity and customs clearance resilience, NSCC based on blockchain technology is attack-resilient and robust. In this sequence, an attacker targets Station A using advanced persistent threat (APT) [56] and backdoor [57] attacks. These attacks are warded off by utilizing the structural benefits of the blockchain.

- **Attack sequence:** The attacker is based at potential attack point C, as shown in Figures 2 and 3. The process before the attack is the same as the basic NSCC process. However, when a train departs, the attacker designates the transit station along the route as a target, launches an APT attack, and simultaneously inserts a backdoor. If the attack is successful, the attacker can control the root authority of transit Station A [58] and modify the transaction data of the blockchain. Subsequently, a discrepancy arises between the data transmitted and received by the train during the verification of customs clearance data with transit Station A.
- **Corresponding sequence:** The train that is refused customs clearance sends its request for permission to pass through to nearby transit Station B. The customs clearance data are checked at transit Station B, which responds with the necessary permission for customs clearance. Furthermore, data sync to transit Station B is requested to restore the blockchain transaction data of transit Station A, which is falsified. To recover the transaction data of transit Station A and conduct its ledger sync process, transit Station B and other transit stations transfer the entire blockchain data to transit Station A, which can re-participate in the customs clearance process.
- **Analysis and discussion:** Root access can be hijacked using numerous methods. Figure 8 shows a straightforward example of backdoor injection via an APT attack. A transit station with social engineering issues is vulnerable to root authority hijacking attacks. This attack falsifies the blockchain data of a transit station node and interferes with customs clearance. Owing to the structural features of the blockchain, data can be restored even if the blockchain data inside one node are altered. All nodes included in the blockchain can participate in the consensus process, as shown in Figure 6. Consequently, transit Station B is required to continue with customs clearance.

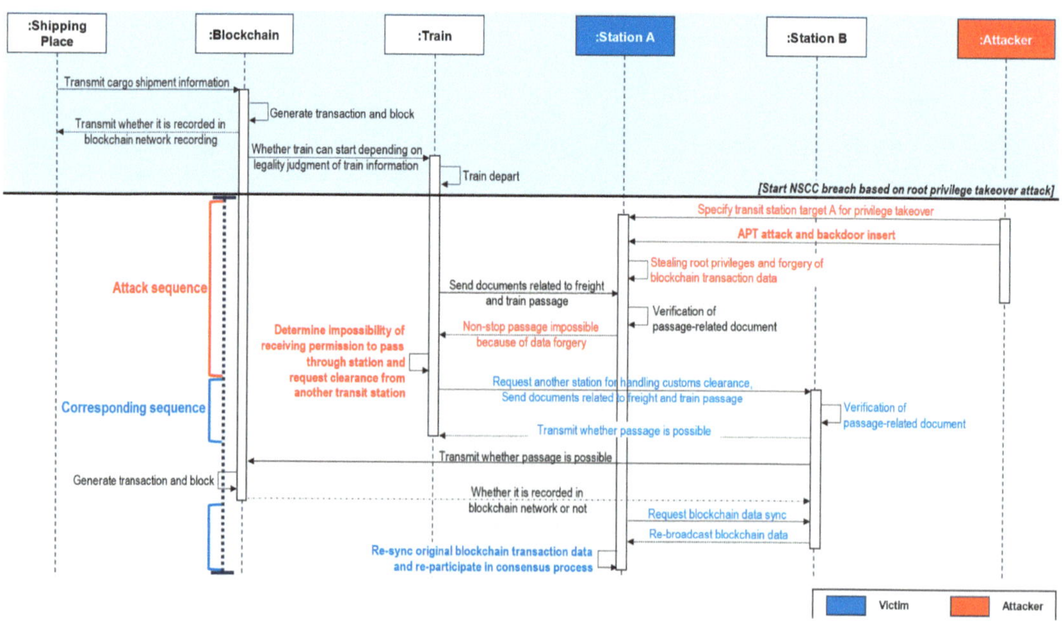

Figure 8. Attack sequence using root access privileges.

5. Experimental Results

In this section, we describe the simulation of the environment in Figure 1 using SUMO, Docker, and Ethereum for experiments, explain the blockchain-based NSCC, and demonstrate the advantages of the approach followed in this study. In addition, the conventional concept of a blockchain is explained, and the quantitative and qualitative metrics used when a blockchain is integrated into the customs clearance system are presented. Table 3 provides information about the software versions and status information of the experimental environment.

Table 3. Simulation and experimental environment.

Category	Description
OS	Windows 11
GPU	RTX 3070 Ti
RAM	16 GB
Docker OS	Ubuntu 20.04
Blockchain environment	Geth v1.10.25
SUMO version	SUMO v1.14.1

5.1. Experiments and Materials

In this study, the NSCC environment was constructed using SUMO, Docker, and Ethereum, with the final experimental results of the NSCC derived based on the experimental results of each component. SUMO makes it possible to simulate the navigation of a given road network by single vehicles in response to a given traffic demand. The simulation is purely microscopic: each vehicle is modeled explicitly, has its own route, and navigates the network individually. Therefore, we used SUMO to derive the travel route and the timing of the train on the railroad. In addition, NSCC was implemented using Docker to configure the train and station as one node, while Geth was used to implement the blockchain network. Figure 9 shows the structure of our experimental environment.

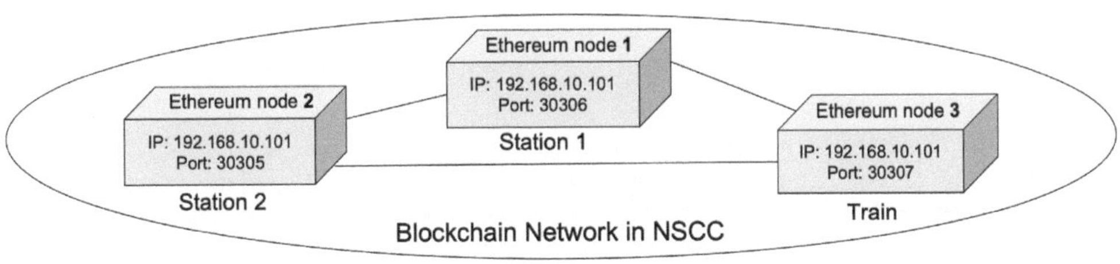

Figure 9. Blockchain network in NSCC using Docker, SUMO, and Ethereum.

Station and Train nodes are implemented as a single Docker container; considering they are implemented on a single local machine for experimentation, they all have the same IP address and different port numbers. However, in a real environment, all nodes have different IP addresses and port numbers. Each node was composed of Geth nodes to connect to the Ethereum network; in the case of the Train node, the SUMO client was run inside the Docker container to serve as a train in this experiment. Each Docker container communicates using the IP address and port number. During the communication process, the Docker container transmits and receives data, interlocks with the Ethereum network, and propagates transactions and blocks. In this study, using SUMO, the NSCC was tested for a simulation involving trains moving between Kazakhstan and Mongolia. Table 4 shows the values and parameters required for the experiment. The train departed from Station 1, a customs clearance station in Mongolia, and traveled to Station 2, a customs clearance station in Kazakhstan, at approximately 150 km/h.

Table 4. Numerical values and parameters required for experiments.

Category	Descriptions of Values and Parameters
Stations and country	Mongolia Station 1 to Kazakhstan Station 2
Train velocity	Approximately 150 km/h
Coordinates of departure station	lat: 44.162919, lon: 80.326560
Coordinates of destination station	lat: 43.632262, lon: 77.647001
Maximum duration of consensus algorithm	Up to 10 min with PoW

Figure 10 shows the SUMO-based simulation environment, with Kazakhstan to the left of the red line and Mongolia to the right. We simulated a train traveling from Station 1 in Mongolia to Station 2 in Kazakhstan. Specific information is listed in Table 4, with the train movement and NSCC procedures tested in this environment. The steps indicated below the figure correspond to Steps 1–4 described in Figure 2. Step 0 involves the train moving from Station 1 to Station 2. While the train is underway, in Step 1, data related to customs clearance are transmitted to the station. In Step 2, the station proceeds with hashing based on the received data. In Step 3, the station compares and verifies the data integrity based on the hash value of the data. In Step 4, the station propagates the final verification result to other station and train nodes. The novelty of this study is that after reducing the speed to a minimum from Step 1 to Step 4, the train passes through the station. In the case of the existing customs clearance procedure, it is necessary to stop at the station to allow the cargo to be inspected. In contrast, the proposed basic methodology for our blockchain-based NSCC enables the train to pass through the station without stopping after cargo integrity is verified.

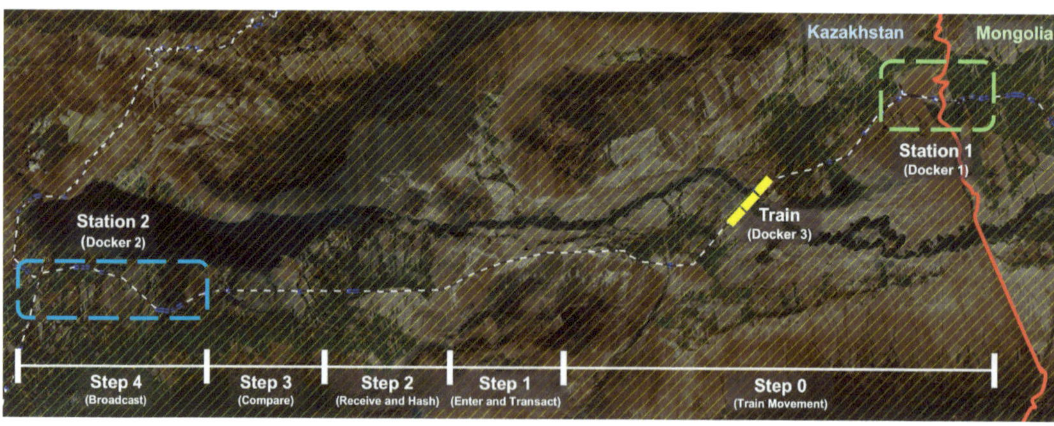

Figure 10. Simulation of NSCC from Mongolia to Kazakhstan using SUMO.

5.2. Results of Blockchain-Based NSCC

This study developed a solution to the problem of time delays caused by the need to obtain customs clearance, consumption of human and material resources, and reliability and integrity of customs clearance of the existing customs clearance system. The system utilizes a Bitcoin-derived blockchain [22]. A blockchain is a peer-to-peer-based system in which distributed nodes share a single ledger [59], and is characterized by integrity, reliability, and traceability. Data integrity can be realized by alerting other nodes when data are forged because nodes are distributed and share the same ledger. Additionally, data are recorded in a setting that ensures integrity. The reliability of previously created data and blocks increases if data are transmitted because transactions and blocks are continuously created by a verification process. Owing to their integrity and reliability, data continually entered into the ledger on a blockchain cannot be falsified. Therefore, the traceability of all data can also be guaranteed.

The average processing times for Step 1 to Step 4 are listed in Table 5. The amount of data allowed per transmission in the process is limited to a maximum of 1000 MB. In Step 1, data are propagated using a VPN environment, which takes a maximum of 10 min based on the minimum data propagation speed over the network. Step 2 takes a maximum of 1 min based on the minimum execution time of SHA-256, which is 20 Mbps in size, to derive a hash value using the SHA-256 hash function after receiving the data. Step 3 takes a maximum of 10 min based on the minimum download speed of 1 Mbps from IPFS to perform the comparison verification based on the data stored in IPFS. Finally, when the verification process is completed based on the customs clearance data, it takes a maximum of 10 min to verify transactions and propagate blocks using the PoW consensus algorithm in the Ethereum-based PoW environment, calculated at 11 TPS. Therefore, the overall process takes a maximum of 31 min to complete if no problems arise. However, considering that connection to the blockchain may involve delays or disconnections depending on the network environment, a maximum of 1 h is considered necessary to enable these issues to be resolved [60].

Table 5. Processing speed and available data size for each NSCC step.

Category	Time for Each Procedure	Speed	Data	Ref.
Step 1: Enter and transact	Up to 10 min with VPN	5~10 Mbps	1000 MB	[61]
Step 2: Receive and hash	Up to 1 min with SHA-256	20 Mbps	1000 MB	[62,63]
Step 3: Compare	Up to 10 min with IPFS	1 Mbps	1000 MB	[46]
Step 4: Broadcast	Up to 10 min with PoW	11 TPS	.	[64]
Total	Up to 31 min	.	.	.

Using these features of a blockchain network and software, we created a "blockchain-based border NSCC" system in this study and tested its resilience against attacks. The main contributions of this study and the differences between the developed NSCC and current customs clearance systems are discussed below.

Table 6 compares the traditional customs clearance system with the blockchain-based NSCC system. The six criteria used for comparison are time, resource, integrity, reliability, transparency, and traceability. The respective criteria are described as follows:

- **Time:** The current customs clearance process is time-consuming because individuals have to directly inspect customs clearance items and cargo. However, with NSCC, customs clearance can be completed in as little as 1 h if the validity of the customs documents is not questionable.
- **Resource:** In the current customs clearance system, people directly participate in customs clearance and personally inspect the goods and cargo. However, resource consumption is minimal because the accuracy of the customs data is verified by machine. Customs clearance is conducted by verifying the integrity using the hash value of the data, which is broadcast to the blockchain network.
- **Integrity:** Data integrity is safeguarded by the distributed ledger technology used in the blockchain. However, data forgery and tampering can occur because documents are stored in a database and written by hand in the current customs clearance system.
- **Reliability:** The current customs clearance system assumes that the people participating in customs clearance are reliable. However, the blockchain-based NSCC system can structurally ensure reliability.
- **Transparency:** The blockchain-based NSCC guarantees that the customs clearance process remains transparent. The participation of each of the member countries in verification and customs clearance enables transparent data management. However, the transparency of the current customs clearance process cannot be ensured because of possible threats by malicious attackers.

- **Traceability:** The current customs system tracks data to documents and databases. However, the blockchain-based NSCC uses a distributed storage and blockchain network to track every step of the continuous customs clearance process from shipment to unloading.

Table 6. Quantitative comparison of the existing customs clearance system and NSCC.

Category	Existing Customs Clearance	Non-Stop Customs Clearance (Ours)
Time to customs clearance	About 1–2 days	Up to 1 h (from Table 5)
Resource	Human and machine	Machine
Integrity	X	O
Reliability	△	O
Transparency	△	O
Traceability	X	O

The comparison in Table 6 shows that blockchain-based NSCC guarantees integrity, reliability, and traceability, reduces the need for human resources, and shortens the time required for customs clearance. In addition, our solution based on the consensus algorithm of the blockchain integrates the interests of existing railway cooperation agreements and organizations, such as OSJD and OTIF. However, the blockchain-based NSCC system is still vulnerable to APT and network attacks aimed at legacy systems, which prompted us to consider the attack resilience of NSCC, as demonstrated in Section 4, using sequence diagrams to provide a structural explanation. Sections 6 and 7 outline the limitations and future research directions.

6. Discussions

In this study, we developed a blockchain-based NSCC system intended as a new customs clearance mechanism with structural robustness and attack resilience. As demonstrated in Section 4, existing customs clearance systems are vulnerable to DoS, APT, and spoofing attacks. The proposed blockchain-based NSCC system includes a method to solve the above-mentioned problems based on the integrity and reliability provided by the blockchain. In addition, the efficiency of the customs clearance process was maximized by reducing the time required for the customs clearance procedure to 1 h. However, real measurements are challenging and significant system resources are required to implement the sequences, as described in Section 4. In the future, each of these sequences can be investigated and additional vulnerabilities to cyberattacks could be considered, with attack sequences tested by modeling and simulation (M&S) [65–67]. In this study, the current railroad customs clearance system was set up as an overlay network for the developed blockchain-based NSCC system. The integrity and reliability offered by a blockchain can be ensured when configured as an overlay. However, because each train and transit station node participates as nodes in the blockchain network, machine resources unnecessary in the conventional customs clearance system are required. Our choice of a PoS-based consensus algorithm minimizes the use of computational resources. We plan to perform M&S for each potential consensus method in our next study.

This study was conducted based on an IoT-based network environment. Further research on IoT and artificial intelligence (AI)-based block seals and smart container capabilities [68] for inspecting cargo integrity is required, which we plan to incorporate in future studies. Concepts of IoT-based block seals and AI can be used to actively inspect cargo integrity in terms of damage and movement. This study is the first step toward proposing a blockchain-based customs clearance procedure. The most important aspect of this study is that we introduced blockchain into the existing railway customs clearance process to maximize the efficiency of the international railway customs clearance process.

7. Conclusions

This study involved developing a blockchain-based NSCC system and structurally demonstrating its resilience to cyberattacks. Cyberattacks aimed at legacy systems can occur because the NSCC system combines a blockchain with the legacy customs clearance system. However, an attack–response sequence diagram was used to demonstrate that cyberattack resilience can be secured by employing the integrity, reliability, and traceability features of the blockchain.

Compared to the current customs clearance methods, the blockchain-based NSCC system excels in terms of integrity, security, and reliability. Reducing the time required for customs clearance can improve the performance of the freight transportation sector using railroads for cross-border trade. Consequently, the developed customs clearance method uses fewer materials and people overall. This study demonstrated the versatility of blockchain technology and its implications for maritime and aviation trade and the customs clearance system for cross-border railroad transport.

This work demonstrated the compatibility of blockchain with traditional systems. Future research could employ trade domains, such as land, sea, and air. The blockchain-based NSCC system proposed in this study can also be improved using IoT and AI-based object recognition systems to verify cargo integrity. Furthermore, M&S of the NSCC system can be conducted based on the created attack–response sequence diagram presented in Section 4 to appropriately apply the environment, such as the consensus algorithm and network protocol. The defense system utilizing MITRE D3FEND can also be extended [69]. Moreover, simulating the connection between the actual train model and IoT equipment based on the experiment conducted in Section 5 could be explored further.

Author Contributions: Conceptualization, S.K. and D.K.; software, S.K.; sequence diagram, S.K. and D.K.; validation, S.K. and D.K.; investigation, S.K.; resources, S.K.; writing—original draft preparation, S.K. and D.K.; writing—review and editing, S.K. and D.K.; attack scenario, S.K. and D.K.; supervision, D.K.; project administration, S.K. and D.K.; funding acquisition, D.K. All authors have read and agreed to the published version of the manuscript.

Funding: This research received no external funding.

Institutional Review Board Statement: Not applicable.

Informed Consent Statement: Not applicable.

Data Availability Statement: Not applicable.

Acknowledgments: This study was supported by a Kyonggi University Research Grant (2022).

Conflicts of Interest: The authors declare no conflict of interest.

References

1. Abramović, B.; Zitricky, V.; Biškup, V. Organisation of railway freight transport: Case study CIM/smgs between Slovakia and Ukraine. *Eur. Transp. Res. Rev.* **2016**, *8*, 27. [CrossRef]
2. Yaga, D.; Mell, P.; Roby, N.; Scarfone, K. Blockchain technology overview. *arXiv* **2019**, arXiv:1906.11078.
3. Zheng, Z.; Xie, S.; Dai, H.N.; Chen, X.; Wang, H. Blockchain challenges and opportunities: A survey. *Int. J. Web Grid Serv.* **2018**, *14*, 352. [CrossRef]
4. Nofer, M.; Gomber, P.; Hinz, O.; Schiereck, D. Blockchain. *Bus. Inf. Syst. Eng.* **2017**, *59*, 183–187. [CrossRef]
5. Li, S.; Xu, L.D.; Zhao, S. The internet of things: A survey. *Inf. Syst. Front.* **2015**, *17*, 243–259. [CrossRef]
6. Rose, K.; Eldridge, S.; Chapin, L. The internet of things: An overview. *Internet Soc.* **2015**, *80*, 1–50.
7. Zikratov, I.; Kuzmin, A.; Akimenko, V.; Niculichev, V.; Yalansky, L. Ensuring data integrity using blockchain technology. In Proceedings of the 20th Conference of Open Innovations Association (FRUCT), St. Petersburg, Russia, 3–7 April 2017; Volume 2017. [CrossRef]
8. Galvez, J.F.; Mejuto, J.C.; Simal-Gandara, J. Future challenges on the use of blockchain for food traceability analysis. *TrAC Trends Anal. Chem.* **2018**, *107*, 222–232. [CrossRef]
9. Lo, S.K.; Xu, X.; Staples, M.; Yao, L. Reliability Analysis for blockchain oracles. *Comput. Electr. Eng.* **2020**, *83*, 106582. [CrossRef]

10. Bach, L.M.; Mihaljevic, B.; Zagar, M. Comparative Analysis of Blockchain Consensus algorithms. In Proceedings of the 41st International Convention on Information and Communication Technology, Electronics and Microelectronics (MIPRO), Opatija, Croatia, 21–25 May 2018; Volume 2018, pp. 1545–1550. [CrossRef]
11. Bamakan, S.M.H.; Motavali, A.; Babaei Bondarti, A.B. A survey of blockchain consensus algorithms performance evaluation criteria. *Expert Syst. Appl.* **2020**, *154*, 113385. [CrossRef]
12. Strom, B.E.; Applebaum, A.; Miller, D.P.; Nickels, K.C.; Pennington, A.G.; Thomas, C.B. Mitre ATT&CK: Design and Philosophy. In *Technical Report*; The MITRE Corporation: Bedford, MA, USA, 2018.
13. Xiong, W.; Legrand, E.; Åberg, O.; Lagerström, R. Cyber security threat modeling based on the MITRE Enterprise ATT&CK Matrix. *Softw. Syst. Model.* **2022**, *21*, 157–177. [CrossRef]
14. Alexander, O.; Belisle, M.; Steele, J. *MITRE ATT&CK for Industrial Control Systems: Design and Philosophy*; The MITRE Corporation: Bedford, MA, USA, 2020; p. 29.
15. Gupta, R.; Tanwar, S.; Kumar, N.; Tyagi, S. Blockchain-based security attack resilience schemes for autonomous vehicles in industry 4.0: A systematic review. *Comput. Electr. Eng.* **2020**, *86*, 106717. [CrossRef]
16. Fernandez, E.; Pelaez, J.; Larrondo-Petrie, M. Attack patterns: A new forensic and design tool. In *Advances in Digital Forensics III, Proceedings of the IFIP International Conference on Digital Forensics, National Centre for Forensic Science, Orlando, Florida, January 28–January 31, 2007*; Springer: New York, NY, USA, 2007; pp. 345–357.
17. SUMO Official Site. Available online: https://sumo.dlr.de/docs/ (accessed on 19 February 2023).
18. Ethereum Official Site. Available online: https://ethereum.org/en/ (accessed on 20 February 2023).
19. Mahjabin, T.; Xiao, Y.; Sun, G.; Jiang, W. A survey of distributed denial-of-service attack, prevention, and mitigation techniques. *Int. J. Distrib. Sens. Netw.* **2017**, *13*, 1550147717741463. [CrossRef]
20. Castro, M.; Liskov, B. Practical byzantine fault tolerance. *OsDI* **1999**, *99*, 173–186.
21. Kaur, M.; Khan, M.Z.; Gupta, S.; Noorwali, A.; Chakraborty, C.; Pani, S.K. MBCP: Performance analysis of large-scale mainstream blockchain consensus protocols. *IEEE Access* **2021**, *9*, 80931–80944. [CrossRef]
22. Nakamoto, S. Bitcoin: A Peer-To-Peer Electronic Cash System Bitcoin. 2009. Available online: https://bitcoin.org/bitcoin.pdf (accessed on 4 January 2023).
23. Kim, H.; Kim, D. Adjusting the block interval in PoW consensus by block interval process improvement. *Electronics* **2021**, *10*, 2135. [CrossRef]
24. Gervais, A.; Karame, G.O.; Wüst, K.; Glykantzis, V.; Ritzdorf, H.; Capkun, S. On the security and performance of proof of work blockchains. In Proceedings of the 2016 ACM SIGSAC Conference on Computer and Communications Security, Vienna, Austria, 24–28 October 2016; pp. 3–16. [CrossRef]
25. De Angelis, S.; Aniello, L.; Baldoni, R.; Lombardi, F.; Margheri, A.; Sassone, V. PBFT vs. proof-of-authority: Applying the CAP theorem to permissioned blockchain. In Proceedings of the Second Italian Conference on Cyber Security, Milan, Italy, 6–9 February 2018.
26. Ekparinya, P.; Gramoli, V.; Jourjon, G. The attack of the clones against proof-of-authority. *arXiv* **2019**, arXiv:1902.10244. [CrossRef]
27. Toyoda, K.; Machi, K.; Ohtake, Y.; Zhang, A.N. Function-level bottleneck analysis of private proof-of-authority ethereum blockchain. *IEEE Access* **2020**, *8*, 141611–141621. [CrossRef]
28. Saleh, F. Blockchain without waste: Proof-of-stake. *Rev. Financ. Stud.* **2021**, *34*, 1156–1190. [CrossRef]
29. Lee, D.R.; Jang, Y.; Kim, H. Poster: A proof-of-stake (PoS) blockchain protocol using fair and dynamic Sharding management. In Proceedings of the 2019 ACM SIGSAC Conference on Computer and Communications Security, London, UK, 11–15 November 2019; pp. 2553–2555. [CrossRef]
30. Li, W.; Andreina, S.; Bohli, J.; Karame, G. Securing proof-of-stake blockchain protocols. In *Data Privacy Management, Cryptocurrencies and Blockchain Technology*; Springer: Cham, Switzerland, 2017; pp. 297–315. [CrossRef]
31. Nguyen, C.T.; Hoang, D.T.; Nguyen, D.N.; Niyato, D.; Nguyen, H.T.; Dutkiewicz, E. Proof-of-stake consensus mechanisms for future blockchain networks: Fundamentals, applications and opportunities. *IEEE Access* **2019**, *7*, 85727–85745. [CrossRef]
32. Yang, F.; Zhou, W.; Wu, Q.; Long, R.; Xiong, N.N.; Zhou, M. Delegated proof of stake with downgrade: A secure and efficient blockchain consensus algorithm with downgrade mechanism. *IEEE Access* **2019**, *7*, 118541–118555. [CrossRef]
33. Wang, Y.; Cai, S.; Lin, C.; Chen, Z.; Wang, T.; Gao, Z.; Zhou, C. Study of blockchains's consensus mechanism based on credit. *IEEE Access* **2019**, *7*, 10224–10231. [CrossRef]
34. OSJD Official Site. Available online: https://en.osjd.org/ (accessed on 4 January 2023).
35. OTIF Official Site. Available online: http://otif.org/en/ (accessed on 4 January 2023).
36. CIM; SMGS. CIT Official Site. Available online: https://www.cit-rail.org/en/freight-traffic/cim-smgs/ (accessed on 4 January 2023).
37. Yadav, T.; Rao, A.M. Technical aspects of cyber kill chain. In *International Symposium on Security in Computing and Communication*; Springer: Cham, Switzerland, 2015; pp. 438–452.
38. Karapantazis, S.; Pavlidou, F.-N. VoIP: A comprehensive survey on a promising technology. *Comput. Netw.* **2009**, *53*, 2050–2090. [CrossRef]
39. Rafique, M.Z.; Akbar, M.A.; Farooq, M. Evaluating dos attacks against SIP-based VoIP systems. In Proceedings of the GLOBECOM IEEE Global Telecommunication Conference, Honolulu, HI, USA, 30 November–4 December 2009; Volume 2009, pp. 1–6. [CrossRef]
40. Docker Documents Official Site. Available online: https://docs.docker.com/ (accessed on 4 January 2023).

41. Go-Ethereum Documents Official Site. Available online: https://geth.ethereum.org/docs (accessed on 4 January 2023).
42. Merkle, R.C. A fast software one-way hash function. *J. Cryptol.* **1990**, *3*, 43–58. [CrossRef]
43. Swaminathan, A.; Mao, Y.; Wu, M. Robust and secure image hashing. *IEEE Trans. Inf. Forensics Secur.* **2006**, *1*, 215–230. [CrossRef]
44. Chang, F.; Dean, J.; Ghemawat, S.; Hsieh, W.C.; Wallach, D.A.; Burrows, M.; Chandra, T.; Fikes, A.; Gruber, R.E. Bigtable: A distributed storage system for structured data. *ACM Trans. Comput. Syst.* **2008**, *26*, 1–26. [CrossRef]
45. Benet, J. Ipfs-content addressed, versioned, p2p file system. *arXiv* **2014**, arXiv:1407.3561.
46. Chen, Y.; Li, H.; Li, K.; Zhang, J. An Improved P2P File System Scheme Based on IPFS and Blockchain. In Proceedings of the 2017 IEEE International Conference on Big Data (Big Data), Boston, MA, USA, 11–14 December 2017; pp. 2652–2657. [CrossRef]
47. IPFS Documents Official Site. Available online: https://docs.ipfs.tech/ (accessed on 4 January 2023).
48. Ferguson, P.; Huston, G. *What Is a VPN?* Technical Report; Cisco Systems: San Jose, CA, USA, 1998.
49. Schurgot, M.R.; Shinberg, D.A.; Greenwald, L.G. Experiments with security and privacy in IoT networks. In Proceedings of the 2015 IEEE 16th International Symposium on A World of Wireless, Mobile and Multimedia Networks (WoWMoM), Boston, MA, USA, 14–17 June 2015; pp. 1–6. [CrossRef]
50. Rachmawati, D.; Tarigan, J.T.; Ginting, A.B.C. A comparative Study of Message Digest 5 (MD5) and SHA256 algorithm. *J. Phys. Conf. Ser.* **2018**, *978*, 012116. [CrossRef]
51. Bertoni, G.; Daemen, J.; Peeters, M.; Van Assche, G. Keccak. In *Advances in Cryptology–EUROCRYPT 2013: 32nd Annual International Conference on the Theory and Applications of Cryptographic Techniques, Athens, Greece, May 26–30, 2013. Proceedings 32*; Springer: Berlin/Heidelberg, Germany, 2013; pp. 313–314.
52. Antonowicz, M. 65 years of OSJD activities in Eurasia. *Probl. Kolejnictwa* **2021**, *65*, 111–120. [CrossRef]
53. Xu, J.; Lee, W. Sustaining availability of web services under distributed denial of service attacks. *IEEE Trans. Comput.* **2003**, *52*, 195–208. [CrossRef]
54. Van der Merwe, J.R.; Zubizarreta, X.; Lukčin, I.; Rügamer, A.; Felber, W. Classification of spoofing attack types. In Proceedings of the 2018 European Navigation Conference (ENC), Gothenburg, Sweden, 14–17 May 2018; pp. 91–99.
55. Wood, G. Ethereum: A secure decentralised generalised transaction ledger. *Ethereum Proj. Yellow Pap.* **2014**, *151*, 1–32.
56. Li, M.; Huang, W.; Wang, Y.; Fan, W.; Li, J. The study of APT attack stage model. In Proceedings of the 2016 IEEE/ACIS 15th International Conference on Computer and Information Science (ICIS), Okayama, Japan, 26–29 June 2016; pp. 1–5. [CrossRef]
57. Li, Y.; Jiang, Y.; Li, Z.; Xia, S.T. Backdoor learning: A survey. *IEEE Trans. Neural Netw. Learn. Syst.* **2022**, 1–18. [CrossRef]
58. Alladi, T.; Chamola, V.; Sikdar, B.; Choo, K.R. Consumer IoT: Security vulnerability case studies and solutions. *IEEE Consum. Electron. Mag.* **2020**, *9*, 17–25. [CrossRef]
59. Rodrigues, R.; Druschel, P. Peer-to-peer systems. *Commun. ACM* **2010**, *53*, 72–82. [CrossRef]
60. Su, Y.; Nguyen, K.; Sekiya, H. A comparison of blockchain recovery time in static and Mobile IoT-blockchain networks. *Future Internet* **2022**, *14*, 330. [CrossRef]
61. Oudah, M.A.; Oudah, M.A. An insight into internet sector in Iraq. *Turk. J. Comput. Math. Educ. TURCOMAT* **2023**, *14*, 22–32.
62. Algredo-Badillo, I.; Morales-Sandoval, M.; Medina-Santiago, A.; Hernández-Gracidas, C.A.; Lobato-Baez, M.; Morales-Rosales, L.A. A SHA-256 hybrid-redundancy hardware architecture for detecting and correcting errors. *Sensors* **2022**, *22*, 5028. [CrossRef] [PubMed]
63. Zhang, P.; Zhang, X.; Yu, J. A parallel hash function with variable initial values. *Wirel. Pers. Commun.* **2017**, *96*, 2289–2303. [CrossRef]
64. Lepore, C.; Ceria, M.; Visconti, A.; Rao, U.P.; Shah, K.A.; Zanolini, L. A survey on blockchain consensus with a performance comparison of PoW, PoS and pure PoS. *Mathematics* **2020**, *8*, 1782. [CrossRef]
65. Guizani, M.; Rayes, A.; Khan, B.; Al-Fuqaha, A. *Network Modeling and Simulation: A Practical Perspective*; John Wiley & Sons: Hoboken, NJ, USA, 2010.
66. Alharby, M.; Van Moorsel, A. Blocksim: A simulation framework for blockchain systems. *SIGMETRICS Perform. Eval. Rev.* **2019**, *46*, 135–138. [CrossRef]
67. Gupta, Y.; Shorey, R.; Kulkarni, D.; Tew, J. The applicability of blockchain in the Internet of things. In Proceedings of the 2018 10th International Conference on Communication Systems & Networks (COMSNETS), Bengaluru, India, 3–7 January 2018; pp. 561–564. [CrossRef]
68. Kupriyanovsky, Y.; Kupriyanovsky, V.; Klimov, A.; Namiot, D.; Dolbnev, A.; Sinyagov, S.; Lipuntsov, Y.; Arsenyan, A.; Evtushenko, S.; Larin, O. Smart container, smart port, BIM, Internet Things and blockchain in the digital system of world trade. *Int. J. Open Inf. Technol.* **2018**, *6*, 49–94.
69. Turtiainen, H.; Costin, A.; Hämäläinen, T. Defensive machine learning methods and the cyber defence chain. In *Artificial Intelligence and Cybersecurity*; Springer: Cham, Switzerland, 2023; pp. 147–163.

MDPI AG
Grosspeteranlage 5
4052 Basel
Switzerland
Tel.: +41 61 683 77 34

Sensors Editorial Office
E-mail: sensors@mdpi.com
www.mdpi.com/journal/sensors

www.ingramcontent.com/pod-product-compliance
Lightning Source LLC
Chambersburg PA
CBHW041356210926
51850CB00013B/3098